Microstructure-Mechanical Properties and Application of Magnesium Alloys

Microstructure-Mechanical Properties and Application of Magnesium Alloys

Editors

Talal Al-Samman
Sangbong Yi
Dietmar Letzig

MDPI • Basel • Beijing • Wuhan • Barcelona • Belgrade • Manchester • Tokyo • Cluj • Tianjin

Editors
Talal Al-Samman
Institute of Physical
Metallurgy and Materials
Physics, RWTH-Aachen
Germany

Sangbong Yi
Institute of Material and
Process Design,
Helmholtz-Zentrum Hereon
Germany

Dietmar Letzig
Institute of Material and
Process Design,
Helmholtz-Zentrum Hereon
Germany

Editorial Office
MDPI
St. Alban-Anlage 66
4052 Basel, Switzerland

This is a reprint of articles from the Special Issue published online in the open access journal *Metals* (ISSN 2075-4701) (available at: https://www.mdpi.com/journal/metals/special_issues/microstructure_mechanical_magnesium_alloys).

For citation purposes, cite each article independently as indicated on the article page online and as indicated below:

LastName, A.A.; LastName, B.B.; LastName, C.C. Article Title. *Journal Name* **Year**, *Volume Number*, Page Range.

ISBN 978-3-0365-4689-6 (Hbk)
ISBN 978-3-0365-4690-2 (PDF)

Cover image courtesy of Christian Schmid

Contents

About the Editors

Talal Al-Samman

Dr.-Ing. Talal Al-Samman is a senior scientist and chief engineer at the Institute of Physical Metallurgy and Material Physics of the RWTH Aachen University, where he also earned his Ph.D. in Engineering Sciences (2008). His research is concerned with advanced structural materials for lightweight applications, focusing on the science and engineering of magnesium alloys and their potential use to lighten automotive structures given their excellent strength to weight ratios. In recent years, he and his group have made influential contributions to understanding deformation mechanisms, recrystallization and grain growth annealing phenomena, and crystallographic texture evolution in HCP metals, which dictate their mechanical performance and operational stability. On the basis of a comprehensive physical understanding of the underlying mechanisms of microstructure and texture evolution, he has introduced various concepts of microstructure and texture engineering in order to improve the cold formability and elevated temperature strength of commercial magnesium alloys.

Sangbong Yi

Dr.-Ing. Sangbong Yi studied Materials Science and Engineering at the Kumoh National Institute of Technology in South Korea and received his doctorate (Dr.-Ing.) on the deformation mechanisms in magnesium alloys from Clausthal University of Technology, Germany, in 2005. He worked as scientist at the Department of Microstructure Physics, Max Planck Institute for Iron Research, Duesseldorf, studying the textures of metallic and biomaterials until 2006. After that, he worked at the Institute of Materials Science and Technology, Clausthal University of Technology on mechanical surface treatments and recrystallization behaviors until 2008. Since that, he has been employed in the Helmholtz-Zentrum Hereon as a senior researcher. His research interest is focused on the process–microstructure–property relationship in lightweight metallic materials. His area of expertise covers alloy and microstructure design, electron and X-ray diffraction, and sheet processing via continuous casting technology.

Dietmar Letzig

Dr. D. Letzig studied physics at the Christian-Albrecht University in Kiel and at the University of Hamburg, and worked as a scientist in the Department of Physical Metallurgy at the Max Planck Institute for Iron Research Duessldorf, Germany. In 1995, he received his PhD on the development of intermetallic phases at the Technical University RWTH Aachen, Germany. After that, he was employed until 2000 as a research scientist in the R&D Centre at Alcan International Limited in Banbury, GB, and developed new 6xxx series aluminum alloys and designed their processing for the European Automotive Manufacturing System. Since then, he has been head of the Department of Wrought Magnesium Alloys at the Helmholtz Research Center, Geesthacht, which was renamed in 2020 as the Helmholtz-Zentrum Hereon. He is the author of many papers in peer-reviewed scientific journals and proceedings of international conferences. His area of expertise is the process–microstructure–property relationship of lightweight materials, and particularly wrought magnesium alloys. He is leading fundamental, as well as application-oriented research projects in metal forming processes, and is an expert in the processing of magnesium sheets via twin roll casting technology and their performance optimization.

Preface to "Microstructure-Mechanical Properties and Application of Magnesium Alloys"

Given that efficiency improvements for internal combustion engines are limited, light-weighting by means of optimized vehicle design combined with the increased use of low-density metals, such as aluminum and magnesium, is an effective tool to tackle CO_2 emissions and produce efficient low-, or even zero-carbon vehicle technology by shifting to electro-mobility. Correspondingly, the global trend towards light-weighting has triggered major international efforts to develop innovative and cost-effective magnesium alloys and processing routes for lightweight structural components. The extensive use of magnesium alloys in transportation is nevertheless hindered, because we still lack a full understanding of their mechanical and electrochemical behavior resulting from the complex interplay between microstructure and alloy chemistry. Compared to steel or aluminum, magnesium alloy research is relatively young, having mostly been published in the past 20 years, which witnessed a significant advancement of high-resolution characterization and atomistic modeling techniques. With this, many new exciting possibilities have emerged that can answer pending questions and stimulate new areas of research.

The aim of this Special Issue is to cover a broad scope of contributions that highlight recent accomplishments, and to provide readers with some perspective on where the research on magnesium alloys is heading in the near future with respect to global challenges. The Special Issue contains papers presenting state-of-the-art methods and research trends regarding the microstructure, properties and industrial application of magnesium alloys. This encompasses advanced material characterization at different length scales, microstructure manipulation using alloying, thermomechanical treatments, and modern material modeling to establish the best composition/processing/microstructure combinations for targeted applications. The contributions in the Special Issue will offer guidance towards the synergetic development and processing of newly developed alloys.

The editors extend their gratitude to all authors for their excellent contributions. Thanks are also given to Betty Jin at MDPI, who provided invaluable administrative support.

We hope that this Special Issue will serve as a useful reference, and will help to incite new ideas in magnesium research not only for the researchers, theoreticians and experimentalists who have been working in this field, but also for those who are new to the field.

Talal Al-Samman , Sangbong Yi, and Dietmar Letzig
Editors

Editorial

Microstructure–Mechanical Properties and Application of Magnesium Alloys

Talal Al-Samman [1,*], Dietmar Letzig [2,*] and Sangbong Yi [2,*]

[1] Institute for Physical Metallurgy and Materials Physics, RWTH Aachen University, 52074 Aachen, Germany
[2] Institute of Materials and Process Design, Helmholtz-Zentrum Hereon, 21502 Geesthacht, Germany
* Correspondence: alsamman@imm.rwth-aachen.de (T.A.-S.); dietmar.letzig@hereon.de (D.L.); sangbong.yi@hereon.de (S.Y.); Tel.: +49-(0)-2418-026871 (T.A.-S.); +49-(0)-4152-871994 (D.L.); +49-(0)-4152-871911 (S.Y.)

Citation: Al-Samman, T.; Letzig, D.; Yi, S. Microstructure–Mechanical Properties and Application of Magnesium Alloys. *Metals* **2021**, *11*, 1958. https://doi.org/10.3390/met11121958

Received: 18 November 2021
Accepted: 24 November 2021
Published: 6 December 2021

Publisher's Note: MDPI stays neutral with regard to jurisdictional claims in published maps and institutional affiliations.

1. Introduction and Scope

Transport is a major contributor to CO_2 emissions and is considered the most urgent global climate problem. Considering the fact that light-weighting is a crucial factor for efficiency improvements for conventional combustion engines as well as electric vehicles, the increased use of low-density metals, such as aluminum and magnesium, is an effective tool to tackle CO_2 emissions and produce an efficient low—or even—zero-carbon vehicle technology. As a result, the global trend towards light-weighting has triggered major international efforts to develop innovative and cost-effective magnesium alloys and processing for lightweight structural components. Extensive use of magnesium alloys in transportation is nevertheless hindered because we still lack a full understanding of their mechanical and electrochemical behavior that results from the complex interplay between the microstructure and alloy chemistry. Compared to steel or even aluminum, magnesium alloy research is relatively young, being mostly published during the past 20 years, where significant advancements in high-resolution characterization and sophisticated atomistic modeling techniques have emerged. This has introduced many new, exciting possibilities to shed light on still pending questions and stimulate new research areas.

The aim of this Special Issue is to cover a broad scope of contributions that highlight current accomplishments, and to provide the readers with some perspectives on the direction of research on magnesium alloys in the near future with respect to global challenges. This Special Issue contains papers presenting the state of the art and the research trends in the relationship among the microstructure, properties and the industrial application of magnesium alloys. The contributions in the Special Issue cover a wide range of research topics, from alloy fabrication to wrought processing, encompassing advanced material characterization at different length scales, microstructure manipulation using alloying, thermo-mechanical treatments, as well as modern material modeling to establish the best composition/processing/microstructure combinations for targeted applications. The contributions included in the Special Issue will provide guidance towards synergetic development in new alloys and in the processing of the newly developed alloys.

2. Contributions

The great success of the Special Issue is evidenced by the 12 high-quality papers covering a wide range of topics, from casting to property optimization to service conditions.

The first article in this Special Issue [1] investigates the interfacial reactions between Mg-40Al and Mg-30Y master alloys at temperatures of 350–400 °C using the diffusion couple method. With a lack of systematic studies of the diffusion kinetics in Mg-Al-Y systems, this study addresses a major gap in determining the interdiffusion interactions of alloying elements and the formation of intermetallic phases at the interface of the diffusion couple. From the interface microstructures, the authors reported that noticeable reaction layers with sub-layers of different contrast were formed at the interface of the diffusion couple. The thickness of these layers was found to increase with the annealing temperature and duration (from 24 h to 72 h). Interestingly, Y in the Mg-30Y matrix did not diffuse into the opposite Mg-40Al matrix, which was attributed to its relatively large atomic radius. The Al2Y intermetallic phase, which is considered a promising grain refiner, was formed only after Al diffused into the Mg-30Y. Thermodynamic calculations showed that this phase is the most stable among other phase compositions in the Al-Y system. Diffusion path analysis along with XRD measurements also confirmed the formation of the Al2Y phase by the interaction between Al and Y. By investigating the growth kinetics of the reaction layers, the authors estimated the diffusion activation energy to be around 90 kJ/mol. They also provided useful diffusion parameters of Al and Y, which were found to diffuse preferably in the layer located in the Mg-30Y matrix. The interdiffusion coefficients of Al at the different temperatures were higher than those of Y. Analogously, the activation energy for the diffusion of Y was higher than that of Al. The results obtained in this study have contributed towards a better understanding of the precipitation behavior in Mg-Al-Y alloys, which are known for their fine microstructure and improved strength properties.

The second contribution, by Z. Yan et al. [2], features a novel severe plastic deformation technique (CEE-AEC) that was used to process an as-cast AZ31B magnesium alloy into extruded plates with a fine grain size and enhanced ductility. This was conducted at 350 °C, with an extrusion rate of 1 mm/s. In the proposed method, cyclic expansion extrusion utilizes an asymmetrical extrusion cavity that introduces shear deformation and thereby modifies the basal texture and increases the Schmid factor for basal slip. The authors used FEM to design the die and billet geometries based on the calculated distribution of effective strain in different deformation zones and the required forces as a function of processing time. Due to a combination of continuous and discontinuous dynamic recrystallization, the resultant microstructure after three passes was fully recrystallized, with a fine grain size of ~10 μm (refinement degree of ~96% relative to the initial as-cast state). The corresponding texture was sharp but showed off-basal peak intensities at 45° from the extrusion direction. With this, the tensile properties of the extruded samples revealed an attractive combination of yield strength (~115 MPa), ultimate tensile strength (~209 MPa) and maximum elongation (~30%). The obtained results clearly demonstrate the feasibility of using this intelligent method to obtain excellent mechanical properties from processed as-cast magnesium alloys readily after three deformation cycles. An additional major advantage is the capability of fabricating large workparts, which enables this severe plastic deformation method to be used on an industrial scale.

The production of highly competitive, thin wires of Mg-Al-Zn (AZ) alloys by means of direct extrusion is discussed in the third contribution, by Nienaber et al. [3]. Wires of magnesium alloys have high specific strength; hence, they are attractive for use as filler materials in joining applications or as biodegradable suture materials in the medical field. Commonly, very thin wires (thicknesses of ~100 μm) are produced through drawing processes of extruded bars employing several passes and intermediate annealing. Alternative production from cast billets via direct extrusion is extremely challenging because of the overwhelmingly high degree of deformation involved. Another challenging aspect is the urgent demand for a high surface quality that has a strong impact on the corrosion behavior of the produced wires. In this study, the authors demonstrated that it is possible to produce 1 mm thin wires of different AZ alloys with various Al content (i.e., AZ31, AZ80 and AZ91) by means of direct extrusion at 325 °C and 0.1 mm/s extrusion speed. The limitation towards reducing the thickness below 1 mm was due to the capacity limit of the extrusion press of 2.5 MN peak force. All Mg alloy wires revealed a homogenous, fine-grained microstructure with an average grain size below 10 μm. The tensile yield strength was between ~180 and ~195 MPa and the ultimate tensile strength was between ~270 and ~300 MPa. With increasing Al content, the maximum texture intensity showed a decrease, accompanied by a more random orientation distribution between the $\langle 10\bar{1}0 \rangle$ and $\langle 11\bar{2}0 \rangle$ poles. This had a positive impact on the ductility, which ranged from ~16%

(fracture strain) for AZ31 to ~21% for AZ91. The surface quality of the produced wires was tested before and after a wrapping test. AZ31 showed the smoothest surface among the other alloys prior to the wrapping test. After the test, the roughness of the wires increased considerably, which depended on the wrapping diameter. Interestingly, the AZ91 alloy with the highest Al content revealed the smallest increase in surface roughness, which highlights the success of the reported process in producing high-performance wires with an excellent profile of mechanical and surface properties.

Damage mechanisms in conventional magnesium alloys exhibiting mechanical twinning are the topic of the fourth paper in this Special Issue [4]. In general, twinning is known to induce strain incompatibility in the microstructure, which needs to be accommodated by slip or interaction twins in adjacent grains; otherwise, it would lead to the nucleation of cracks. Investigations of twin accommodation effects at grain and twin boundaries are therefore highly important for understanding damage initiation in magnesium and other HCP metals. In an effort to achieve this, the authors performed thorough microstructural and fractographical analyses on the interactions between twins and other deformation modes and microstructural defects. They employed electron microscopy combined with in-situ electron backscatter diffraction characterization on plane-strain extruded AM30 magnesium alloy with a double fiber texture. This offered the possibility of tracking deformation through various neighboring grain orientations with favorable and unfavorable activation of $\{10\bar{1}2\}$ and $\{10\bar{1}1\}$ twinning. The results revealed various types of twin interactions with slip dislocations, grain boundaries and other twins, which led to local cracking. Nucleated twins of a very low macroscopic Schmid factor were less tolerant to damage initiation through these interactions. An interesting observation was made in the case of $\{10\bar{1}2\}$ twins in favorably oriented grains, where twin variants that were able to nucleate easily at grain boundaries showed favorable growth compared to other variants. It was suggested that this could have been related to the enhanced nucleation of glissile (mobile) disconnections at the intersection between grain and twin boundaries rather than between two twin boundaries. Another interesting finding was that $\{10\bar{1}1\}$ twins were found to nucleate easier under $\langle c \rangle$-axis contraction than what is reported in the literature under $\langle c \rangle$-axis compression. Although this aspect has not yet been investigated in detail, the authors propose that it is related to the complex stress state in the case of $\langle c \rangle$-axis contraction that could facilitate the atomic shuffle associated with $\{10\bar{1}1\}$ twinning. With these important findings, the study provides valuable insights towards a better understanding of twin–microstructure interactions and their effect on twin formation and damage initiation in magnesium alloys. It also offers beneficial input for advancing our present crystal plasticity simulation models.

The paper written by Bian, M., Huang, X. and Chino, Y. [5] describes the development of a new type of precipitation-hardenable magnesium sheet alloy. The authors chose the magnesium silver system with the addition of calcium and explored the feasibility of developing precipitation-hardenable Mg sheet alloys. Based on the Pandat software, they calculated the Mg-Ag phase diagram containing Ca content fixed at 0.1 wt%. They prepared three alloys with different Ag contents (1.6 wt%, 6 wt% and 12 wt%), extruded them to sheets and rolled them to a final gauge. They investigated the age-hardening response at 170 °C by measuring the Vickers hardness and tensile properties and analyzed the microstructure using various electron microscopy techniques.

In a T4-treated condition, the TYS of Mg-1.5Ag-0.1Ca alloy sheet is only 85 MPa, 57 MPa and 47 MPa along the RD, 45° and TD, respectively. With an Ag content of 12 wt%, the tensile yield strength (TYS) is increased in a T4-treated condition to 193 MPa, 130 MPa and 117 MPa along the RD, 45° and TD direction. Artificial aging at 170 °C for 336 h (T6) further increases the TYS of Mg-12Ag-0.1Ca alloy sheet to 236 MPa, 163 MPa and 143 MPa along the RD, 45° and TD direction. The microstructure characterization reveals that $AgMg_4$ are responsible for the strength improvement, and, in a high-resolution HAADF-STEM image analyzing the FFT patterns, the orientation relationship between the α-Mg matrix and $AgMg_4$ is confirmed by the authors to be $(0001)_\alpha \parallel (0001)AgMg_4$, $[-2110]_\alpha$

‖ [10-10]AgMg$_4$. These findings show that it is possible to improve the age-hardening response and accelerate the aging kinetics of Mg-Ag-Ca alloys by microstructural design.

The contribution, written by Ostapovets, A., Kushnir, K., Máthis, K. and Šiška, F. [6] is focused on the numerical analysis of the interaction between the growing tensile twin and obstacles in magnesium. The study is based on a multiscale approach as the twin growth is simulated using an atomistic model, while the overall stress state is evaluated using a finite element method. As is known, abundant twinning can lead to the formation of structure inhomogeneity during the plastic deformation of magnesium and its alloys. Such inhomogeneity can be a serious restriction for engineering applications. Suppression of the twinning by twin interactions with obstacles can also be considered as a means of decreasing the plastic anisotropy and improving the mechanical properties of the material. The obstacle for twin boundary migration was inserted into the simulation block. Two types of obstacles were considered. One type was a void, and the other type of obstacle was obtained by freezing atoms inside the selected volume. The size of the obstacle was the same as the size of the void. The external tensile stress was applied in the X-direction on the vertical sides, which represents tension in the crystallographic c-direction. The results revealed an increase in critical resolved shear stress, which was higher for the passage of the twin boundary through a row of voids than for the interaction with non-shearable obstacles. Two basal stacking faults were nucleated during the detachment of the boundary from the obstacle. These stacking faults were trailed by the migrating boundary and grew together with the twin. It could be seen that this stress is dependent on the type of obstacle as well as on the initial density of disconnections. It is interesting that voids serve as a stronger barrier for twin boundary migration than non-shearable obstacles.

In the next fundamental study, the authors Mouhib, F., Sheng, F., Mandia, R., Pei, R., Korte-Kerzel, S. and Al-Samman, T. [7] aimed to further the understanding of the texture development in a binary Mg-1 wt%Er and in a ternary Mg-1wt %Er-1wt%Zn alloy during recrystallization and grain growth under the influence of solutes and second-phase precipitates. The rolling texture of the binary alloy showed a typical basal component with a pole spread towards the rolling direction (RD) and moderate intensity. In contrast, the ternary alloy exhibited a weaker and much softer rolling texture characterized by two off-basal components at $\pm20°$ RD that bear a significant pole spread in the transverse direction $\pm40°$ TD. The discussion addresses the microstructure and texture development in both binary and ternary versions of the alloy during recrystallization and grain growth under the influence of solutes and second-phase precipitates and the different recrystallization kinetics. Therefore, both alloys underwent characterization of the microstructure and their evolution, their texture development, as well as EBSD-assisted slip trace analysis in order to understand the activation of deformation modes during tension to 5% strain. The authors report that the addition of Zn alters the substitutional solute chemistry of the Mg-1%Er alloy, which would obviously lead to the complex interaction of solute species within the matrix. From an energetic perspective, Zn and Er can cluster in the lattice to relieve the misfit strains arising from the solute size mismatch. They would also cosegregate to defects in the microstructure and are therefore likely to have a stronger interaction with grain boundaries and dislocations, leading to strikingly different slip system activation during deformation and boundary migration characteristics during annealing. The authors showed a remarkable enhancement in yield strength, strain hardening capability and failure ductility due to precipitates, solute strengthening effects and a favorable soft texture, and they also identified the main mechanism for the $\pm20°$ RD $\rightarrow \pm40°$ TD of the deformed Mg-1%Er-1%Zn alloy.

The modification of the microstructure to adjust the mechanical properties and the reduction of the asymmetric yield behavior and anisotropy of the extruded magnesium alloys Mg2Nd and Mg2Yb is reported in the article of Schmidt, J., Beyerlein, I., Knezevic, M. and Reimers, W. [8]. For their study, they extruded the binary magnesium alloys Mg2 wt%Nd and Mg2 wt%Yb and investigated their deformation behavior at room temperature. To modify the microstructure and thus the mechanical properties, they used variations in extrusion parameters. In addition due to the low solubility of Nd and ytterbium (Yb) in Mg, subsequent heat treatments were used to further increase the strength. The extruded bars were investigated by electron microscopy (SEM, TEM), laboratory X-ray texture measurements and mechanical testing (compression, tension) to determine the mechanical properties, the microstructure and their changes during deformation. To gain further insight into the different deformation behavior, the authors used a combination of in situ energy-dispersive X-ray synchrotron diffraction and applied simulations with the elastoplastic self-consistent (EPSC) model. In the Mg2Nd alloy, subsequent heat treatments lead to the formation of fine precipitates, generating a significant hardening effect. In the case of the Mg2Yb alloy, heat treatments for precipitation hardening were less effective. The authors report that, by reducing the extrusion temperature, a decrease in the grain size of the Mg2Yb and an advantageous texture could be achieved. This was accompanied by a significant increase in YS. Since critical resolved shear stresses for tension twinning (CRSSttw) in particular are very sensitive to grain size, the compressive yield strength (CYS) increases more than the tensile yield strength (TYS). By adjusting the process parameters, including heat treatments, the grain size can be reduced and the texture improved as well, so that, despite differences in plastic deformation, the yield strengths in compression are almost equal to the yield strengths in tension. This resulted in high strengths and low strength differential effects (SDE). As a result, the Mg2Nd series, which had high yield strengths and an SDE close to 0, could be extruded. By adjusting the extrusion parameters, the grain size of the Mg2Nd alloy can be gradually reduced, thereby significantly increasing the strength. The Mg2Yb series tend to have larger grain sizes and less pronounced rare earth textures. This leads to intensive tension twinning (TTW-ing) under compressive stress, resulting in high SDEs. Subsequent heat treatments can increase the YS. Since the effect on the slip systems is stronger than on the formation of TTWs, this leads to further increased SDEs. However, by reducing the extrusion temperature, the grain size of the Mg2Yb alloy is also reduced and an advantageous texture can be achieved. This is accompanied by a significant increase in YS. Since CRSSttw, in particular, is very sensitive to grain size, the CYS increases more than the TYS. Therefore, it is also possible to obtain an SDE close to 0 for the Mg2Yb alloy.

The active deformation mechanisms during the hot rolling of magnesium single crystals, with different initial orientations, are reported in the contribution of Estrada-Martinez, J., Hernandez-Silva, D. and Al-Samman, T. [9]. In this study, the texture evolution during the rolling of the single crystals with two different initial orientations is evaluated, in terms of the prismatic axis aligned in the compression direction (sheet normal direction, ND), while the extension in the c-axis is allowed by the c-axis laid in the rolling direction (RD) in both cases. The systematic investigations using XRD and EBSD revealed the high activities of extension twinning and the concomitant reorientation of original orientations. The orientations formed by the extension twinning are determined by the active twin variants. When the $[10\bar{1}0]$ direction is parallel to the ND, the reorientation by extension twinning results in the c-axis being parallel to the ND. In contrast, the active twin variants by compression in the $[11\bar{2}0]$ axis lead the twin orientations with the c-axis to be aligned at an angle of 30° from the ND. Further deformation of the fully reoriented matrix induces the contraction twins, in which the basal slip occurs relatively easily. The recrystallization triggered at the contraction twins brings about continuously recrystallized grains with non-basal orientations. By using the single crystal and the deformation in a specific orientation, the active deformation mechanism and its contribution to the final texture could be successfully described with respect to the initial orientation.

The tenth contribution, by Jo, S., Letzig, D. and Yi, S. [10], deals with the microstructure and texture control of a non-flammable Mg alloy containing Y and Ca simultaneously. The results regarding the interrelationship between the Al content, texture weakening and mechanical properties provide an efficient guideline towards the improvement of sheet formability, which is important to widen the industrial application of magnesium alloys. By decreasing the Al content, the amount of secondary phases decreases, while the amount of solute atoms, especially Ca, dissolved in the matrix concomitantly increases. A higher amount of solute atoms enhances the activity of non-basal deformation modes and retards the recrystallization such that the texture weakening and the accompanying formation of the non-basal type components are observed in the alloy sheets with smaller Al amounts. The thermodynamic calculation complementarily supports the experimental observations regarding the formation of secondary phases and the solute amount. This study contributes to establishing an alloy design strategy facilitating the development of high-performance magnesium alloy sheets.

The superplastic behavior of Al-free ZK60 alloy having a fine and homogeneous grain structure, which was processed by indirect extrusion, was investigated in the contribution of Palacios-Trujillo, C., Victoria-Hernandez, J., Hernandez-Silva, D., Letzig, D. and Garcia-Bernal, M. [11]. From the tensile tests at various testing conditions, the superplastic behavior was found at 250 °C and 10^{-4}/s, resulting in an elongation to failure of 464%. The tensile sample deformed under a superplastic regime showed a dynamic recrystallized grain structure, while significant texture weakening was observed in the deformed sample. These results imply that the superplastic behavior of the ZK60 alloy at intermediate temperature is brought about by the combined effect of dynamic recrystallization and the grain boundary sliding, while the increase in the strain rate or temperature reduction causes the change in the deformation mechanism to dislocation glide. The activation energy, based on the Zener–Hollomon parameter, of 446 kJ/mol through the range of strain rates and temperature indicates that the grain boundary sliding is responsible for the superplasticity of the examined ZK60 alloy. The most important findings of this study indicate that the superplastic behavior can be obtained at an intermediate temperature even in commercial magnesium alloys processed without employing severe plastic deformation.

The contribution of R. Yamada, S. Yoshihara and Y. Ito examined the effect of equal channel angular pressing (ECAP) on the fatigue properties of AZ31B alloy [12]. The experimental results comprising the tensile properties, fatigue behavior and residual stress indicate that the strengthening brought about by the grain refinement after one ECAP pass effectively improves the fatigue property, while the sample after eight ECAP passes shows a deteriorated fatigue property despite its improved ductility. Moreover, a lower compressive residual stress in the ECAP samples after the fatigue test was measured in comparison to that in the annealed sample. The authors inferred from the different residual stresses after the fatigue tests that the ECAP-treated samples should have a higher tensile residual stress before the fatigue test, which is harmful for the fatigue properties. The experimental results of this study clearly show that strength improvement, especially yield strength, is a more effective means of enhancing the fatigue properties than a ductility increase.

The editors are grateful to all authors for their excellent contributions. Thanks are due to Betty Jin at MDPI, who provided invaluable administrative support.

3. Conclusions and Outlook

We hope that this Special Issue will serve as a useful reference and will help to incite new ideas in magnesium research, not only for those researchers, theoreticians and experimentalists who have been working in this field but also for those who are new to the field.

Conflicts of Interest: The authors declare no conflict of interest.

References

1. Dai, J.; Jiang, B.; Xie, H.; Yang, Q. Interfacial Reactions between Mg-40Al and Mg-30Y Master Alloys. *Metals* **2020**, *10*, 825. [CrossRef]
2. Yan, Z.; Zheng, J.; Zhu, J.; Zhang, Z.; Wang, Q.; Xue, Y. High Ductility with a Homogeneous Microstructure of a Mg–Al–Zn Alloy Prepared by Cyclic Expansion Extrusion with an Asymmetrical Extrusion Cavity. *Metals* **2020**, *10*, 1102. [CrossRef]
3. Nienaber, M.; Yi, S.; Kainer, K.U.; Letzig, D.; Bohlen, J. On the Direct Extrusion of Magnesium Wires from Mg-Al-Zn Series Alloys. *Metals* **2020**, *10*, 1208. [CrossRef]

4.   Russell, W.; Bratton, N.; Paudel, Y.; Moser, R.; McClelland, Z.; Barrett, C.; Oppedal, A.; Whittington, W.; Rhee, H.; Mujahid, S.; et al. In Situ Characterization of the Effect of Twin-Microstructure Interactions on {1 0 1 2} Tension and {1 0 1 1} Contraction Twin Nucleation, Growth and Damage in Magnesium. *Metals* **2020**, *10*, 1403. [CrossRef]

5.   Bian, M.; Huang, X.; Chino, Y. Microstructures and Mechanical Properties of Precipitation-Hardenable Magnesium–Silver–Calcium Alloy Sheets. *Metals* **2020**, *10*, 1632. [CrossRef]

6.   Ostapovets, A.; Kushnir, K.; Máthis, K.; Šiška, F. Interaction of Migrating Twin Boundaries with Obstacles in Magnesium. *Metals* **2021**, *11*, 154. [CrossRef]

7.   Mouhib, F.-Z.; Sheng, F.; Mandia, R.; Pei, R.; Korte-Kerzel, S.; Al-Samman, T. Texture Selection Mechanisms during Recrystallization and Grain Growth of a Magnesium-Erbium-Zinc Alloy. *Metals* **2021**, *11*, 171. [CrossRef]

8.   Schmidt, J.; Beyerlein, I.; Knezevic, M.; Reimers, W. Adjustment of the Mechanical Properties of Mg2Nd and Mg2Yb by Optimizing Their Microstructures. *Metals* **2021**, *11*, 377. [CrossRef]

9.   Estrada-Martínez, J.; Hernández-Silva, D.; Al-Samman, T. Hot Rolling of Magnesium Single Crystals. *Metals* **2021**, *11*, 443. [CrossRef]

10.   Jo, S.; Letzig, D.; Yi, S. Effect of Al Content on Texture Evolution and Recrystallization Behavior of Non-Flammable Magnesium Sheet Alloys. *Metals* **2021**, *11*, 468. [CrossRef]

11.   Palacios-Trujillo, C.; Victoria-Hernández, J.; Hernández-Silva, D.; Letzig, D.; García-Bernal, M. Superplasticity at Intermediate Temperatures of ZK60 Magnesium Alloy Processed by Indirect Extrusion. *Metals* **2021**, *11*, 606. [CrossRef]

12.   Yamada, R.; Yoshihara, S.; Ito, Y. Fatigue Properties of AZ31B Magnesium Alloy Processed by Equal-Channel Angular Pressing. *Metals* **2021**, *11*, 1191. [CrossRef]

Article

Interfacial Reactions between Mg-40Al and Mg-30Y Master Alloys

Jiahong Dai [1],*, Bin Jiang [2],*, Hongmei Xie [1] and Qingshan Yang [3]

[1] College of Materials Science and Engineering, Yangtze Normal University, Chongqing 408100, China; xhm1983@126.com
[2] State Key Laboratory of Mechanical Transmissions, College of Materials Science and Engineering, Chongqing University, Chongqing 400044, China
[3] School of Metallurgy and Materials Engineering, Chongqing University of Science and Technology, Chongqing 401331, China; qsyang@cqu.edu.cn
* Correspondence: daijiahong@yznu.edu.cn (J.D.); jiangbinrong@cqu.edu.cn (B.J.); Tel.: +86-159-0236-1594 (J.D.); +86-135-9419-0166 (B.J.)

Received: 14 May 2020; Accepted: 18 June 2020; Published: 20 June 2020

Abstract: Interfacial reactions between Mg-40Al and Mg-30Y master alloys were investigated at intervals of 25 °C in the 350–400 °C by using a diffusion couple method. Noticeable reaction layers were formed at the interfaces of the diffusion couples. The concentration profiles of the reaction layers were characterized. The diffusion path of the diffusion couple at 400 °C is constructed on the Mg-Al-Y ternary isothermal temperature phase diagram. The phases of the reaction layer were characterized by X-ray diffraction. The interfacial reaction thermodynamics of diffusion couples were studied. These results indicate that Al_2Y is the only new formed intermetallic phase in the reaction layers. The growth constants of the reaction layers were calculated. In the reaction layer II, the integrated interdiffusion coefficients of Al are higher than Y, the diffusion activation energy of Y is higher than that of Al.

Keywords: magnesium alloys; interface reaction; diffusion; intermetallic phases

1. Introduction

Magnesium alloys are important structural materials for automotive, aircraft, and aerospace lightweighting, due to their lower densities compared with steel and aluminum [1,2]. Currently, Mg-Al alloys such as AM60 and AZ91 are the most widely used commercial magnesium alloys. However, the applications of Mg-Al alloys are limited to room or near room-temperature because of inferior creep resistance and poor tensile properties at elevated temperatures above 120 °C. Adding Y to precipitate intermetallic phases in Mg-Al alloys is an efficient method to refine grain and improve the mechanical properties [3–7]. Therefore, a complete knowledge of the precipitation of intermetallic phases for the Mg-Al-Y system is crucial to a better understanding the role of Y in Mg-Al alloys.

In recent years, a lot of studies have been carried out on the in-situ formation of Al_2Y in Mg-Al alloys by adding a Y element [3–6]. Al_2Y is a very promising grain refiner. The Al_2Y phase transformation temperatures during solidification process were determine by thermal analysis combined with microstructural and EDX analysis. The nucleation crystallography and wettability of Al_2Y on the Mg grains have been investigated. However, no systematic diffusion kinetics studies have been performed for the Mg-Al-Y ternary system so far.

In recent decades, there have been thermodynamic and phase diagram studies on the Mg-Al-Y alloy systems. Shakhshir et al. [8] have established the Mg-Al-Y ternary thermodynamic model based on the extrapolation of the Mg-Al, Al-Y, Mg-Y binary subsystems by the CALPHAD approach. The thermodynamic properties and the calculated phase diagrams are consistent with the literature.

Huang et al. [9] have calculated the excess free-energy, enthalpies of formation, excess entropies and activity values of all components of Mg-Al-Y ternary alloy via the Miedema formation enthalpy model. The results show that enthalpies of formation, excess free-energy and excess entropies of the ternary alloy are negative in the whole content range. Dri et al. [10] presented Mg-Al-Y system isothermal sections between 300–500 °C for a vertical section at 80 wt % Mg. Zar et al. [11] and Odi et al. [12] constructed an isothermal section at 400 °C and confirmed the existence of a ternary compound Al₄MgY. In addition, Das et al. [13,14] have researched diffusion kinetic in the Mg-Al and Mg-Y systems. The impurity diffusion coefficients of Al and Y in Mg were determined. The interdiffusion coefficients and growth constants of the intermetallic phases were calculated. However, interfacial reactions and systematic diffusion kinetics studies in Mg-Al-Y systems are not fully reported. Therefore, studies are required to evaluate the interdiffusion interactions of elements and intermetallic phases formation in the interfacial reactions between Mg-Al and Mg-Y.

In this article, the interfacial reactions of (Mg-40Al)/(Mg-30Y) master alloys have been investigated at temperatures of 350–400 °C by diffusion couple method. The formation of the intermetallic phases at the interface of diffusion couples has been studied. The growth constants of the reaction layers have been analyzed by appropriate theories. The interdiffusion coefficients of Al and Y in the diffusion reaction lays have been determined.

2. Analytical Framework for Diffusion

2.1. Growth of Reaction Layers

The growth constants of the reaction layers were determined using a parabolic trend with time due to the diffusion-controlled process. The growth constant, k, can be calculated by:

$$x = \sqrt{kt} \tag{1}$$

where the thickness of reaction layer is x, and the annealing time is t. An Arrhenius relationship:

$$k = k_0 \exp\left(\frac{-Q}{RT}\right) \tag{2}$$

where the k_0 is pre-exponential factor, R is the gas constant, Q is the activation energy, and the annealing temperature is T.

2.2. Interdiffusion

Interdiffusion flux for each component was calculated according to the method proposed by Dayananda [15]:

$$\widetilde{J_i} = \frac{1}{2t} \int_{C_i^{-\infty}}^{C_i(x)} (x - x_0) dC_i, \, (i = Mg, Al, Y) \tag{3}$$

where $C_i(x)$ is the concentration of i, and t is the anneal time, and x_0 is the location of the Matan plane. For each reaction layer, the integrated interdiffusion coefficient $\widetilde{D}_i^{Int, \, layer}$ is calculated by the cumulative interdiffusion fluxes of individual components:

$$\widetilde{D}_i^{Int, \, layer} = \int_{x_1}^{x_2} \widetilde{J_i}(x) dx, \, (i = Mg, Al, Y) \tag{4}$$

where x_1 and x_2 are the position correspond to the intermetallic phase boundaries.

While the average effective interdiffusion coefficient, $\widetilde{D}_i^{eff}$, of each relevant reaction layer is calculated by [16]:

$$\widetilde{D}_i^{eff} = \frac{\int_{x_1}^{x_2} \widetilde{J_i}(x) dx}{\Delta C_i}, \, (i = Mg, Al, Y) \tag{5}$$

where ΔC_i is the difference of solute concentration at the end of i reaction layer.

3. Experimental

The terminal alloys of the diffusion couples were prepared from commercial Mg-40Al (Mg-40 wt % Al) and Mg-30Y (Mg-30 wt % Y) master alloy. The Mg-40Al master alloy is brittle and fragile, and its melting point (460 °C) is lower than that of Mg-30Y master alloy (595 °C). It is not feasible to make solid diffusion couples by the jig method. So a solid–liquid contact method was employed to produce the (Mg-40Al)/(Mg-30Y) diffusion couples. The method of preparing the (Mg-40Al)/(Mg-30Y) diffusion couples are the same as the previous works [17–19]. The Mg-40Al master alloy was melted in a crucible. Then, a piece of Mg-30Y master alloy with surface oxide removed was immediately submerged into Mg-40Al master alloy melt, and the crucible was taken from the furnace. An intimate contact between Mg-40Al and Mg-30Y master alloys was formed during solidification and the diffusion couple samples were obtained. Then the diffusion couples were isothermally heat-treated in a resistance furnace. The diffusion couples were carried out 350, 375 and 400 °C for 24, 48 and 72 h, respectively. Water-quenching after annealing is completed.

The diffusion couple samples were ground using 200–1000 emery papers. The microstructure of the cross-section was observed by scanning electron microscopy (SEM, TESCAN VEGA 2, TESCAN Co., Brno, Czech). Concentrations of Mg, Al and Y at the interfaces of diffusion couples were determined by energy dispersive spectroscopy (EDS, Oxford Inca, Oxford Instrument Technology Co., Ltd., Oxford, UK). Each EDS result is the average concentration of the elements for 5 μm × 150 μm rectangular map scanning, and the step length of the scanning is 5 μm. The intermetallic phases in the reaction layers were determined by X-ray diffractometer (XRD, D/MAX-2500PC, Dandong Fangyuan Instrument Co., Ltd., Dandong, China) and EDS. The thickness of the reaction layer is obtained by averaging the thickness of 10 random position measurement of the reaction layer.

4. Results and Discussion

4.1. Interfaces Microstructure

Typical back-scattered electron (BSE) micrographs of the cross-section for (Mg-40Al)/(Mg-30Y) diffusion couples annealing at 350, 375 and 400 °C for 24, 48 and 72 h, respectively, shown in Figure 1. As is seen in these micrographs, it is found that reaction layers are formed at the interfaces of diffusion couples. The intermetallic phases are formed in the reaction layers, and the reaction layer has obvious sub-layers. Thicknesses of the reaction layers increases with the increase in annealing temperature and time. As shown in Figure 1g, the sub-layer with a darker contrast located on the Mg-40Al matrix is named as layer I, and the intermetallic compound sub-layer with a brighter contrast located on the Mg-30Y matrix is named as layer II. As shown in Figure 1i, most of the newly formed intermetallic phases in the reaction layer distribute along the morphology of $Mg_{24}Y_5$ before diffusion in region 1. However, the morphology of the newly formed intermetallic phases near the Mg-30Y matrix is filamentous in region 2. In region 3, a dark stripe is formed at the end of the reaction layer near the Mg-30Y matrix.

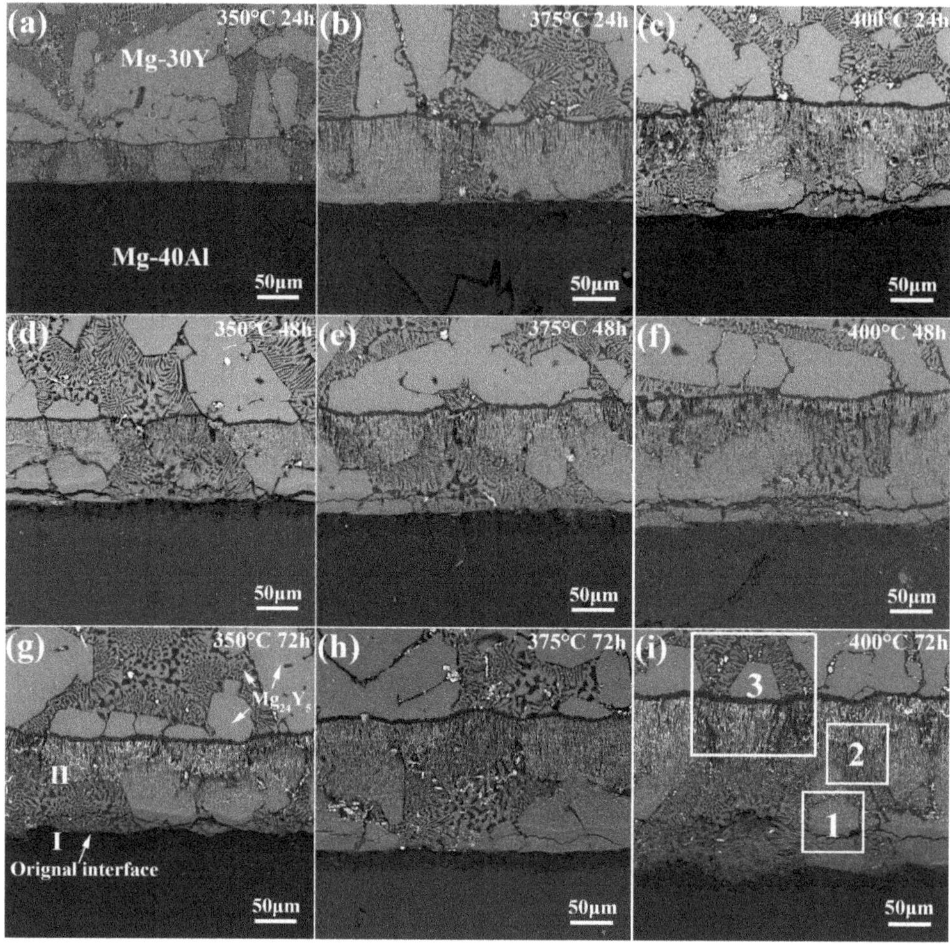

Figure 1. Back-scattered electron (BSE) micrographs of the diffusion couples annealing at 350, 375 and 400 °C for 24, 48 and 72 h. (**a**) at 350 °C, (**b**) at 375 °C and (**c**) at 400 °C for 24 h; (**d**) at 350 °C, (**e**) at 375 °C and (**f**) at 400 °C for 48 h; (**g**) at 350 °C, (**h**) at 375 °C and (**i**) at 400 °C for 72 h.

4.2. Concentration Profiles at Interfaces

Figure 2 shows the concentration profiles that were performed across the reaction lays of diffusion couples annealing at 350, 375 and 400 °C for 72 h, respectively. It is found that Y in Mg-30Y matrix hardly diffused into Mg-40Al matrix, but Al in Mg-40Al matrix diffused into Mg-30Y matrix. The concentration of Al in the reaction layers sharply falls from about 36 at.% to about 10 at.%, the concentration is essentially constant in the reaction layer I. then rises about 25 at.% when entering layer II, and finally decrease gradually are shown Figure 2a,c. Therefore, the sub-layer I is Al depletion compared, which is mainly Mg(Al) solid solution. The Al-Y intermetallic phases are formed when Al from Mg-40Al matrix diffused into Mg-30Y matrix. There is no significant fluctuation in the concentration of Y element in the sub-layer II. The result showed that Al and Y diffused differently in Mg.

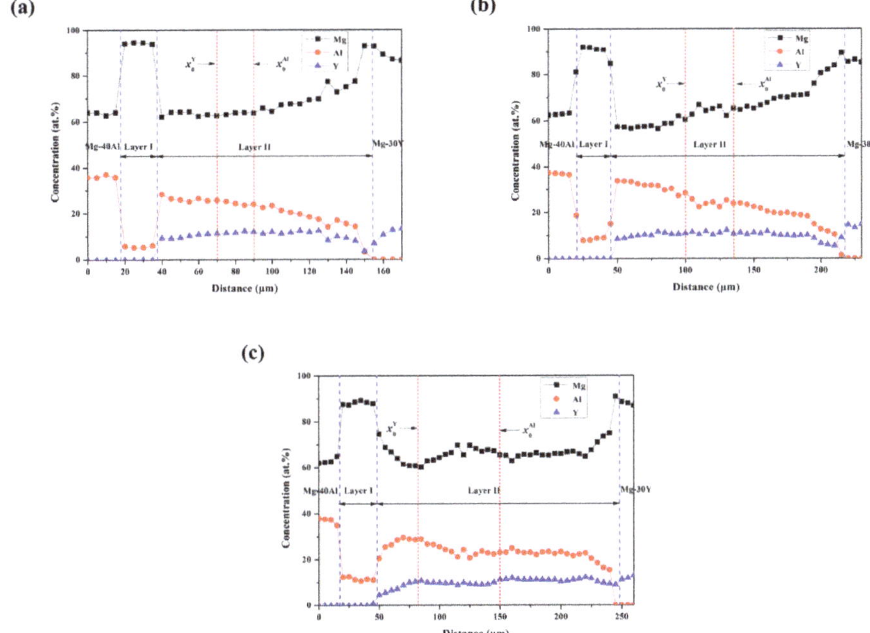

Figure 2. Concentration profiles Mg, Al and Y on the cross-section along the direction of diffusion for the diffusion couples annealed for 72 h at temperatures of (**a**) 350, (**b**) 375 and (**c**) 400 °C, respectively. x_0^{Al} and x_0^{Y} is the location of Matano plane for Al and Y, respectively.

4.3. Diffusion Path

The microstructure prediction of the diffusion process and the qualitative information on the relative diffusion behavior of the components can be realized by diffusion path. Any time on the concentration curve in the diffusion process can be mapped to a ternary isotherm [20,21]. Mg-Al-Y ternary isothermal temperature phase diagram of 400 °C has been determined [22]. Figure 3 shows the concentration curve is mapped onto Mg-Al-Y ternary phase diagram as diffusion path at 400 °C for 72 h. As shown in Figure 3b, the phase regions through which the diffusion path are as follow: Mg-40Al (terminal alloy) → Al_2Y + $\gamma(Mg_{17}Al_{12})$ + (Mg) → Al_2Y + (Mg) → Al_2Y + (Mg) + $Mg_{24}Y_5$ → Mg-30Y (terminal alloy). These concentration profiles in layer I are concentrated at Mg-10at%Al of the Mg-Al-Y ternary phase diagram. However, these concentration profiles in layer II are concentrated in Al_2Y + $\gamma(Mg_{17}Al_{12})$ + (Mg) and Al_2Y + (Mg) regions. Mg is the base of the terminal alloys, $Mg_{17}Al_{12}$ and $Mg_{24}Y_5$ exist in Mg-40Al and Mg-30Y alloys, respectively. The shape of the diffusion path is S-shaped. This agrees with the results of the diffusion path in the single-phase region reported in the literature [21]. So Al_2Y is the only newly formed phase in the diffusion path.

Figure 4 shows a schematic diagram of sample preparation process and the XRD result of the diffusion couple annealing for 72 h at 400 °C. The XRD pattern shows the existence of $Mg_{24}Y_5$ and Al_2Y phases. The $Mg_{24}Y_5$ phase already exists in the Mg-30Y matrix. The Al_2Y phases are formed by the interaction between Al and Y.

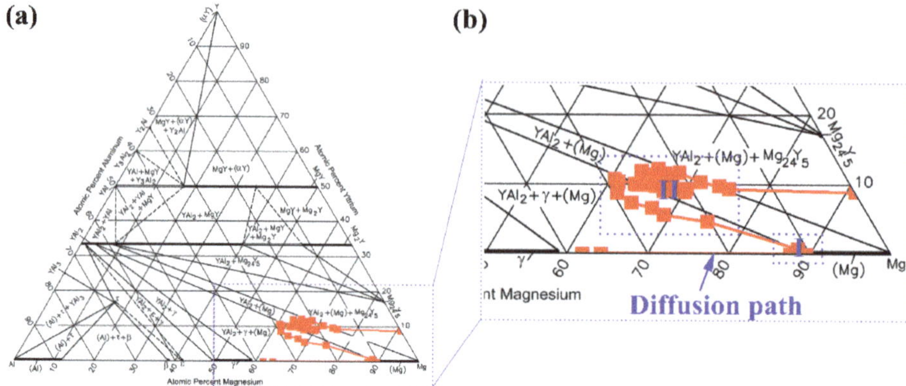

Figure 3. (**a**) The diffusion path for the diffusion couple on isothermal section at 400 °C of the Mg-Al-Y system, (**b**) a larger version of the dotted box in Figure 3a.

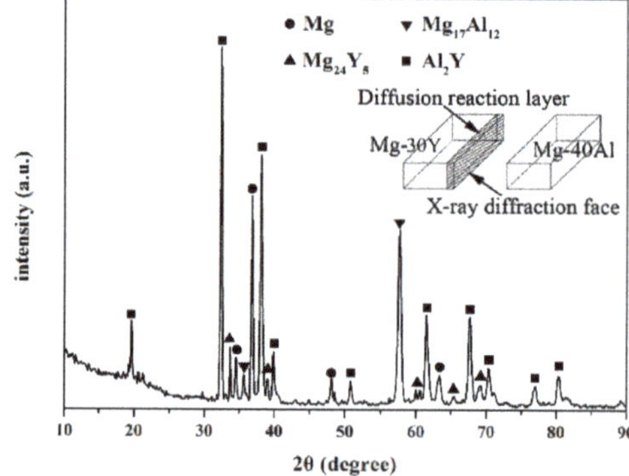

Figure 4. XRD pattern for the diffusion couple annealing for 72 h at 400 °C.

4.4. Thermodynamic Analysis

The standard enthalpy of formation (ΔH) can predict the binary alloy formation. According to the thermodynamic parameters of Mg, Al and Y in the literature [23], ΔH can be calculated using Miedema model [24]. Figure 5 shows ΔH of formation for Mg-Al, Mg-Y and Al-Y systems. It is clear that ΔH of formation for Mg-Al, Mg-Y and Al-Y systems are all negative, ΔH for Mg-Y system are closer to the Mg-Al system than the Al-Y system, and ΔH for Mg-Al and Mg-Y systems is far higher than Al-Y system. It indicates that the stable Al-Y intermetallic phases are first formed in the Mg-Al-Y system. This is consistent with the thermodynamic calculation results in the literature [9]. In addition, the Al$_4$MgY phase is the only ternary phase reported in the Mg-Al-Y system [22]. The melting point of the Al-Y intermetallic phases are higher than the Al$_4$MgY phase [8], therefore, during the existence of the Al-Y intermetallic phase in the Mg-Al-Y system, the Al$_4$MgY phase is difficult to form in this study.

The Gibbs energy of formation (ΔG_f) can be an effective prediction of intermetallic phase formation. Moreover, the reaction with the lowest ΔG_f tends to occur among all the possible reactions. ΔG_f of the Al-Y intermetallic phases are described by a linear function of the temperature T: $\Delta G_f = a + bT$. The thermodynamic parameters of a and b can be evaluated using phase diagram and thermodynamic data. Optimized thermodynamic parameters for the Al-Y intermetallic phases are listed in Table 1 [8]. As shown in Figure 6, ΔG_f of Al$_2$Y is lower than those of Al$_3$Y, AlY, Al$_2$Y$_3$ and AlY$_2$, which indicates that that Al$_2$Y is the most stable phase of the Al-Y system. The variation of ΔG_f for the Al-Y intermetallic phases is consistent with that reported in literature [25,26].

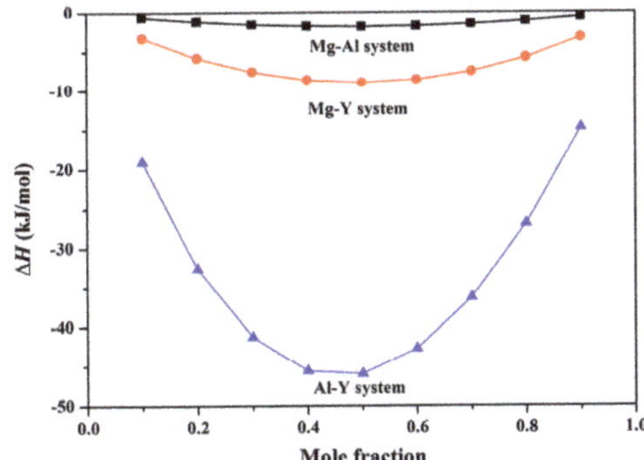

Figure 5. Standard enthalpy of formation of Mg-Al, Mg-Y and Al-Y systems.

Table 1. Optimized thermodynamic parameters for Al-Y intermetallic phases [8].

Intermetallic Phase	a (J mol^{-1})	b (J (mol K)$^{-1}$)
Al_3Y	−39,727.972	8.036
Al_2Y	−50,410.046	10.230
AlY	−48,074.303	11.536
Al_2Y_3	−45,347.395	12.364
AlY_2	−38,200.000	10.568

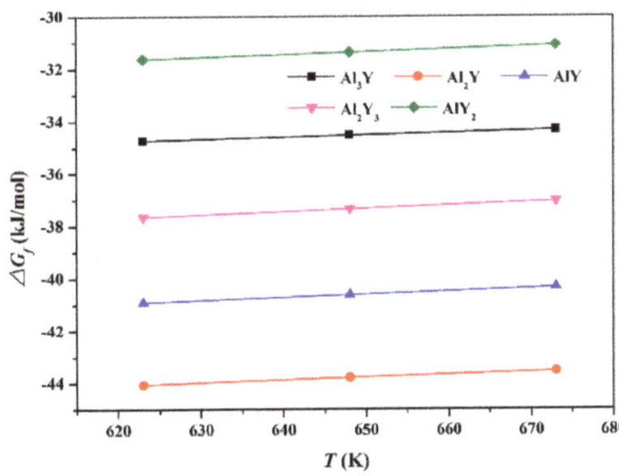

Figure 6. Gibbs energy of formation of Al-Y intermetallic phases in the temperature range of this experiment.

4.5. EDS Characterizing Intermetallic Phases

BSE images of regions 1 and 2 in Figure 1i are shown in Figure 7. In Figure 7a, it is found that after Al atoms diffused into the Mg-30Y matrix, the Al-Y compounds basically formed along the morphology of the previous $Mg_{24}Y_5$, which were the dense block phases of close to the Mg-40Al matrix and the acicular phase of far from the Mg-40Al matrix. In region 2, a large number of lamellar intermetallic phases are apparent. EDS point analysis was carried out to identify the composition of the reaction lay and their constituents. Table 2 lists the results of EDS analysis. Many studies in the literature have reported that Al_2Y is the most common Al-Y intermetallic phase in Mg alloys [3–6,27]. A Mg-Al-Y vertical section at 80 wt % Mg computed by Shakhshir et al. [8] is compared. It is found that in addition to $Mg_{24}Y_5$ and $Mg_{17}Al_{12}$, only in Al-Y intermetallic phases, Al_2Y, was formed at 80 wt % Mg. Combined with the EDS, diffusion path, thermodynamics analysis and XRD results, the phase composition of each point is determined and shown in Table 2.

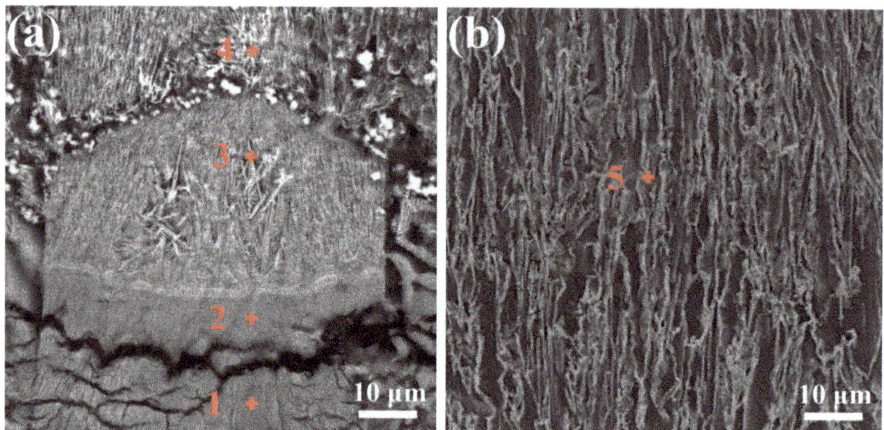

Figure 7. BSE images of (**a**) region 1 and (**b**) region 2 for the Figure 1i.

Table 2. EDS results of the reaction layer as denoted in Figure 7.

Point	Mg (at.%)	Al (at.%)	Y (at.%)	Corresponding Phase
1	74.71	20.56	4.72	$Al_2Y + \gamma + (Mg)$
2	66.87	26.42	6.71	$Al_2Y + \gamma + (Mg)$
3	60.78	28.68	10.54	$Al_2Y + (Mg)$
4	62.91	25	12.09	$Al_2Y + (Mg)$
5	65.7	22.95	11.35	$Al_2Y + (Mg)$

The line scan in Figure 8 shows that a lean Y region is formed at the end of the reaction layer near the Mg-30Y matrix. It is possible that Y in Mg-30Y matrix diffused into the reaction layer to form Al_2Y compounds. Yu et al. [28,29] showed that Y tends to diffused towards the interface.

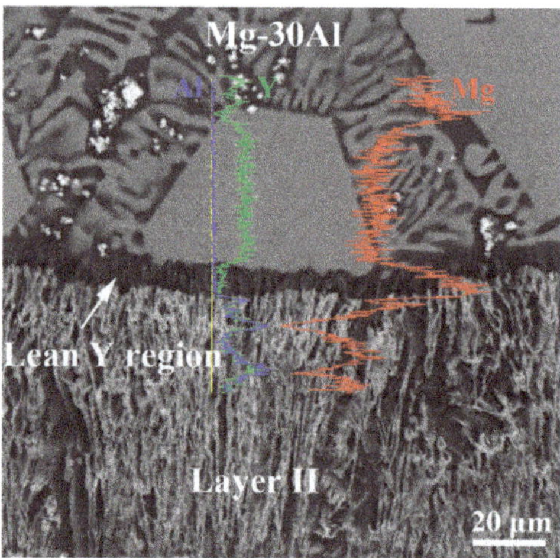

Figure 8. BSE/EDS line of the region 3 in Figure 1i.

4.6. Growth Kinetics of Reaction Layers

Figure 9 shows the thickness reaction layer plotted against the square root of time. It is found that the linear extrapolation of the experimental data intersects through the origin, which proves that the growth constant of the reaction layer and the thickness of the diffusion layer have a parabolic trend with time. Therefore, the growth of reaction layers is controlled by diffusion mechanisms. The growth constants of the reaction layer are calculated from Equations (1) and (2). Figure 10 presents an Arrhenius plot of the reaction layers. The growth constants of the reaction layer are listed in Table 3. The magnitudes of the growth constants for the reaction layers are consistent with the previous study on the interfacial reaction of Mg alloys [17–19]. Especially at 400 °C, the growth constant of the reaction layer is $k_{(Mg-40Al)/(Mg-30Y)} > k_{(Mg-40Al)/(Mg-30Nd)} > k_{(Mg-40Al)/(Mg-20Ce)} > k_{(Mg-40Al)/(Mg-20Ca)}$. This is exactly the same as the law of the radius of Y, Nd, Ce, and Ca atoms ($r_Y > r_{Nd} > r_{Ce} > r_{Ca}$). Therefore, the growth constants of reaction layers may be related to the atomic size of alloying elements in magnesium alloys. The activation energy of the reaction layer is larger than we studied before [17–19].

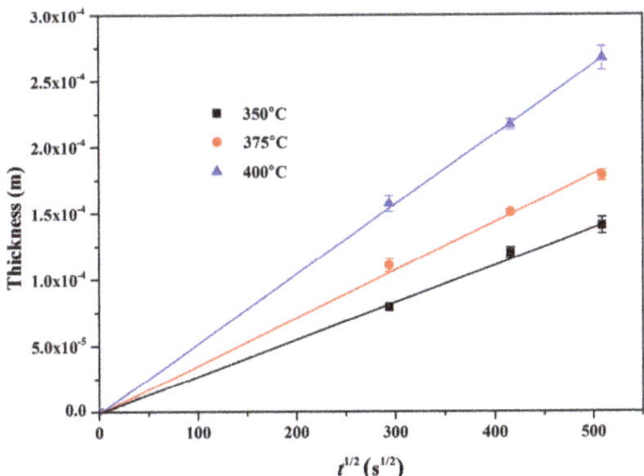

Figure 9. Thickness of reaction layer and square root of the time in the temperature range of 350–400 °C.

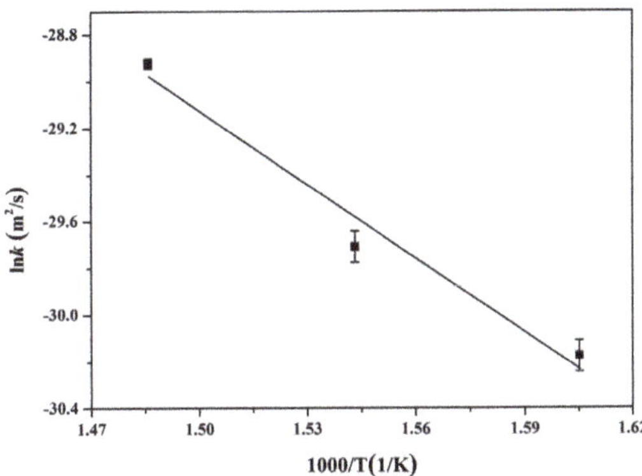

Figure 10. Growth constants for the reaction layers as a function of temperature.

Table 3. The growth constants of reaction layer of the diffusion couples.

T (°C)	k (m^2 s^{-1})	k_0 (m^2 s^{-1})	Q (kJ mol^{-1})
350	$(7.84 \pm 0.13) \times 10^{-14}$		
375	$(1.25 \pm 0.08) \times 10^{-13}$	$(1.48 \pm 0.10) \times 10^{-6}$	87.09 ± 0.73
400	$(2.75 \pm 0.05) \times 10^{-13}$		

4.7. Interdiffusion Coefficients

In this study, Mg is the matrix, so the interdiffusion coefficient of Mg is not calculated. Layer I is formed by the Al element in Mg-40Al matrix diffused across the original interface to Mg-30Y matrix, but Y in Mg-30Y matrix does not diffused to Mg-40Al matrix. However, both Al and Y diffused in layer II. Therefore, the interdiffusion coefficients of Al and Y in layer II are calculated. In addition, ΔC_i could not be determined in this study. Thus, the average effective interdiffusion coefficients were not calculated. The integrated interdiffusion coefficients, $\widetilde{D}_i^{Int,II}$, of Al and Y are calculated from Equation (4) and given in Table 4. The magnitude of $\widetilde{D}_{Al}^{Int,II}$ agree well with Brennan et al. [30] and our previous study [18]. In layer II, $\widetilde{D}_{Al}^{Int,II}$ are higher than $\widetilde{D}_{Y}^{Int,II}$, and $\widetilde{Q}_{Y}^{Int,II}$ is higher than $\widetilde{Q}_{Al}^{Int,II}$. In the solid diffusion process, the element diffusion is proportional to the melting point of the alloy and inversely proportional to the atomic radius. The melting points of Mg-30Y master alloy (595 °C) are higher than Mg-40Al master alloy (460 °C). In addition, Al (143 pm) has a smaller atomic radius than Y (227 pm). Therefore, Al atoms are more easily diffused than Y atoms in magnesium alloys.

Table 4. The diffusion parameters of the Al and Y in the layer II.

i	T (°C)	$\widetilde{D}_i^{Int,II}$ (m^2/s)	$\widetilde{D}_0^{Int,II}$ (m^2 s^{-1})	$\widetilde{Q}_i^{Int,II}$ (kJ mol^{-1})
Al	350	1.44×10^{-14}		
	375	2.98×10^{-14}	2.20×10^{-6}	97.61
	400	5.83×10^{-14}		
Y	350	5.13×10^{-15}		
	375	1.24×10^{-14}	2.40×10^{-5}	115.29
	400	2.68×10^{-14}		

5. Conclusions

1. Two sub-layers were formed in the diffusion couples at 350–400 °C. Most of the newly formed intermetallic phases in the reaction layer distribute along the morphology of Mg$_{24}$Y$_5$ before diffusion.

2. During the whole interfacial reactions process, Y in Mg-30Y matrix hardly diffused into Mg-40Al matrix, but Al from Mg-40Al matrix diffused into Mg-30Y matrix. The Al-Y intermetallic phases were formed when Al diffused into Mg-30Y matrix. Diffusion path, XRD, thermodynamics analysis and EDS results show that Al$_2$Y is the only newly formed phase at the interface of diffusion couple.

3. The growth of reaction layers is controlled by diffusion mechanisms. The diffusion activation energy for the reaction layers is calculated to be (87.09 ± 0.73) kJ/mol. The $\widetilde{D}_{Al}^{Int,II}$ are higher than $\widetilde{D}_{Y}^{Int,II}$ in the temperature range of 350–400 °C. The diffusion activation energy of Y is higher than that of Al.

Author Contributions: Conceptualization, J.D. and B.J.; methodology, J.D.; software, Q.Y.; investigation, H.X.; writing—original draft preparation, J.D.; writing—review and editing, J.D. and B.J. All authors have read and agreed to the published version of the manuscript.

Funding: This research was sponsored by National Natural Science Foundation of China (11847077, 51701033, U1764253), Chongqing Science and Technology Commission (cstc2018jcyjAX0760), Research Foundation of Yangtze Normal University (2016RYQD15, 2018QNRC06).

Conflicts of Interest: The authors declare no conflict of interest.

References

1. Karakulak, E. A review: Past, present and future of grain refining of magnesium castings. *J. Magnes. Alloy.* **2019**, *7*, 355–369. [CrossRef]

2. Song, J.; She, J.; Chen, D.; Pan, F. Latest research advances on magnesium and magnesium alloys worldwide. *J. Magnes. Alloy.* **2020**, *8*, 1–41. [CrossRef]

3. Qiu, D.; Zhang, M.-X. The nucleation crystallography and wettability of Mg grains on active Al$_2$Y inoculants in an Mg-10 wt% Y Alloy. *J. Alloys Compd.* **2014**, *586*, 39–44. [CrossRef]

4.   Qiu, D.; Zhang, M.-X.; Kelly, P.M. Crystallography of heterogeneous nucleation of Mg grains on Al_2Y nucleation particles in an Mg-10 wt.% Y alloy. *Scr. Mater.* **2009**, *61*, 312–315. [CrossRef]

5.   Chang, H.-W.; Qiu, D.; Taylor, J.; Easton, M.; Zhang, M.-X. The role of Al_2Y in grain refinement in Mg-Al-Y alloy system. *J. Magnes. Alloy.* **2013**, *1*, 115–121. [CrossRef]

6.   Jiang, Z.; Feng, J.; Chen, Q.; Jiang, S.; Dai, J.; Jiang, B.; Pan, F. Preparation and Characterization of Magnesium Alloy Containing Al_2Y Particles. *Materials* **2018**, *11*, 1748. [CrossRef]

7.   Tang, Q.; Sun, H.; Zhou, M.; Quan, G. Effect of Y Addition on the Semi-Solid Microstructure Evolution and the Coarsening Kinetics of SIMA AZ80 Magnesium Alloy. *Metals* **2017**, *7*, 416. [CrossRef]

8.   Shakhshir, S.A.; Medraj, M. Computational thermodynamic model for the Mg-Al-Y system. *J. Ph. Equilib. Diffus.* **2006**, *27*, 231–244. [CrossRef]

9.   Huang, W.; Yan, H. Calculation of thermodynamic parameters of Mg-Al-Y alloy. *J. Wuhan Univ. Technol.* **2014**, *29*, 374–378. [CrossRef]

10.   Drits, M.E.; Padezhnova, E.M.; Dobatkina, T.V. Phase Equilibria in Magnesium-Yttrium-Aluminum Alloys. *Metall* **1979**, *3*, 223–227.

11.   Zarechnyuk, O.S.; Drits, M.E.; Rykhal, R.M.; Kinzhibalo, V.V. Examination of the Mg-Al-Y System (0-33 at% Y) at 400 °C. *Metall* **1980**, *5*, 242–244.

12.   Odinev, K.O.; Ganiev, I.N.; Kinzhibalo, V.V.; Kurbanov, K.K. Phase Equilibria in Aluminum-Magnesium-Yttrium and Aluminum-Magnesium-Cerium Systems at 673 K. *Tsvetn. Metall* **1989**, *4*, 75–77.

13.   Das, S.K.; Kim, Y.; Ha, T.; Gauvin, R.; Jung, I.-H. Anisotropic Diffusion Behavior of Al in Mg: Diffusion Couple Study Using Mg Single Crystal. *Metall. Mater. Trans. A* **2013**, *44*, 2539–2547. [CrossRef]

14.   Das, S.K.; Kang, Y.-B.; Ha, T.; Jung, I.-H. Thermodynamic modeling and diffusion kinetic experiments of binary Mg-Gd and Mg-Y systems. *Acta Mater.* **2014**, *71*, 164–175. [CrossRef]

15.   Dayananda, M.A. An analysis of concentration profiles for fluxes, diffusion depths, and zero-flux planes in multicomponent diffusion. *Metall. Mater. Trans. A* **1983**, *14*, 1851–1858. [CrossRef]

16.   Dayananda, M.A. Average effective interdiffusion coefficients and the Matano plane composition. *Metall. Mater. Trans. A* **1996**, *27*, 2504–2509. [CrossRef]

17.   Dai, J.; Jiang, B.; Li, X.; Yang, Q.; Dong, H.; Xia, X.; Pan, F. The formation of intermetallic phases during interdiffusion of Mg-Al/Mg-Ce diffusion couples. *J. Alloys Compd.* **2015**, *619*, 411–416. [CrossRef]

18.   Dai, J.; Shen, S.; Jiang, B.; Zhang, J.; Yang, Q.; Jiang, Z.; Dong, H.; Pan, F. Interfacial reaction in (Mg-37.5Al)/(Mg-6.7Nd) diffusion couples. *Met. Mater. Int.* **2016**, *22*, 1–6. [CrossRef]

19.   Dai, J.; Xiao, H.; Jiang, B.; Xie, H.; Peng, C.; Jiang, Z.; Zou, Q.; Yang, Q.; Pan, F. Diffusion behavior and reactions between Al and Ca in Mg alloys by diffusion couples. *J. Mater. Sci. Technol.* **2018**, *34*, 291–298. [CrossRef]

20.   Kirkaldy, J.S.; Brown, L. Diffusion behaviour in ternary, multiphase systems. *Can. Metall. Quart.* **1963**, *2*, 89–115. [CrossRef]

21.   Morral, J.E. Diffusion path theorems for ternary diffusion couples. *Metall. Mater. Trans. A* **2012**, *43*, 3462–3470. [CrossRef]

22.   Raghavan, V. Al-Mg-Y (aluminum-magnesium-yttrium). *J. Ph. Equilib. Diff.* **2007**, *28*, 477–479. [CrossRef]

23.   De Boer, F.R.; Boom, R.; Mattens, W.C.M.; Miedema, A.R.; Niessen, A.K. *Cohesion in Metals: Transition Metal Alloys*; Elsevier: Amsterdam, The Netherlands, 1988.

24.   Miedema, A.; De Chatel, P.; De Boer, F. Cohesion in alloys-fundamentals of a semi-empirical model. *Physica B + C* **1980**, *100*, 1–28. [CrossRef]

25.   Liu, S.; Du, Y.; Chen, H. A thermodynamic reassessment of the Al-Y system. *Calphad* **2006**, *30*, 334–340. [CrossRef]

26.   Ran, Q.; Lukas, H.L.; Effenberg, G.; Petzow, G. A thermodynamic optimization of the Al-Y system. *J. Less Common Met.* **1989**, *146*, 213–222. [CrossRef]

27.   Peng, Z.Z.; Shao, X.H.; Guo, X.W.; Wang, J.; Ma, X.L. Atomic-Scale Insight into Structure and Interface of Al_2Y Phase in an Mg-Al-Y Alloy. *Adv. Eng. Mater.* **2018**, *20*. [CrossRef]

28.   Yu, X.; Jiang, B.; Yang, H.; Yang, Q.; Xia, X.; Pan, F. High temperature oxidation behavior of Mg-Y-Sn, Mg-Y, Mg-Sn alloys and its effect on corrosion property. *Appl. Surf. Sci.* **2015**, *353*, 1013–1022. [CrossRef]

Metals **2020**, *10*, 825

29. Yu, X.; Jiang, B.; He, J.; Liu, B.; Jiang, Z.; Pan, F. Effect of Zn addition on the oxidation property of Mg-Y alloy at high temperatures. *J. Alloys Compd.* **2016**, *687*, 252–262. [CrossRef]

30. Brennan, S.; Bermudez, K.; Kulkarni, N.S.; Sohn, Y. Interdiffusion in the Mg-Al System and Intrinsic Diffusion in β-Mg_2Al_3. *Metall. Mater. Trans. A* **2012**, *43*, 4043–4052. [CrossRef]

Article

High Ductility with a Homogeneous Microstructure of a Mg–Al–Zn Alloy Prepared by Cyclic Expansion Extrusion with an Asymmetrical Extrusion Cavity

Zhaoming Yan [1], Jie Zheng [1], Jiaxuan Zhu [2], Zhimin Zhang [1,*], Qiang Wang [1] and Yong Xue [1,*]

[1]   School of Material Science and Engineering, North University of China, Taiyuan 030051, China; zmyan1027@126.com (Z.Y.); cqzhengjie@163.com (J.Z.); qingwangnuc@126.com (Q.W.)

[2]   College of Mechatronics Engineering, North University of China, Taiyuan 030051, China; nucjxzhu@126.com

*   Correspondence: zhangzhimin@nuc.edu.cn (Z.Z.); yongxuenuc@126.com (Y.X.); Tel.: +86-351-3921778 (Z.Z.)

Received: 14 July 2020; Accepted: 13 August 2020; Published: 14 August 2020

Abstract: In the current work, cyclic expansion extrusion with an asymmetrical extrusion cavity (CEE-AEC), as a relatively novel severe plastic deformation method, was applied to fabricate an AZ31B magnesium alloy plate with a size of $50 \times 100 \times 220$ mm, and the resultant microstructure, texture development, and mechanical properties were systematically investigated. A refined and homogeneous grain structure was achieved after three passes of deformation due to dynamic recrystallization. The grain refinement degree in comparison to as-cast alloys was more than ~96%. With the increasing number of CEE-AEC passes, a basal inclination texture was gradually formed, with the basal planes inclined ~45° from the transverse direction to the extrusion direction, which could be attributed to the introduction of an asymmetrical extrusion cavity that led to an increasing Schmid factor for the activation of basal <a> slip systems. The comprehensive mechanical properties were improved by successive multi-passes of CEE-AEC processing, especially due to the ductility reaching to $30.0 \pm 1.3\%$ after three passes of deformation. The competition between the grain refinement and texture modification were the main strengthening mechanisms.

Keywords: cyclic expansion extrusion with asymmetrical extrusion cavity; AZ31B alloy; microstructure; texture; mechanical properties

---

1. Introduction

Magnesium (Mg) and its alloys, which have the advantages of a low density, a high specific strength, easy recyclability, etc., have broad application prospects in national defense, aerospace, automobile, and 3C communication [1–4]. However, due to their poor strength and low ductility at room temperature, the development and application of Mg alloys are still limited [5]. Thus, research into improving the strength and toughness of Mg alloys is of great importance to promote the development of Mg alloys and Mg industries [6]. Grain refinement and texture modification have been proven to be effective ways to improve the ductility of Mg alloys [1,7]. In last few decades, severe plastic deformation (SPD) has attracted more and more attention in the Mg alloy field, because it is believed to be a practical and promising technology to prepare high strength/toughness Mg alloys by modifying their microstructures and textures.

Many pervious SPD methods with various feathers have been proposed and applied to prepare Mg alloys with fine grain structures and excellent properties [8]. Equal channel angular pressing (ECAP) and high-pressure torsion (HPT) are two of the most famous techniques [9,10]. Kim et al. [11] investigated that ECAP-processed AZ61 alloys and demonstrated a significant grain refinement and improvement in both strength and ductility. Stráská et al. [12] showed that the HPT-processed AZ31 alloys with grain sizes of approximately 150–205 nm exhibited high microhardness levels.

The advantage of these SPD methods is the introduction of high plastic strain to achieve dramatic grain refinement and texture modification [13]. However, their disadvantages are deformed, small sized billet and high equipment requirements. Thus, these methods are still limited in laboratory research, and they are rarely used in industrial production.

We propose a novel SPD method entitled cyclic expansion extrusion with an asymmetrical extrusion cavity (CEE-AEC) to fabricate thick plate 50 × 100 × 220 mm (thickness × length × height) Mg alloys. Besides the deformed billet with large size, the introduction of shear strain by attaching an asymmetrical extrusion cavity is the core advantage of this technology. HCP metals are known to form a (0001) fiber texture that is parallel to the extrusion direction after axisymmetric deformation. Therefore, for axisymmetric extruded Mg alloys, slipping on the basal plane is difficult. From our early studies, we prepared the Mg–Gd–Y–Zn–Zr alloys with improved mechanical properties by refining their microstructure and optimizing their basal texture via CEE-AEC [14]. We attributed these improvements to the introduction of an asymmetrical extrusion cavity that could effectively modify the basal texture in a way that corresponded to the increase of the Schmid factor (SF). Twist extrusion (TE) has proven that shear deformation can improve the uniformity of metals and obtain materials with gradients [15]. Chang et al. [16] and Xu et al. [17] also illustrated the significant effect of asymmetrical extrusion for the improvement of the ductility of Mg alloys.

In the present work, the multi-pass isothermal CEE-AEC process was conducted on AZ31B. The microstructure, texture, and strengthening mechanisms were analyzed and are discussed. Furthermore, the mechanical properties were also investigated and correlated to the grain refinement and texture modification.

2. Experimental Procedures

2.1. Materials and Process

The material chosen in this study was commercial AZ31B with a composition of 3% Al (wt %), 1% Zn (wt %), 0.3% Mn (wt %), and balance Mg, which was supplied in form of cast rod that was 220 mm in diameter and 300 mm in length. The as-cast rod was machined into 50 × 100 × 220 mm (thickness × length × height) billets to prepare for the CEE-AEC process. Due to the serious segregation and defects in as-cast alloys, homogenization treatment was carried out under conditions of 400 °C/12 h; the microstructures are shown in Figure 1. It can be seen that the $Mg_{17}Al_{12}$ phases were dissolved in the matrix after homogenization, and the alloy showed an inhomogeneous grain structure with an average grain size of more than 300 μm and an irregular (0001) basal texture. The CEE-AEC process was conducted on the die structure that consisted of a punch, an upper bottom die, and a lower bottom die. H13 steel was used to fabricate the die parts. The schematic diagram is shown in Figure 2a. The cuboid billet 1 was first put into the channel to achieve the expansion. Then, billet 2 was added in the channel to complete the extrusion process of billet 1, followed by a new cycle of CEE-AEC, as shown in Figure 2b. The extrusion rate was 1 mm/s, and the deformation temperature was set to 350 °C. The oil-based graphite was used to reduce the friction between the die structure and the billets. The complete forming process and processed objects are shown in Figure 2c. A comparison between axisymmetric deformation and CEE-AEC on the effect of crystallographic orientation was made, and it was noticed that the grain inclined after CEE-AEC.

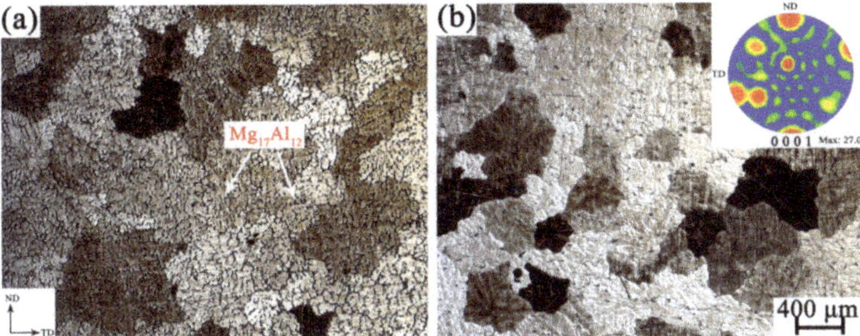

Figure 1. Microstructures of (**a**) as-cast and (**b**) as-homogenized AZ31B alloys. (Two figures have the same scale bars).

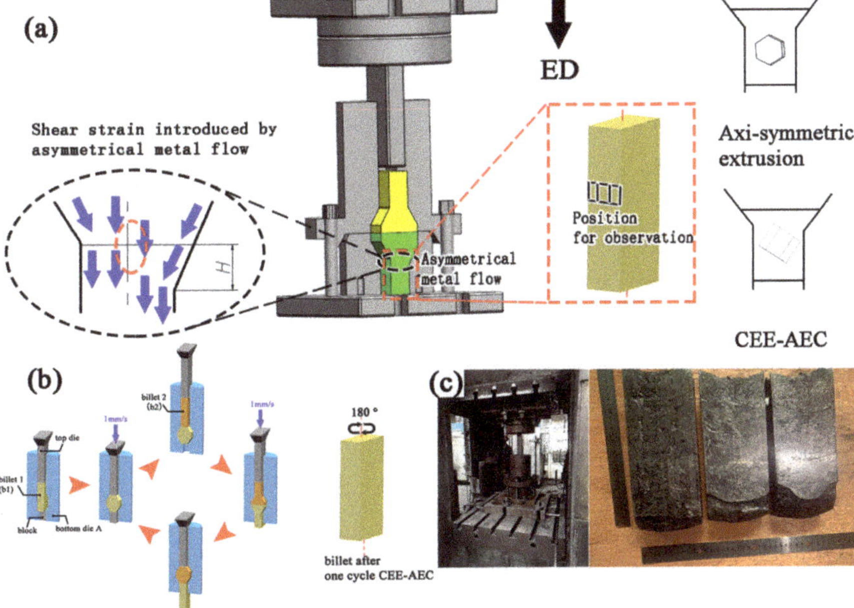

Figure 2. Schematic illustration of the cyclic expansion extrusion with an asymmetrical extrusion cavity (CEE-AEC) process: (**a**) deformation molds, (**b**) one-cycle deformation, and (**c**) specific dies and processed objects.

The microstructures of the CEE-AEC-processed samples were analyzed using an optical microscope (OM, Carl Zeiss A2m, Oberkochen, Germany). Electron backscatter diffraction (EBSD, EDAX Inc., Mahwah, NJ, USA) interfaced with a field emission gun and a scanning electron microscope (SEM, Hitachi SU5000, Tokyo, Japan) were used to reveal the distribution of grain size, grain orientations and crystallographic texture. The OM samples with longitudinal sections along the extrusion direction were polished and etched in a solution of 2.1 g of picric acid, 5 mL of acetic acid, 5 mL of water, and 35 mL of ethanol. The EBSD measurements were implemented at 20 kV and a step size of 1.0 μm over an area of 460×585 μm^2, and all the data were processed with the TSL OIM^{TM} software (version 7.3, EDAX Inc., Mahwah, NJ, USA). The tensile samples were machined to a dog-bone shape with a gage length of 15 mm, a width of 4 mm, and a thickness of 2 mm. The tensile experiments were carried out at room temperature using an Instron 3382 testing machine with a quasi-static strain rate of 1.0×10^{-3} s^{-1}.

2.2. Finite Element Method

The finite element method (FEM) is an effective way to give constructive suggestions for physical experiments by investigating the deformation behavior of the material [18]. The die parts and billet geometries were designed as shown in Figure 2. DEFORM-3D V 11.2 (SFTC Co., Santa Fe, NM, USA) was used to analyze the deformation behavior of the first pass of CEE-AEC. The stress–strain curves of the AZ31B alloy were imported into the system, and the required database was customized. The molds were built by Unigraphics NX (version 10.0, SIMENS, Berlin, Germany) with the same dimensions as the processed ones, and they were imported through third-party interfaces. The punch and die were set to be rigid bodies, and the billets were defined as plastic objects with 6000 four-node elements. The friction factor was defined as 0.3, the speed of the top die was 1mm/s, and the deformation temperature was 350 °C.

3. Results

3.1. FEM Analysis

Figure 3 shows the FEM results of one pass of CEE-AEC, as well as the microstructures of different zones. It can be seen in Figure 3a that the materials first expanded and filled the chamber, and then the extrusion channel was opened. Three points were selected in outer and central areas to study the strain homogenization. For expansion, materials in the central area showed a higher effective strain than the outer areas. However, the pattern of the effective strain after extrusion changed, as shown in Figure 3b. The highest amounts of effective strain happened in the outer areas. Meanwhile, with the extrusion process continuing, the effective strain in Position 1 (P1) was gradually higher than in P3. This was because that the introduction of the asymmetrical cavity in CEE-AEC changed the distribution of the effective strain. Thus, the asymmetrical deformation area where P1 was located provided strong deformation behavior. Figure 3d–f shows the microstructures corresponding to P1, P2, and P3, respectively. It was clearly seen that the morphologies at different deformation areas showed great differences. More DRX grains nucleated and grew in the outer areas, and P1 showed a more homogeneous grain structure than P3 due to a higher effective strain. In P2, more coarse grains existed together with fine grains and formed a "necklace" structure. Twins were not observed during CEE-AEC due to the high deformation temperature of 350 °C. The critical resolved shear stress of the non-basal slip systems, such as prismatic and pyramidal slips, were easier to activate than the twinning mechanism [19–21]. The highest deformation force occurred in the extrusion process, as shown in Figure 3c. According to the plastic forming law, the peak value of the extrusion force appeared when the material flew through the die land.

3.2. Microstructure Evolutions

Figure 4 shows the OM microstructures of the CEE-AEC-processed samples with different numbers of passes. After two passes of CEE-AEC, a significant grain refinement was obtained, and the differences at different areas obviously decreased. At the asymmetrical cavity and non-asymmetrical cavity zones, which are represented the P1 and P3, respectively, in Figure 3, there was no existence of coarse grains, and a large number of fine grains (lower than 10 μm) were observed. Such was expected from the DRX that occurred in high strain area due to multi-pass severe plastic deformation. At the center zone, more coarse grains occupied the scanned area, and the heterogeneous microstructure was attributed to the lower accumulative strain than outer areas. Furthermore, a serrated grain boundary was observed along the coarse grain shown in Figure 4b, and it could have been the site for nucleation of dynamic grain through bulging [22]. By increasing CEE-AEC process, the grain structure was further refined and distributed in a relatively homogeneous fashion. The problem of the heterogeneity of a material processed by CEE-AEC was overcome due to the multi-pass deformation. We concluded that the condition of three passes of CEE-AEC was the simplest deformation pattern for obtaining a material with a fine grain structure.

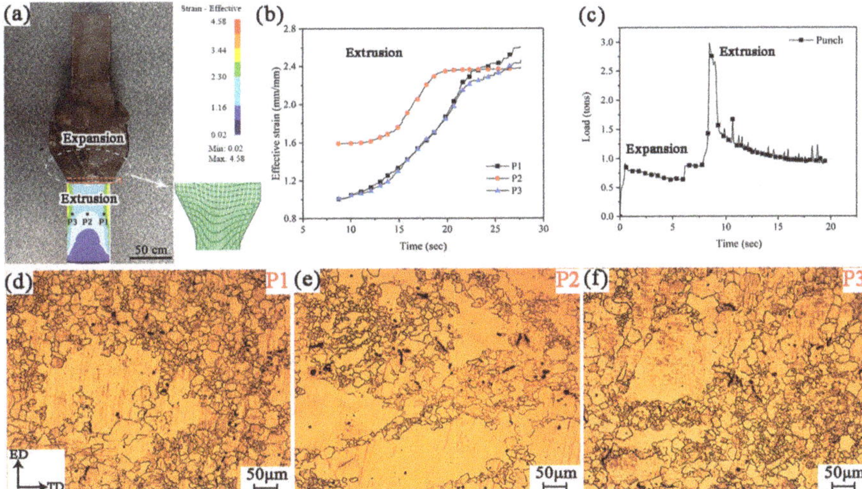

Figure 3. (**a**) Distribution of effective strain; (**b**) plot of effective strain of different deformation zones (Position (P) 1, 2, and 3); (**c**) finite element method (FEM) calculated force versus processing time; and (**d,e,f**) microstructures of 1 pass of CEE-AEC in P1, P2, and P3, respectively.

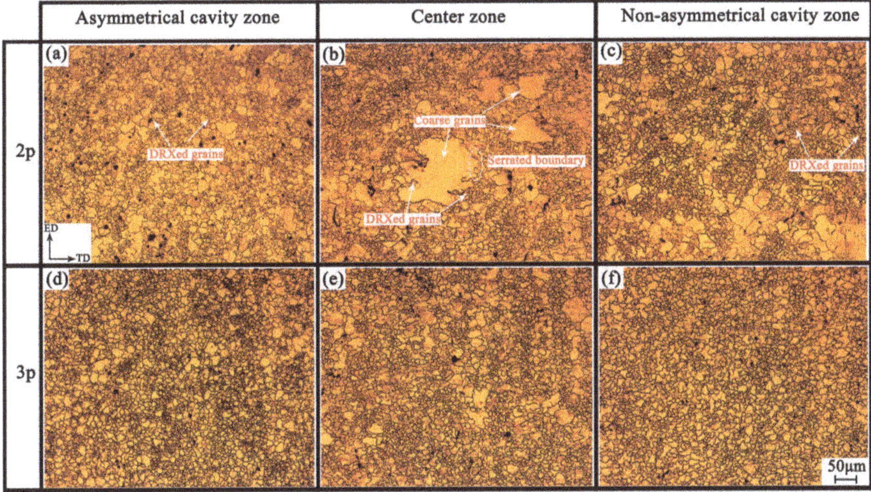

Figure 4. Microstructures of AZ31B alloys processed by CEE-AEC: (**a,b,c**) grain structures of 2 passes of CEE-AEC at asymmetrical cavity zone, center zone and non-asymmetrical cavity zone, respectively; (**d,e,f**) grain structures of 3 passes of CEE-AEC at asymmetrical cavity zone, center zone and non-asymmetrical cavity zone, respectively. (All the figures have the same scale bars).

A fully CEE-AEC-processed grain structure based on EBSD technology after one pass is shown in Figure 5. The sample location for analysis is demonstrated in Figure 5c. The EBSD-based microstructure was relatively similar to what we observed with the OM (Figure 3). Coarse grains were mixed together with fine grains to form the heterogeneous structure. A few fine DRX grains formed along the prior coarse grain boundaries, as marked with the blue circle, G1, shown in Figure 5a, resulting in a "necklace" structure. The average grain size was calculated to be 37.9 ± 1.8 μm. A high peak of misorientation angle distribution was seen at 30°, which is a typical characteristic for recrystallization of Mg [23], as shown in Figure 5b. Galiyev et al. [24] and Yi et al. [25] found that DRX grains of Mg alloys nucleated and grew at 30° around the c-axis of the crystal. There were numerous low angle grain boundaries (LAGBs) in the CEE-AEC-processed material. The LAGBs were primarily distributed in the coarse grains, and many subgrains and fine DRX grains could be observed inside coarse grains, as highlighted by the red arrows in Figure 5d. The high angle grain boundaries (HAGBs) were mainly composed of DRX grain boundaries and some coarse grain boundaries. Figure 5d shows the DRX behavior of the deformed grain G1. It can be noticed that G1 mainly consisted of yellow- and orange-colored regions. The line profile of the point-to-point and point-to-origin along the black line AB indicates that the accumulative misorientation, $\Sigma\theta$, gradually increased up to 20°, thus showing the continuous change of orientation occurring in G1, as shown in Figure 5e. It can be suggested from Figure 5f,g that the basal plane orientations of the two regions had a gradual change of orientation. This can be further confirmed by the three-dimensional diagram of crystallography along AB in Figure 5d. As usually occurs, the LAGBs trapped mobile dislocations and evolved into HAGBs, eventually achieving fine DRX grains. This is a typical continuous dynamic recrystallization (CDRX) mechanism [26]. Additionally, the subgrains were isolated by sub-GBs, and the coarse grains bowed and bulged corresponding to fine DRX grains that formed the serrated grain boundaries, as shown in Figure 5d. This is defined as discontinuous dynamic recrystallization (DDRX).

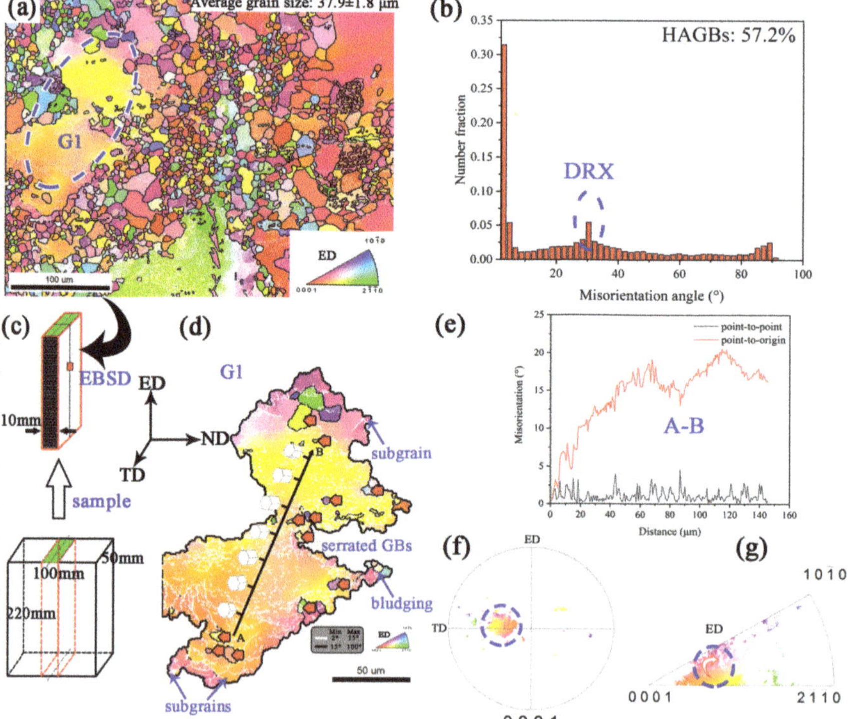

Figure 5. (**a**) Inverse pole figure map of the 1 pass of the CEE-AEC-processed sample, (**b**) misorientation distribution, (**c**) schematic diagram of the location of electron backscatter diffraction (EBSD) measurement within the sample, (**d**) inverse pole figure map of the coarse grain G1, (**e**) line profile of misorientation angle along AB in (**d**), (**f**) the crystallographic orientations in G1, and (**g**) their corresponding inverse pole figures.

Figure 6 shows EBSD-based microstructure of the sample after two passes of CEE-AEC. The grain structure was further refined with an average grain size of 24.7 ± 1.8 μm. The fine recrystallization grains of about 10 μm with HAGBs were evolved near and along coarse grain boundaries. The number fraction of HAGBs was increased to 62.3%. Due to insufficient accumulative strain, orientation gradients were also observed in coarse grain. Herein, some sub-GBs that contained high density dislocations were successfully transformed into HAGBs, resulting in the formation of new GBs, as indicated by the ellipse in Figure 6d. It can be seen that orientation gradient was distinct in coarse grain G2, revealing a high dislocation density in coarse grains of the two passes of deformation; on the other hand, this suggests that CDRX also occurred during the 2 passes of CEE-AEC. The crystallographic orientation of the basal planes of G2 were more inclined than that of the first pass, as shown in Figure 6e,f. This can be attributed to the shear strain that was introduced by attaching an asymmetrical extrusion cavity.

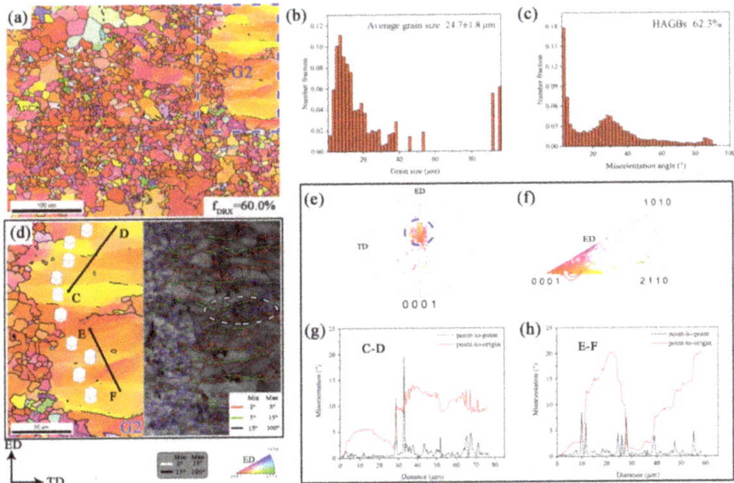

Figure 6. EBSD results of 3 passes of the CEE-AEC-processed sample: (**a**) inverse pole figure map; (**b**) grain size distribution; (**c**) misorientation angle distribution; (**d**) inverse pole figure map of the region in (**a**), G1, and the corresponding crystallographic orientations shown in the (0001) pole figure of (**e**) and the inverse pole figure of (**f**); and (**g**,**h**) line profile of misorientation angle along CD and EF, respectively.

Figure 7 shows the EBSD results of the three passes of the CEE-AEC-processed sample. The DRX fraction markedly increased to 91.0%, but the grain structures were relatively homogeneous, though not equiaxed, as shown in Figure 7a,b. Thus, we know that the DRX process that occurred during the three passes of CEE-AEC did not fully take place, especially in the recrystallized region. For a better understanding of the DRX behavior in the recrystallized region, typical regions R1 and R2 are selected in Figure 7a, and the results of their processing are shown in Figure 7c. It can be seen that fine DRX grains nucleated at serrated GBs and at the triple junctions of the previous DRX grains, indicating the consecutive occurrence of DDRX in the recrystallized region. The DDRX grains had evidently deviated orientations, as can be seen by them surrounding recrystallized grains without preferred selection. Thus, we can conclude that the DDRX played dominant role in the microstructural evolution during three passes of CEE-AEC.

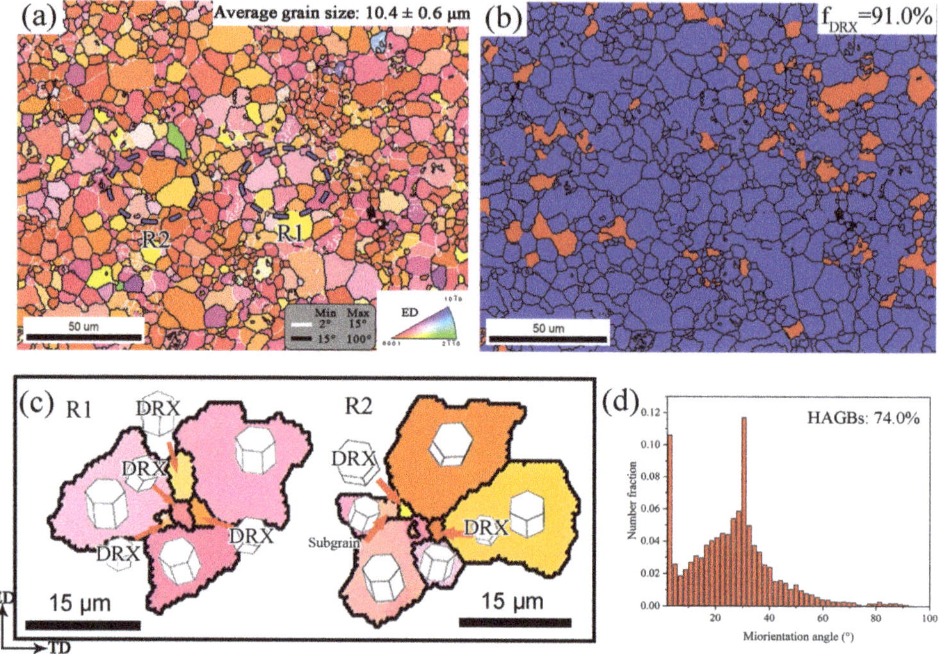

Figure 7. Microstructure of the sample after 3 passes of CEE-AEC: (**a**) inverse pole figure map; (**b**) grain orientation spread map; (**c**) the selected regions R1 and R2 in Figure 7a; and (**d**) misorientation angle distribution.

Table 1 summarizes the evolution of grain size, the DRX fraction, HAGBs, and LAGBs. A significant grain refinement was observed from initial grain size of more than 300 to 10.4 ± 0.6 µm after three passes. Meanwhile, the DRX fraction markedly increased to 91.0% with the increasing number of CEE-AEC passes. Furthermore, the HAGB fraction was observed to continuously increase to 74.0% by the end of the three passes of CEE-AEC, corresponding to a continuous decrease in LAGBs.

Table 1. The changes of main indexes during the CEE-AEC process, including grain size, DRX fraction, high angle grain boundaries (HAGBs), and low angle grain boundaries (LAGBs).

Pass	Grain Size (µm)	DRX Fraction (%)	HAGBs (%)	LAGBs (%)
1	37.9 ± 1.8	42.0	57.2	42.8
2	24.7 ± 1.8	60.0	62.3	37.7
3	10.4 ± 0.6	91.0	74.0	26.0

Figure 8 shows the kernel average misorientation (KAM) maps of the AZ31B samples after different numbers of CEE-AEC passes. Usually, KAMs are used to explain the local residual strain concentration and dislocation density in a processed alloy, and different colors that transition from blue to red represent the increasing degree of strain and dislocation concentration [27]. After one pass of CEE-AEC, as illustrated in Figure 8a, it can be noticed that yellow and green regions mainly occupied the coarse deformed grains and their grain boundaries. Due to the accumulative strain from CEE-AEC, residual strain and dislocations began to generate and accumulate. Meanwhile, the newly formed DRX grains, which are presented in blue, indicated the lower dislocation accumulation compared to that of the deformed grains. The volume fraction of DRX increased with the increasing number of CEE-AEC passes, and a large number of dislocations was consumed in the recrystallization process. A comparatively homogeneous microstructure was observed after three passes of deformation, and the scanned area was occupied by blue areas, thus indicating a much lower strain concentration compared to the first CEE-AEC pass. Thus, the CEE-AEC-processed alloy remained in a steady state, which was good for the following process.

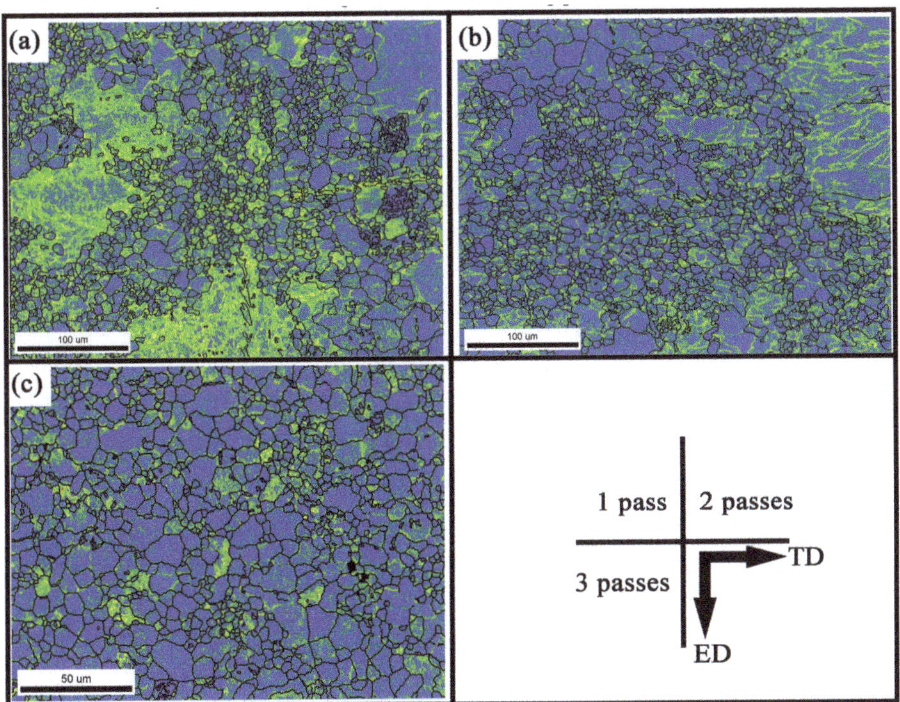

Figure 8. The kernel average misorientation (KAM) maps of (**a**) 1 pass, (**b**) 2 passes, and (**c**) 3 passes of the CEE-AEC-processed sample.

3.3. Texture Evolutions

The (0001) and (10$\bar{1}$0) pole figures of samples based on EBSD results after different numbers of CEE-AEC passes are shown in Figure 9. After one pass of CEE-AEC, as shown in Figure 9a, an inclined basal texture with 0–20° spreading from TD to ED was observed. This was attributed to the shear deformation introduced by the asymmetrical extrusion cavity. The appearance of this non-basal texture component was investigated by Chang et al. [16] and Xu et al. [17,18]. Furthermore, due to the occurrence of DRX, the concentrated basal poles began to spread over a large angular from ED to TD. With further CEE-AEC passes, as shown in Figure 9b,c, the samples showed a more obvious inclined texture, with the (0001) planes of most grains being parallel and inclined ~45° to ED. Moreover, the basal poles were more concentrated, and the maximum intensity increased with the increasing number of CEE-AEC passes.

In order to avoid the limitation of micro-area EBSD texture for understanding the orientation behavior of the AZ31B alloys during the CEE-AEC process, a further investigation of macro-area (0002) pole figure from XRD measurements was carried out, and the results are shown in Figure 10. In general, the global texture presented a similar situation to that of the EBSD-based micro-area texture. The homogenized sample showed a random texture, as shown in Figure 10a. After CEE-AEC, the grains presented a preferred orientation, which was different from the typical fiber texture processed by conventional axisymmetric extrusion. The c-axes of most grains inclined at an angle of 5–45° with the ED. Moreover, due to the increasing volume fraction of DRX, the spreading of the basal planes of most grains and a weak intensity were observed after two passes of CEE-AEC, as shown in Figure 10c. However, the high accumulative strain after three passes of deformation resulted in a further increase of the titling angle and intensity, as shown in Figure 10d.

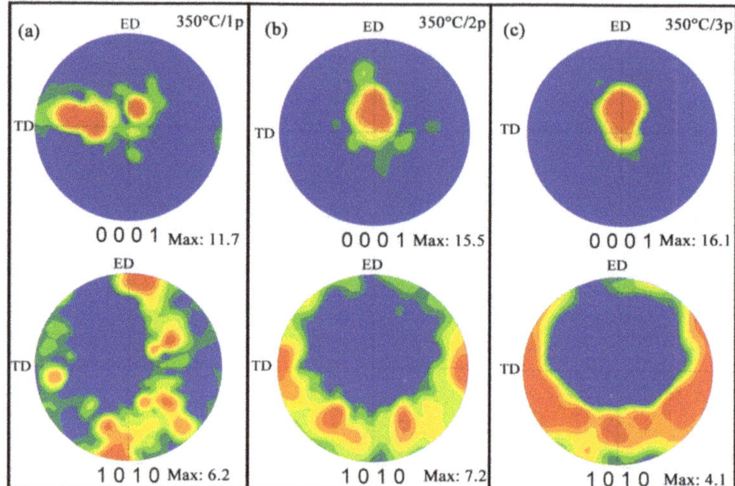

Figure 9. The pole figures of samples processed by CEE-AEC measured by EBSD analysis: (**a**) 1 pass, (**b**) 2 passes, and (**c**) 3 passes.

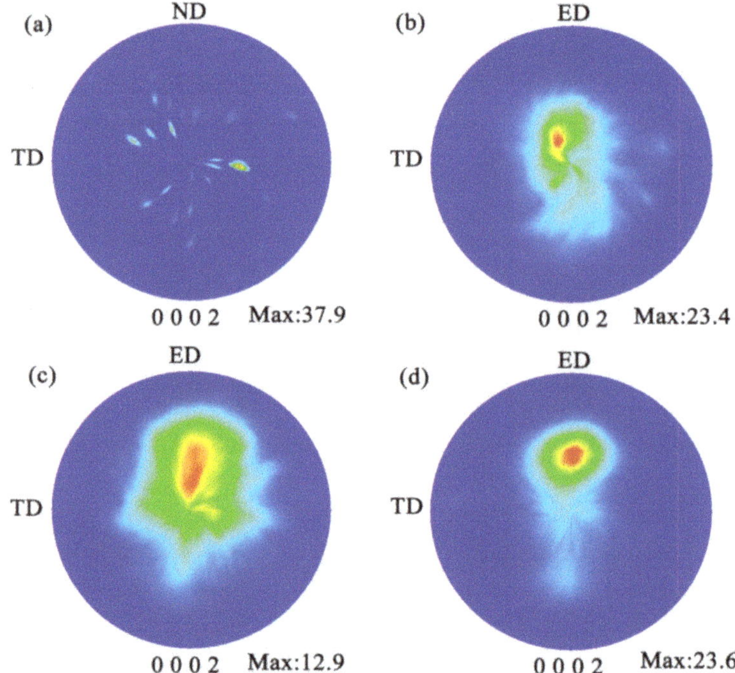

Figure 10. (0002) pole figure based on XRD results of (**a**) the homogenized alloy, (**b**) the alloy after 1 pass, (**c**) the alloy after 2 passes, and (**d**) the alloy after 3 passes of CEE-AEC.

3.4. Mechanical Properties

The room temperature mechanical properties of the as-cast and CEE-AEC-processed samples are shown in Figure 11, and the values of ultimate tensile strength (UTS), tensile yield strength (TYS), and elongation to failure (EF) are listed in Table 2. The abbreviations of "asy," "cen," and "non-asy" represent the selected areas for tensile tests, and they are "asymmetrical area," "center area," and "non-asymmetrical area," respectively. The as-cast alloy showed the lowest tensile properties, both in strength and ductility. The UTS, TYS, and EF were 148 ± 5 MPa, 65 ± 4 MPa, and 15.7 ± 0.8%, respectively. After two passes of CEE-AEC, the mechanical properties showed an obviously increasing tendency. In particular, after one pass, the values of UTS and TYS in the center zone of the CEE-AEC-processed sample presented a dramatic increase to 194 ± 2 and 94 ± 2 MPa, respectively, and the mechanical anisotropy that resulted from the heterogeneous microstructure showed better tensile properties in the asymmetrical area than in the non-asymmetrical area; the center area had the worst tensile properties. After three passes of CEE-AEC, the UTS and TYS remained steady in comparison those after two passes, while the EF increased rapidly to 30.0 ± 1.3%. From the observation of microstructure and the calculation of average grain size, it was concluded that the superior mechanical properties corresponded to the fine and uniform grain structure. The specific mechanical mechanisms are discussed in the following section.

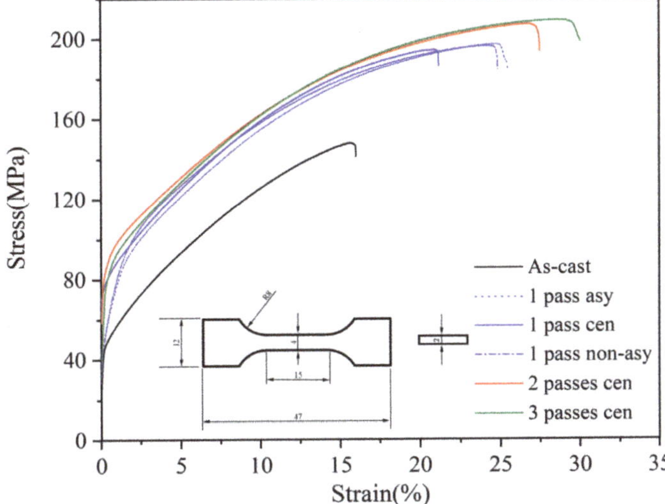

Figure 11. Engineering stress–strain plots for AZ31B alloys with and without CEE-AEC processing at room temperature (The different lines represent different sampling positions and states of AZ31B alloys processed by CEE-AEC, and the dog-bone shaped tensile sample is shown in the bottom of the figure).

Table 2. Room temperature tensile properties of AZ31B alloys without and with CEE-AEC passes. Asy: asymmetrical area; cen: center area; non-asy: non-asymmetrical area; UTS: ultimate tensile strength; TYS: tensile yield strength; EF: elongation to failure.

State	UTS (MPa)	TYS (MPa)	EF (%)	Grain Size (μm)
As-cast	148 ± 5	65 ± 4	15.7 ± 0.8	>300
1 pass asy	198 ± 5	97 ± 3	25.2 ± 1.6	28.6 ± 1.4
1 pass cen	194 ± 2	94 ± 2	21.3 ± 1.2	37.9 ± 1.8
1 pass non-asy	196 ± 3	96 ± 3	24.5 ± 1.3	33.2 ± 1.7
2 passes cen	206 ± 5	112 ± 3	27.2 ± 1.7	24.7 ± 1.8
3 passes cen	209 ± 2	115 ± 4	30.0 ± 1.3	10.4 ± 0.6

4. Discussion

The experiment results of the CEE-AEC provide important information concerning the effects of grain refinement and texture modification on the alloy's mechanical properties. Hence, understanding the relationship and competitive mechanisms of the microstructure and texture evolution of Mg alloys prepared by CEE-AEC is of significant importance for designing a test to obtain samples with superior mechanical properties in the future.

4.1. Grain Refinement during CEE-AEC

Several mechanisms for an Mg alloy to achieve grain refinement have been reported in the literature. Li et al. [27] proposed that the grain structure evolution of Mg–Gd–Y–Zn–Zr alloys during ECAP is due to the transformation from LAGBs into HAGBs, as well as the particle-stimulated nucleation (PSN) mechanism. Miura et al. [28] explained that the grain refinement mechanisms of AZ61 alloys fabricated by multidirectional forging (MDF) were determined by temperature. Twinning and CDRX were combined to refine grain structures during MDF processing with a decreasing temperature. Zhang et al. [29] investigated the DRX behavior of an Mg–Zn–Y–Zr alloy using hot compression tests. They found that DDRX firstly occurred on the original grain boundaries. As further deformation was carried out, CDRX took place inside the original grains. Ma et al. [30] reported twinning-induced dynamic recrystallization in an Mg alloy.

In our work, the dynamic recrystallization mechanism was the dominant mechanism in grain refinement. Twinning and deformation bands were rarely found in the evolved microstructures due to the high deformation temperature.

The dynamic recrystallization mechanisms in Mg alloys can be divided into two groups: CDRX and DDRX. The continuous absorption of dislocations in sub-GBs is the main characteristic of CDRX, while DDRX grains nucleate and grow via the migration of HAGBs. As discussed above, it can be concluded that CDRX and DDRX play important roles in grain refinement during CEE-AEC. Figure 12 shows DRX behavior in different stages.

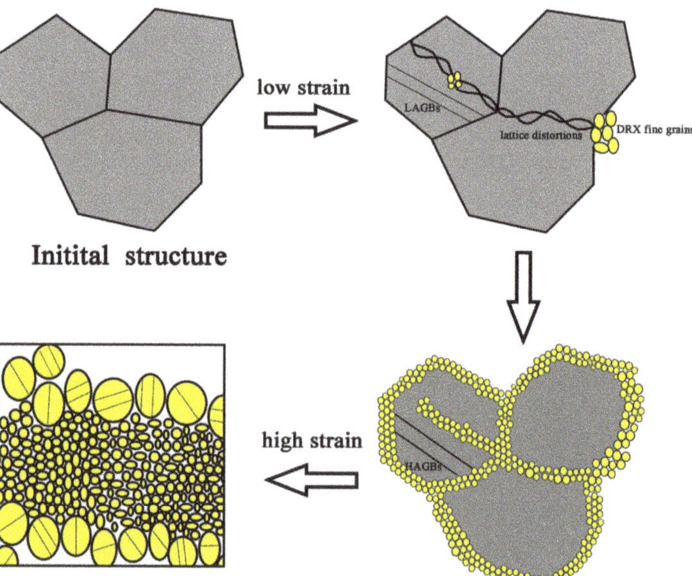

Figure 12. Schematic diagram of grain refinement during the CEE-AEC process.

In first stage of the CEE-AEC process, DRX fine grains with grain size smaller than 10 μm were observed, both within the coarse grains and along serrated grain boundaries. Discontinuous dynamic recrystallized grains showed no distinct gradients, either in recrystallized grains or in the original grains, while continuous dynamic recrystallized grains resulted from the progressive rotation of subgrains adjacent to preexisting grain boundaries. Thus, it can be concluded that the dominant mechanism of Mg alloys prepared by CEE-AEC to achieve grain refinement at low and medium strains is a combination of CDRX and DDRX.

At second stage of the CEE-AEC process with higher accumulative strain (three passes of CEE-AEC), the coarse grains were consumed by recrystallized grains, but they were not equiaxial. Fine grains could also be observed at recrystallized regions, especially at the triple junctions of the DRX grains. Thus, it can be known that the CDRX mechanism is not a dominant force refining grain structure, and DDRX determines further grain refinement.

4.2. Texture Evolution Mechanism

The ductility of an Mg alloy at room temperature is mainly related to its activated slips. In Mg alloys, basal <a> slip systems predominate, although the prismatic slip operated for grains is oriented to suppress the basal slip. The modification of the texture component during the SPD process can affect the activation of basal <a> slips. Kim et al. [11,31] and Tong et al. [32] found that the shear deformation provided from ECAP could inform the basal inclination texture, which was proven to be helpful in improving the ductility.

In the present work, the introduction of an asymmetrical extrusion cavity changed the crystallographic orientation of most grains, and a mixed texture component was obtained. Figure 13 shows a schematic image of the texture evolution during CEE-AEC. The red color represents a more concentrated orientation distribution, and the blue area represents a dispersion texture. As seen in Figure 13a, the initial structure was mainly composed of random orientated grains, and the random texture was remarkably changed after one pass of CEE-AEC, as shown in Figure 13b. An obvious mixed texture component occurred, with basal planes parallel and inclined from TD to ED, due to the shear deformation at the asymmetrical extrusion cavity. The phenomenon of TD spread has been reported in many studies and may be related to the DRX mechanism, which intensifies the dispersion of basal textures [14,22,27,33]. After three passes, as shown in Figure 13c, the orientations of most grains were redistributed, and the basal planes were inclined ~45° to ED.

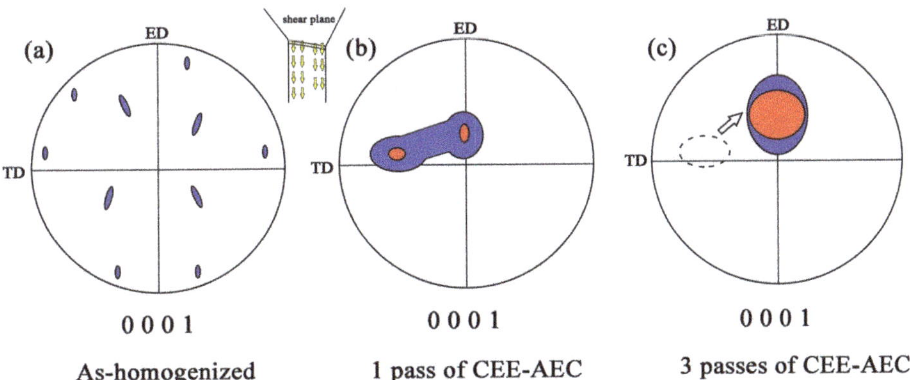

Figure 13. Schematic illustration of the dominant textures during CEE-AEC: (**a**) the as-homogenized; (**b**) 1 pass of CEE-AEC; (**c**) 3 passes of CEE-AEC.

4.3. Strengthening Mechanisms

In Figure 14, the tensile yield strength is plotted against $d^{-1/2}$ for the as-cast and CEE-AEC-processed samples. The solid circle indicates the data for the cast material, and the solid squares represent the points for the CEE-AEC-processed samples. The standard Hall–Petch relationship with a positive slope was valid in the data for the samples from the state of homogenization to two passes of CEE-AEC. Meanwhile, these points were well-correlated to a single line. For the three-pass sample, however, the slope of tensile yield stress versus $d^{-1/2}$ obviously declined. Kim et al. [11] found that the tensile stress of Al alloys subject to ECAP is mainly determined by grain size, but Mg alloys subject to ECAP did not behave in the same way. In subsequent studies, they found that the tensile stress of Mg alloys subject to ECAP was notably lower than that of the extruded alloys with the same grain size; texture modification was responsible for these results.

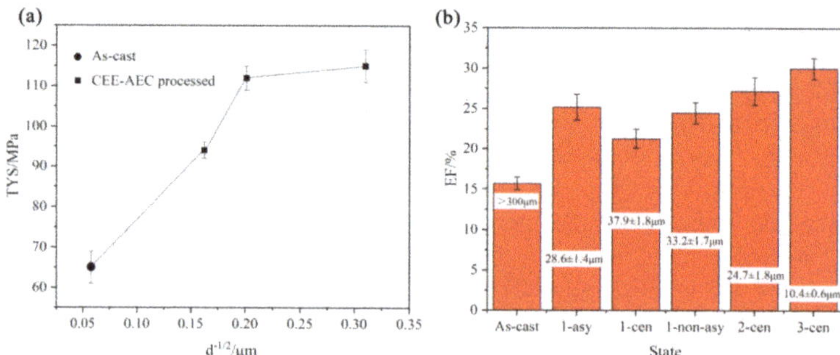

Figure 14. Schematic illustration of (**a**) the relationship between TYS and d-1/2, and (**b**) the EF of different sampling positions with different passes.

In general, the room temperature mechanical properties of the AZ31 alloys in the present research were mainly dependent on the grain size and texture modification. From Table 2 and Figure 14, it can be seen that the TYS increased and the average grain size decreased with the increasing number of CEE-AEC passes. The appearance of non-standard Hall–Petch relationship demonstrated the competition of grain refinement strengthening and texture modification.

During the two-pass deformation of CEE-AEC, an obvious linear increase relation in yield strength and a linear decrease relation in grain sizes were observed, as shown in Figure 14. Moreover, the sample after two passes of CEE-AEC presented an embryonic form of the basal inclination texture. This suggested that the determination of TYS in the first two passes of CEE-AEC was predominately controlled by grain refinement strengthening. Additionally, another piece of evidence that proves this conclusion was the relationship between TYS and grain sizes in different processing areas. As shown in Figure 11, the finer grain structure in asymmetrical areas showed better properties than the non-asymmetrical and center areas.

The CEE-AEC-processed AZ31B alloys, after three passes with a grain size of 10.4 ± 0.6 μm, had almost the same TYS as the sample that went through two passes. This was because the basal texture of the CEE-AEC-processed AZ31 alloy showed an inclination of ~45° from TD to ED. The basal inclination texture made grains with soft orientations for the operation of basal <a> slips at room temperature. Figure 15 shows the EBSD analysis of the Schmid factor for the (0001) $<11\bar{2}0>$ slip system of the CEE-AEC-processed samples loading along the ED. The higher Schmid factor value meant a lower tensile yield strength in the same conditions. This was because a high Schmid factor leads to a low stress for the activation of basal slip and material yielding [34]. From the quantitative analysis of Schmid factor during CEE-AEC, as shown in Figure 15, it can be seen that an increasing value of Schmid factor from 0.22 to 0.32 was obtained due to the shear deformation, which redistributed the texture component and formed a typical texture with inclined c-axis of most grains towards ED. The sample after three passes of CEE-AEC was more favorable for the dislocation glide on (0001) plane.

Thus, it can be concluded that the determination of TYS of the sample after three passes of deformation was mainly controlled by texture modification rather than grain refinement strengthening.

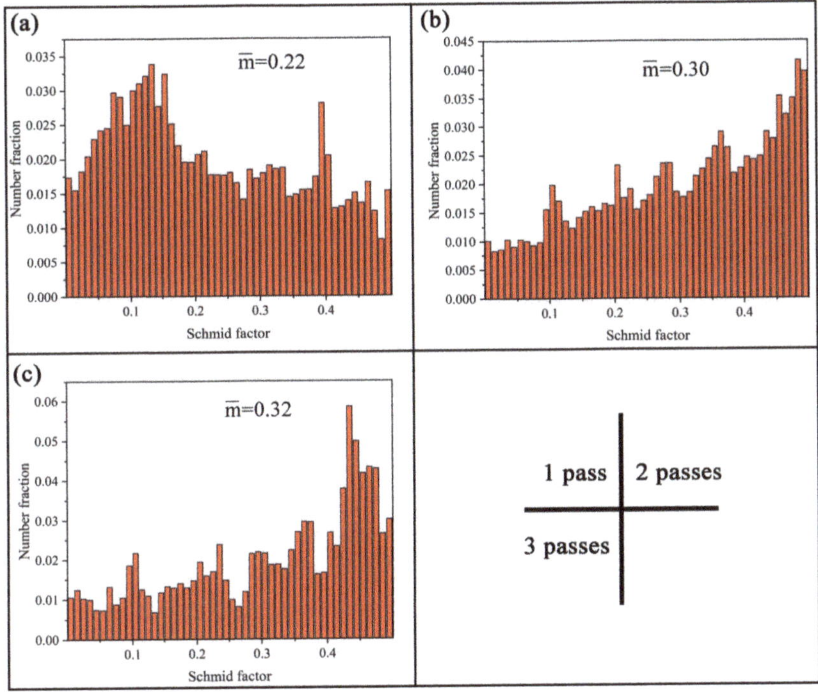

Figure 15. Distribution of the Schmid factor of basal slip of samples after different numbers of CEE-AEC passes: (**a**) 1 pass, (**b**) 2 passes, and (**c**) 3 passes.

It was clearly seen that the elongation the AZ31B alloys processed by CEE-AEC was remarkably improved, achieving a good balance of strength and ductility. A fractography analysis is shown in Figure 16. A lot of cleavage steps and tear ridges were observed on the surface of the as-cast alloy, implying that the main fracture mechanism was cleavage damage, which resulted in a poor ductility, as shown in Figure 16a. After one pass of CEE-AEC, the refined grain structure and the increasing volume fraction of the fine DRX grains effectively hindered the cleavage damage. The number of cleavage steps and tear ridges obviously decreased, and many dimples were observed on the fracture surfaces, representing cleavage mixed with ductile fracture behavior, as shown in Figure 16b. With the increasing passes, the coarse grains were consumed, and a more homogeneous grain structure with a basal inclination texture was obtained. Meanwhile, the increasing Schmid factor value promoted the activation of the basal <a> slip system. As shown in Figure 16c,d, the fracture surfaces were occupied by a large number of dimples, thus showing a typical ductile fracture behavior. The improvement of the ductility of the CEE-AEC-processed alloy was related to grain refinement and the formation of a basal inclination texture. The increasing volume fraction grain boundary restricted the required energy of crack growth, and the activation of the basal slip system promoted the slide of the high density of dislocation formed during the tensile experiments.

The CEE-AEC-processed AZ31B alloys were found to achieve a dramatic grain refinement from >300 to 10.4 ± 0.6 μm for the as-cast to the three-passed alloys, respectively. The refinement degree was more than ~96%. Moreover, the TYS and ductility were increased by ~95% and ~100%, respectively. The excellent elongation of the sample after three passes of deformation (30.0 ± 1.3%) was mainly dominated by a refined and homogeneous grain structure, as well as the formation of a basal inclination texture. The introduction of an asymmetrical extrusion cavity effectively improved the homogeneity of the microstructure and modified the texture component.

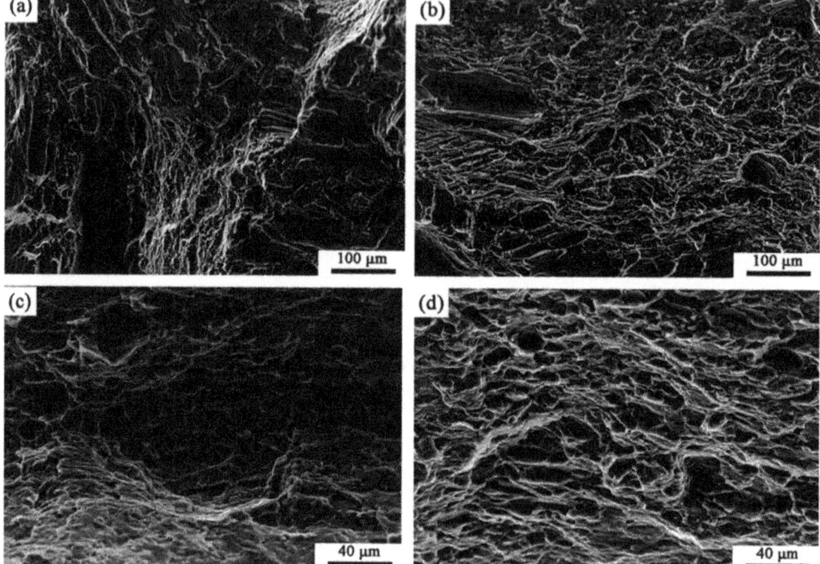

Figure 16. The fracture morphologies of the AZ31B alloys: (**a**) as-cast, CEE-AEC-processed for 1 pass (**b**), 2 passes (**c**) and 3 passes (**d**).

5. Conclusions

In present research, AZ31B alloys with homogeneous grain structures and enhanced ductility were successfully fabricated by a novel SPD technology: CEE-AEC. The corresponding DRX behavior, texture evolution and strengthening mechanisms were systematically discussed. The following main observations occurred in this study.

(1). CEE-AEC could effectively refine grain structure and improve microstructure homogeneity, mainly via DRX. CDRX and DDRX acted on the first two passes of deformation, and DDRX further refined the microstructure after three passes. Finally, a homogenous microstructure with an average grain size of 10.4 ± 0.6 μm was observed, and the grain refinement degree in comparison with the as-cast alloys was more than ~96%.

(2). The texture characterization after CEE-AEC revealed a basal inclination texture, with (0001) planes inclined ~45° to the ED. With further CEE-AEC processing, the typical basal texture rotation resulted in an asymmetric texture corresponding to an increasing Schmid factor.

(3). The tensile properties of the CEE-AEC-processed samples were remarkably improved, with the UTS, TYS, and EF of the as-cast alloys being 148 ± 5 MPa, 65 ± 4 MPa, 15.7 ± 0.8%, respectively, and the same qualities of three passes being 209 ± 2 MPa, 115 ± 4 MPa, and 30.0 ± 1.3%, respectively. The formed basal inclination texture facilitated the activation of basal <a> slip systems, and the TYS of the three pass-deformation showed a slight improvement in comparison with the two pass-deformation.

(4). Grain refinement strengthening and texture modification were the main strengthening mechanisms. The competition between the two mechanisms influenced the whole CEE-AEC process. During the first two passes of deformation, grain refinement was the dominant way of improving the tensile strength. Texture modification determined the high ductility after three passes of deformation.

6. Patents

In order to protect the intellectual property rights, the CEE-AEC technology is patented with the number of CN201910324749.4 of China.

Author Contributions: Z.Z. and Y.X. designed the experiments; Z.Y. and J.Z. (Jie Zheng) performed the experiments and collected the data; Z.Y., Q.W., and J.Z. (Jiaxuan Zhu) analyzed the data; Z.Y. wrote the paper. All authors have read and agreed to the published version of the manuscript.

Funding: The present study was supported by the National Natural Science Foundation of China (NSFC) under Grant no. 51775520 and no. 51675492, the Natural Science Foundation of Shanxi Province under Grant no. 201801D121106, the Key R&D program of Shanxi Province (International Cooperation) under grant No. 201903D421036, and the Scientific and Technological Innovation Programs of Higher Education Institutions in Shanxi under grant No. 2018002.

Conflicts of Interest: The authors declare that they have no conflict of interest.

References

1. Foley, D.; Al-Maharbi, M.; Hartwig, K.T.; Karaman, I.; Kecskes, L.J.; Mathaudhu, S. Grain refinement vs. crystallographic texture: Mechanical anisotropy in a magnesium alloy. *Scr. Mater.* **2011**, *64*, 193–196. [CrossRef]
2. Zheng, J.; Yan, Z.; Yu, J.; Zhang, Z.; Fan, H.; Xu, K.; Xue, Y. Microstructure and mechanical properties of Mg-Gd-Y-Zn-Zr alloy by cyclic expansion-extrusion with an asymmetrical extrusion cavity (CEE-AEC). *Mater. Res. Express* **2019**, *6*, 1065c8. [CrossRef]
3. Yan, Z.; Fang, M.; Lian, Z.; Zhang, Z.; Zhu, J.; Zhang, G.; Wang, Y. Research on AZ80 + 0.4%Ce (wt%) Ultra-Thin-Walled Tubes of Magnesium Alloys: The Forming Process, Microstructure Evolution and Mechanical Properties. *Metals* **2019**, *9*, 563. [CrossRef]
4. Xue, Y.; Chen, S.; Liu, H.; Zhang, Z.; Ren, L.; Bai, B. Effect of Cyclic Expansion-Extrusion Process on Microstructure, Deformation and Dynamic Recrystallization Mechanisms, and Texture Evolution of AZ80 Magnesium Alloy. *Adv. Mater. Sci. Eng.* **2019**, *2019*, 1–10. [CrossRef]
5. Zhang, Z.; Yan, Z.; Du, Y.; Zhang, G.; Zhu, J.; Ren, L.; Wang, Y. Hot Deformation Behavior of Homogenized Mg-13.5Gd-3.2Y-2.3Zn-0.5Zr Alloy via Hot Compression Tests. *Materials* **2018**, *11*, 2282. [CrossRef] [PubMed]
6. Yan, Z.M.; Zhang, Z.M.; Du, Y.; Zhang, G.S.; Ren, L.Y. Effect of homogenization treatment on microstructure and mechanical properties of Mg-13Gd-3.5Y-2Zn-0.5Zr magnesium alloy. *J. Mater. Eng.* **2019**, *47*, 93–99.
7. You, S.; Huang, Y.; Kainer, K.U.; Hort, N. Recent research and developments on wrought magnesium alloys. *J. Magnes. Alloy.* **2017**, *5*, 239–253. [CrossRef]
8. Valiev, R.Z.; Islamgaliev, R.; Alexandrov, I. Bulk nanostructured materials from severe plastic deformation. *Prog. Mater. Sci.* **2000**, *45*, 103–189. [CrossRef]
9. Suh, J.; Victoria-Hernández, J.; Letzig, D.; Golle, R.; Volk, W. Effect of processing route on texture and cold formability of AZ31 Mg alloy sheets processed by ECAP. *Mater. Sci. Eng. A* **2016**, *669*, 159–170. [CrossRef]
10. Lee, H.-J.; Lee, S.K.; Jung, K.H.; Lee, G.A.; Ahn, B.; Kawasaki, M.; Langdon, T.G. Evolution in hardness and texture of a ZK60A magnesium alloy processed by high-pressure torsion. *Mater. Sci. Eng. A* **2015**, *630*, 90–98. [CrossRef]
11. Kim, W.; Hong, S.; Kim, Y.; Min, S.; Jeong, H.; Lee, J. Texture development and its effect on mechanical properties of an AZ61 Mg alloy fabricated by equal channel angular pressing. *Acta Mater.* **2003**, *51*, 3293–3307. [CrossRef]
12. Stráská, J.; Janeček, M.; Gubicza, J.; Krajňák, T.; Yoon, E.Y.; Kim, H.S. Evolution of microstructure and hardness in AZ31 alloy processed by high pressure torsion. *Mater. Sci. Eng. A* **2015**, *625*, 98–106. [CrossRef]
13. Iwahashi, Y.; Wang, J.; Horita, Z.; Nemoto, M.; Langdon, T.G. Principle of equal-channel angular pressing for the processing of ultra-fine grained materials. *Scr. Mater.* **1996**, *35*, 143–146. [CrossRef]
14. Yan, Z.; Zhang, Z.; Li, X.; Xu, J.; Wang, Q.; Zhang, G.; Zheng, J.; Fan, H.; Xu, K.; Zhu, J.; et al. A novel severe plastic deformation method and its effect on microstructure, texture and mechanical properties of Mg-Gd-Y-Zn-Zr alloy. *J. Alloys Compd.* **2020**, *822*, 153698. [CrossRef]
15. Beygelzimer, Y.; Kulagin, R.; Estrin, Y.; Toth, L.S.; Kim, H.S.; Latypov, M.I. Twist Extrusion as a Potent Tool for Obtaining Advanced Engineering Materials: A Review. *Adv. Eng. Mater.* **2017**, *19*, 1600873. [CrossRef]

16. Chang, L.; Wang, Y.; Zhao, X.; Huang, J.C. Microstructure and mechanical properties in an AZ31 magnesium alloy sheet fabricated by asymmetric hot extrusion. *Mater. Sci. Eng. A* **2008**, *496*, 512–516. [CrossRef]

17. Xu, J.; Jiang, B.; Song, J.; He, J.; Gao, P.; Liu, W.; Yang, T.; Huang, G.; Pan, F. Unusual texture formation in Mg–3Al–1Zn alloy sheets processed by slope extrusion. *Mater. Sci. Eng. A* **2018**, *732*, 1–5. [CrossRef]

18. Xu, J.; Yang, T.B.; Jiang, B.; Song, J.F.; He, J.J.; Wang, Q.H.; Chai, Y.F.; Huang, G.S.; Pan, F.S. Improved mechanical properties o Mg-3Al-1Zn alloy sheets by optimizing the extrusion die angles: Microstructural and texture evolution. *J. Alloys Compd.* **2018**, *762*, 719–729. [CrossRef]

19. Liu, X.-Y.; Lu, L.; Sheng, K.; Zhou, T. Microstructure and Texture Evolution during the Direct Extrusion and Bending–Shear Deformation of AZ31 Magnesium Alloy. *Acta Metall. Sin. (Engl. Lett.)* **2018**, *32*, 710–718. [CrossRef]

20. Beausir, B.; Suwas, S.; Toth, L.S.; Neale, K.W.; Fundenberger, J.-J. Analysis of texture evolution in magnesium during equal channel angular extrusion. *Acta Mater.* **2008**, *56*, 200–214. [CrossRef]

21. Su, C.W.; Lu, L.; Lai, M.O. Mechanical behaviour and texture of annealed AZ31 Mg alloy deformed by ECAP. *Mater. Sci. Technol.* **2007**, *23*, 290–296. [CrossRef]

22. Zhang, Z.; Meng, Y.; Yan, F.; Gao, Z.; Yan, Z.; Zhang, Z. Microstructure and texture evolution of Mg-RE-Zn alloy prepared by repetitive upsetting-extrusion under different decreasing temperature degrees. *J. Alloys Compd.* **2020**, *815*, 152452. [CrossRef]

23. Lei, W.; Liang, W.; Wang, H.; Sun, Y. Effect of annealing on the texture and mechanical properties of pure Mg by ECAP at room temperature. *Vacuum* **2017**, *144*, 281–285. [CrossRef]

24. Galiyev, A.; Kaibyshev, R.; Gottstein, G. Correlation of plastic deformation and dynamic recrystallization in magnesium alloy ZK60. *Acta Mater.* **2001**, *49*, 1199–1207. [CrossRef]

25. Yi, S.-B.; Zaefferer, S.; Brokmeier, H.-G. Mechanical behaviour and microstructural evolution of magnesium alloy AZ31 in tension at different temperatures. *Mater. Sci. Eng. A* **2006**, *424*, 275–281. [CrossRef]

26. Jiang, M.; Xu, C.; Yan, H.; Fan, G.; Nakata, T.; Lao, C.; Chen, R.; Kamado, S.; Han, E.-H.; Lu, B. Unveiling the formation of basal texture variations based on twinning and dynamic recrystallization in AZ31 magnesium alloy during extrusion. *Acta Mater.* **2018**, *157*, 53–71. [CrossRef]

27. Li, B.; Teng, B.; Chen, G. Microstructure evolution and mechanical properties of Mg-Gd-Y-Zn-Zr alloy during equal channel angular pressing. *Mater. Sci. Eng. A* **2019**, *744*, 396–405. [CrossRef]

28. Miura, H.; Yu, G.; Yang, X. Multi-directional forging of AZ61Mg alloy under decreasing temperature conditions and improvement of its mechanical properties. *Mater. Sci. Eng. A* **2011**, *528*, 6981–6992. [CrossRef]

29. Zhang, Y.; Zeng, X.; Lu, C.; Ding, W. Deformation behavior and dynamic recrystallization of a Mg–Zn–Y–Zr alloy. *Mater. Sci. Eng. A* **2006**, *428*, 91–97. [CrossRef]

30. Ma, Q.; Li, B.; Marin, E.B.; Horstemeyer, S.J. Twinning-induced dynamic recrystallization in a magnesium alloy extruded at 450 °C. *Scr. Mater.* **2011**, *65*, 823–826. [CrossRef]

31. Kim, W.J.; Kim, J.K.; Park, T.Y.; Hong, S.I.; Kim, D.I.; Kim, Y.S.; Lee, J.D. Enhancement of strength and superplasticity in a 6061 Al alloy processed by equal-channel-angular-pressing. *Metall. Mater. Trans. A* **2002**, *33*, 3155–3164. [CrossRef]

32. Tong, L.; Chu, J.; Jiang, Z.; Kamado, S.; Zheng, M. Ultra-fine grained Mg-Zn-Ca-Mn alloy with simultaneously improved strength and ductility processed by equal channel angular pressing. *J. Alloys Compd.* **2019**, *785*, 410–421. [CrossRef]

33. Tang, L.; Liu, C.; Chen, Z.; Ji, D.; Xiao, H. Microstructures and tensile properties of Mg–Gd–Y–Zr alloy during multidirectional forging at 773K. *Mater. Des.* **2013**, *50*, 587–596. [CrossRef]

34. Mostaed, E.; Fabrizi, A.; Dellasega, D.; Bonollo, F.; Vedani, M. Microstructure, mechanical behavior and low temperature superplasticity of ECAP processed ZM21 Mg alloy. *J. Alloys Compd.* **2015**, *638*, 267–276. [CrossRef]

Article

On the Direct Extrusion of Magnesium Wires from Mg-Al-Zn Series Alloys

Maria Nienaber *, Sangbong Yi, Karl Ulrich Kainer, Dietmar Letzig and Jan Bohlen *

Magnesium Innovation Centre (MagIC), Helmholtz-Zentrum Geesthacht, Max-Planck-Str. 1, 21502 Geesthacht, Germany; sangbong.yi@hzg.de (S.Y.); karl.kainer@hzg.de (K.U.K.); dietmar.letzig@hzg.de (D.L.)
* Correspondence: Maria.Nienaber@hzg.de (M.N.); Jan.Bohlen@hzg.de (J.B.)

Received: 14 August 2020; Accepted: 7 September 2020; Published: 9 September 2020

Abstract: Wires of magnesium alloys possess a high potential, e.g., as filler materials, for joining applications but also for biodegradable applications, such as suture materials. While the typical process of producing wires is based on a wire drawing process, direct extrusion by using adjusted dies to deal with high degrees of deformation allows a one-step manufacturing of wires to some extent. In this work, the extrusion of wires with a thickness of 1 mm and even lower is shown feasible for pure magnesium and three Al-containing magnesium alloys (AZ31, AZ80, AZ91). The surface quality and the mechanical properties are improved with increasing Al content. It is shown that, despite the large difference in the degrees of deformation, the properties and their development are similar to those of extruded round bars. Wrapping tests were carried out as an exemplary more complex forming procedure, and the behavior is correlated to the microstructure and texture of the extruded wires.

Keywords: magnesium wire; extrusion; characterization; mechanical properties; wrapping test; AZ-series

---

1. Introduction

Thin magnesium alloy wires have been used in the form of welding wire as filler materials for enabling same- and multi-material joining of parts [1]. In such cases, the mechanical properties of the wires may not be of great interest as long as the processing of the wires during the joining procedure can be ensured. A growing interest in the use of wires from magnesium alloys is driven by the potential application in the biomedical sector, e.g., in the form of sutures for wound closure purposes [2]. In such cases, the implementation of this class of materials is emphasized where removal operations can be avoided due to the natural degradation properties of magnesium alloys, as well as their compatibility with the body. Then, the mechanical performance, especially ductility and formability, of the wires is prerequisite for the suitable application of the wires.

A commonly used method to produce thin wires is the application of drawing processes to thin extruded bars [3–8]. Typically, several drawing passes are involved, including intermediate annealing, i.e., softening, of the wires. This schedule allows the preparation of wires with thicknesses even below 0.1 mm [9,10]. In this process, single processing steps can be adjusted to the individual needs of the respective alloys, including the establishment of clean and homogeneous surfaces of the wires.

Few studies so far have investigated the direct extrusion of wires to the required product thickness [11–13]. The challenge in such a procedure is the very high degree of deformation, which is applied to the wire from a typical sized cast billet used for extrusion. This will especially limit the ability to reach very small thicknesses for the wires. Furthermore, the direct establishment of a smooth surface depends directly on the forming behavior of the respective alloys at the preset extrusion parameters. This surface condition has a profound impact on the resulting corrosion behavior, also associated with Fe inclusions on the surfaces during processing [14]. However, the advantage of this method is the manufacturing of the final product in one single processing step. Alternatively, such extruded wires may serve well as feedstock for a concurrent wire drawing process, therefore easing the following processing effort.

With respect to the forming process, the extruded wire has a qualitatively comparable deformation history like an extruded round bar and will therefore behave in a comparable way, despite the high degree of deformation. The extrusion of round bars from magnesium alloys has been studied widely [15]. Property relevant findings include the process parameter related grain structure development, i.e., the grain size and the crystallographic texture. For example, Liu et al. [16] demonstrated the influence of the extrusion parameters, e.g., temperature T and speed v, on the properties of the extruded bars for AZ31 alloy. It was found that the mechanical properties of extruded bars are more influenced by the extrusion speed rather than by the temperature. In addition, higher speeds are associated with a decline of the strength and an increased elongation of the bars. The influence of the extrusion ratio was investigated by Shahzad et al. [17] for AZ80 alloy where a higher degree of recrystallization is shown at a higher ratio, which also leads to a weakening of the texture. Using recycled AZ91, Hu et al. [18] investigated the influence of the degree of deformation on the mechanical properties. They were able to show that a higher degree of deformation leads to higher strength, which has been attributed to a finer grained microstructure.

Microstructures of magnesium alloy extrusions are mostly the result of partly or full recrystallization due to the massive deformation applied during processing. The corresponding grain structure largely depends on the impact of dynamic recrystallization and grain growth. As this impact is obviously temperature dependent, it also changes with the applied extrusion speed. An increase of the extrusion speed results in higher deformation related heating and, therefore, higher forming temperatures and corresponding grain growth [19–21].

The texture development typically results in a fiber texture with a prismatic $<10\bar{1}0>$ component parallel to the extrusion direction (ED) [22,23]. This texture is especially distinct in cases with partly recrystallized microstructure, where the unrecrystallized fraction reveals the deformation texture [21,24,25]. Fully recrystallized microstructures tend to correspond to a tilting component of the above with a rotation of up to 30° around the c-axis. Then, the orientations are concentrated around the $<2\bar{1}\,\bar{1}0>$ pole parallel to ED [26], or they often seen as an intensity distribution between the two mentioned poles. In all cases, these textures represent a preference for basal planes aligned parallel to ED. A general variation of this type of texture development has been found with alloys containing rare earth elements or Ca [27]. In such cases, enhanced non-basal slip and retarded recrystallization allow textures with distinct tilt component for the basal planes [21,24,28–31]. Then, enhanced basal slip and corresponding higher strain hardening ability allows a significant enhancement of the ductility of the extruded alloys. However, the typical texture with aligned basal planes parallel to the extrusion direction limits the ability of strain hardening in tension and favors twinning in compression along the extrusion direction. The latter leads to decreasing yield stresses due to its preferred activation, as well as distinctly higher strain hardening rates and therefore limited ductility [32]. Thus, a distinct anisotropy of the mechanical properties is very typical of such magnesium alloy extrusions.

For extruded Al-containing bars, it has been demonstrated that the aluminium content and the processing (degree of deformation and process parameters) have an influence on the microstructure and/or texture development and consequently also on the mechanical properties [16,17]. It has also been shown that there is a correlation between strength and grain size [33,34].

In the case of extruded wires, the degree of deformation is several times higher and, consequently, also the resulting exit speeds as a function of the ram speed during extrusion. As a result the processing window is even more restricted since hot cracks can easily occur, as shown by Atwell and Barnett [20].

Based on these considerations, it was the aim of this paper to show the extrusion of magnesium alloy wires feasible directly in only one processing step with a wire thickness of 1 mm or smaller. The alloy specific correlation of the process to the developing microstructures and to excellent mechanical properties and surface conditions is revealed.

2. Materials and Methods

For the extrusion of wires, billets were prepared from cast ingots with a length of 150 mm and a diameter of 49 mm to fit the container with 50 mm diameter. The billets were homogenized for 16 h at 400 °C prior to extrusion to maintain the alloying elements, especially Al, in solid solution as far as possible. The chemical composition for pure magnesium and the three used Mg-Al-based alloys, AZ31, AZ80, and AZ91, was analyzed by X-ray micro fluorescence (µXRF, M4 Tornado, Bruker, Billerica, MA, USA). The results are listed in Table 1.

Table 1. Chemical analysis in wt% (Mg in balance).

Alloy	Al	Zn	Mn
AZ31	2.88	1.04	0.22
AZ80	8.01	0.35	0.23
AZ91	8.67	0.70	0.22
Mg	-	-	-

Wires with a thickness of 1 mm and in case of AZ31 0.6 mm were produced by direct extrusion, respectively. The corresponding significant reduction of the resulting material flow (speed) in AZ31 did not allow the continuation of thinner wire extrusion for AZ80 and AZ91. For all experiments, extrusion was performed using a 2.5 MN automatic extrusion press (theoretical maximum force limit) from Müller Engineering (Müller Engineering GmbH & Co. KG, Todtenweis/Sand, Germany). The billets were preheated for 60 min to the extrusion temperature. The processing temperature was 325 °C and the extrusion speed setting (ram speed) was 0.1 mm/s, which translates into a profile exit speed of 3.75 m/min for the 1 mm thick wire or 10.42 m/min for the 0.6 mm thick wire (the different units are used to distinguish between ram speed and profile exit speed). Note, that this maximum level is not reached throughout the experiments (see Table 2). A schematic illustration of the wire extrusion process and the die used in this case with the most important dimensions is shown in Figure 1. The extrusion experiments were carried out using a die with 4 nozzles positioned equi-distant from the center position and with the respective diameters, 1 mm (resulting in an extrusion ratio of 1:625) or 0.6 mm (extrusion ratio 1:1736). Thus, 4 wires are extruded same time and the corresponding extrusion ratios are thereby reduced. The wires were coiled by using an associated coiler coupled to the extrusion press. Spools for coiling had a diameter of 80 mm. An offset per rotation of the spools of 1.1 mm was applied to adjust the winding pattern. A low initial tension of maximum 350 N was set during coiling to avoid early fracture of the wires. The wires were analyzed in this as-extruded condition without any further heat-treatment.

Table 2. Extrusion peak force and exit speed of the wire.

Wire	Real Exit Speed (m/min)	Peak Force (MN)
Mg-1 mm	3.3 ± 0.10	1.36 ± 0.02
AZ31-1 mm	2.8 ± 0.10	2.50 ± 0.02
AZ31-0.6 mm	0.4 ± 0.04	2.61 ± 0.02
AZ80-1 mm	0.9 ± 0.17	2.53 ± 0.01
AZ91-1 mm	1.6 ± 0.15	2.53 ± 0.02

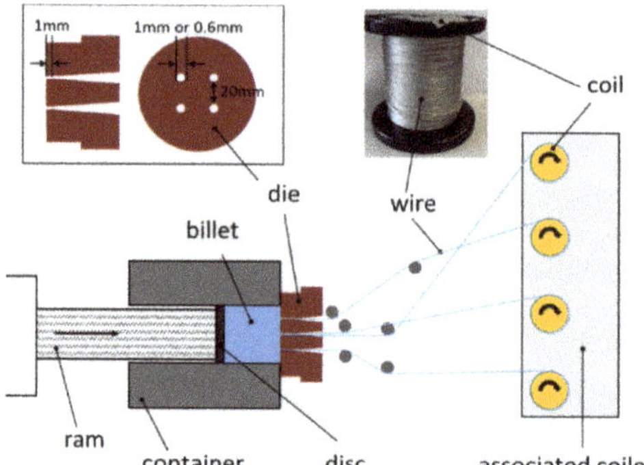

Figure 1. Schematic illustration of the wire extrusion process and the die.

The surface characterization of the wires was performed using a confocal laser-scanning microscope VK-1000 (Keyence, Osaka, Japan). The surface roughness (mean arithmetic height (Sa)) was measured according to the standard DIN EN ISO 25178-2 [35]. This parameter (Sa) is the extension of the line roughness parameter Ra (arithmetic mean) into the area. It is the amount of the height difference of each point compared to the arithmetic mean of the surface. The filters for the evaluation of the measurement results were selected according to DIN EN ISO 25178-3 [36]. The low pass filter (S filter) was set to 0.8 µm. The high-pass filter (L-filter) should be at least 5 times as large as the scale of the largest structure. Consequently, a value of 0.05 mm was chosen. Three areas of 100 µm × 100 µm each were measured.

The grain structure and texture of the wires were analyzed using a field emission scanning electron microscope (SEM) in combination with electron backscatter diffraction (EBSD), Zeiss Crossbeam 550L (SEM, Ultra 55, Carl Zeiss AG, Oberkochen, Germany) equipped with an EDAX-TSL OIM^{TM}; system (AMETEK Inc., Berwyn, PA, USA). An acceleration voltage of 15 kV and a step size of 0.35 µm were used at 300× magnification. SEM images in backscatter contrast (BSE) were received in 500× magnification with again 15 kV with a Vega3 SB SEM from TESCAN (TESCAN, Brünn, Czech Republic). The samples were ground with fine SiC paper (#4000) in longitudinal direction up to midplane and then polished with OPS (oxid polishing suspension) and 1 µm diamond suspension. For the EBSD measurements, the samples were additionally electro-polished for 25 s at a voltage of 10 V by using −20 °C cooled Struers AC2^{TM} solution followed by rinsing with nitric acid to remove the oxide layer. For receiving the area fraction of precipitates, as well as their average size an image analysis tool "analysis pro 5.0 (Olympus Soft Imaging Solutions GmbH, Münster, Germany)", has been employed based on a color contrast approach. For the area fraction, an area of 0.2 mm^2 and, for the particle size, at least 100 precipitates were analyzed.

Mechanical properties of the wires were characterized by tensile and wrapping tests. Tensile tests were performed at room temperature with a gauge length of 50 mm and at a strain rate of 10^{-3} s^{-1} using a universal testing machine (MTS Acumen^{TM}, 3 kN, Eden Prairie, MN, USA). At least 3 specimens were tested.

Wrapping tests were manually carried out according to ISO 7802: 2013 [37] at room temperature on defined diameters from 4 mm to 0.6 mm. At least 6 windings were wrapped.

3. Results and Discussion

Processing: Table 2 shows the resulting extrusion parameters, the measured exit speeds of the wires during extrusion calculated from the gradient of time versus the ram position multiplied by the extrusion ratio, as well as the maximum extrusion force (peak force) which is related to the same extrusion parameter settings (T = 325 °C and v = 0.1 mm/s) for all experiments.

Considering the speed and extrusion force, distinct differences can be observed for varying alloy composition and wire thickness. For the wires with 1 mm thickness, Mg has the lowest peak force of 1.36 MN, which increases to the capacity limit of the extrusion press at ca. 2.5 MN for AZ31. The peak force represents the beginning of the material flow and depends on the flow stress of the material in its respective condition and a force relevant component due to the friction between billet and container wall. Since all billets have the same initial length, the friction component is considered comparable. In all other cases, a corresponding decrease of the extrusion force was not found, leaving the extrusion experiments clearly at the press limit for AZ80 and AZ91, as well as for the 0.6 mm thick wire of AZ31. Correspondingly, the specified speed could not be reached because the force range was exhausted and the resulting extrusion speeds shown in Table 2 are much lower than for Mg and AZ31 (1 mm). For the 1-mm wires, Mg shows the highest exit speed with 3.3 m/min followed by AZ31 with 2.8 m/min. AZ80 and AZ91 show significantly lower speeds with 0.9 and 1.6 m/min, respectively. The 0.6 mm AZ31 wire has by far the lowest exit speed with 0.4 m/min. The addition of the alloying elements and in particular aluminium obviously leads to a strengthening of the material between pure Mg and AZ31, which continues with the further increase of the aluminium content.

Thus, although extrusion was carried out close to the processing limit with these parameter settings (especially in the form of a low extrusion temperature), it is shown feasible. It is suggested that the processing temperature for AZ-series alloys can also be higher than the 325 °C of this study which would further improve the forming ability but not reach a processing limit, e.g., in the form of hot cracking.

Microstructure and Texture Development: The grain structures and corresponding textures of the wires measured from longitudinal sections by EBSD are shown in Figure 2. High angle grain boundaries (HAGB, with misorientation angles larger than 12°) are marked in black. The corresponding average grain sizes are listed in Table 3.

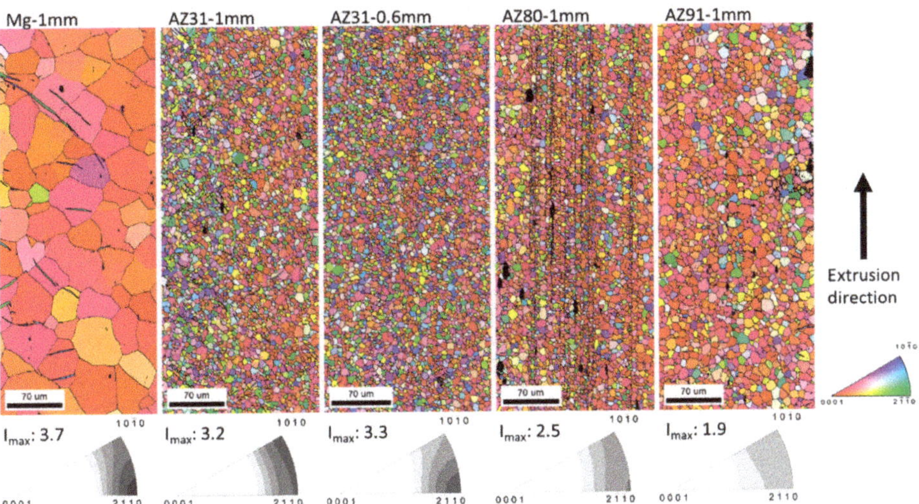

Figure 2. Orientation maps from electron backscatter diffraction (EBSD) measurements on longitudinal section of the extruded wires (extrusion direction (ED) vertical) and the corresponding inverse pole figures in the ED (Level: 1,2, ... 5 multiple random degree, m.r.d.).

Table 3. Grain size calculated from the electron backscatter diffraction (EBSD) measurement and mechanical properties from tension tests parallel to ED (tensile yield stress (TYS), ultimate tensile stress (UTS)) with standard deviations.

Wire	Grain Size (µm)	TYS (MPa)	UTS (MPa)	Fracture Strain (%)
Mg-1mm	42.6 ± 1.0	90 ± 9	169 ± 1	11.2 ± 0.3
AZ31-1mm	6.2 ± 0.1	179 ± 7	268 ± 2	16.2 ± 0.8
AZ31-0.6mm	5.3 ± 0.2	168 ± 10	272 ± 1	18.2 ± 1.5
AZ80-1mm	6.2 ± 0.1	195 ± 6	316 ± 2	18.7 ± 0.6
AZ91-1mm	9.5 ± 0.1	180 ± 1	302 ± 1	21.5 ± 0.6

All wires reveal a completely recrystallized microstructure with equiaxed grains. Vertical lines parallel to the extrusion direction are associated with particle stringers from precipitates. The Mg wire has by far the coarsest grain structure with an average grain size of 43 µm. The visible twins in the microstructure are remains from sample preparation, which is sometimes hard to avoid especially in large grains. The Al-containing wires with 1 mm thickness show considerably finer grained homogeneous microstructures. Although there is no visible difference between AZ31 (6.2 µm) and AZ80 (6.2 µm), the AZ91 wire is somewhat coarser grained at an average of 9.5 µm. It is noteworthy, that the same grain size of AZ31 and AZ80 with their different aluminium contents corresponds to a much lower resulting extrusion exit speed of the wires. On the contrary, the larger grain size of the AZ91 wire corresponds to a rather small variation in the alloy composition compared to AZ80, but a higher extrusion exit speed for AZ91 compared to AZ80. The thinner wire of AZ31 (0.6 mm) is also slightly finer grained at an average grain size of 5.3 µm, despite the significantly higher degree of deformation. However, the resulting extrusion speed is also the lowest in this study, which corresponds to this finest grained material. It is the hypothesis that the resulting lower extrusion speed limits deformation related heating, which leaves the recrystallized microstructure finer grained whereas the higher degree of deformation does not have a significant impact.

The very high degree of deformation imposed to the billet during extrusion results in considerable dynamic recrystallization during the forming process. The major impact on the kinetics of this dynamic recrystallization is due to temperature and deformation rate, often combined in the form of a temperature compensated strain rate, the Zener Hollomon parameter [38–41]. The impact of the imposed strain as a result of the extrusion ratio is indirectly included as the strain rate—seen as the extrusion exit speed—which results from a constant extrusion ram speed. It increases with increasing extrusion ratio. Furthermore, a higher strain rate during extrusion will result in higher deformation related heating, therefore increasing the relevant temperature for dynamic recrystallization [42]. The above results on the grain size development do not allow a quantitative analysis regarding the kinetics of recrystallization but confirm a well-known grain refinement of alloying with aluminium in Mg. Furthermore, grain growth during recrystallization appears to increase with the profile exit speed and determines the microstructure development besides of alloying effects. Especially in the AZ80 wire, it is furthermore visible that a fraction of smaller grains developed in the vicinity of the stringer particles. This indicated a particle related growth restriction of grains during recrystallization. Earlier works have emphasized two important mechanisms in this regard: a particle related restriction of the grain boundary mobility due to the particles [43] or a stimulation of recrystallization by particles which increases the nucleation rate of recrystallized grains [44,45]. As a result, of the fully recrystallized microstructures in cannot be distinguished which mechanisms dominated in this regard, however, a local effect on the developing texture cannot be revealed (not shown).

The inverse pole figures in the extrusion direction (ED) are also shown in Figure 2 to represent the texture of the wires. As being typical for extruded round bars, a distinct preference for alignment of the basal planes parallel to the ED results. This is visible from the highest intensities in the pole figures along the arc between the <10$\bar{1}$0>- and the <2$\bar{1}$ $\bar{1}$0> poles [46]. Even further, in some cases the completely recrystallized microstructures correspond with high intensity at the <2$\bar{1}$ $\bar{1}$0> pole [26]. This is especially pronounced for the Mg wire which has the most distinct texture with a maximum intensity of 3.7 m.r.d. With increasing content of Al from AZ31 to AZ91, the significance of the texture and the maximum intensity decrease, i.e., the orientation distribution has a more random character. AZ80 shows a maximum intensity of 2.5 m.r.d. and AZ91 of 1.9 m.r.d. The higher extrusion ratio of the thin 0.6 mm thick AZ31 wire corresponds to a more distinct development of intensity at the <2$\bar{1}$ $\bar{1}$0> pole compared to the wire with 1 mm thickness. This finding again corresponds to enhanced dynamic recrystallization despite the smaller grain size. In this case, a lower temperature increase and comparably retarded recrystallization during extrusion with the higher extrusion ratio is hypothetically balanced out by a further development of the recrystallized grain structure due to the higher imposed strain.

Especially in the case of the AZ80 wire, an inhomogeneity in the grain structure is visible due to lined-up sub-structures along the extrusion direction. Figure 3 collects corresponding SEM images to reveal second phase particles in BSE-contrast (backscattered electrons). Additionally, the average particle size and the area fraction of the particles, measured on the represented image, are shown. Energy dispersive X-ray spectroscopy (EDX) reveals that all the particles measured consist of a combination of Al with Mn (not shown). No particles consistent with the stable $Mg_{17}Al_{12}$-phase could be identified throughout the alloys. Especially large particles are found in AZ80, not in AZ91. Note that the smaller average particle size corresponds to the obviously advanced grain growth in this alloy. On the other hand, the area fraction of particles increases continuously with the aluminium content. Furthermore, fine structures aligned in extrusion direction as particle stringers can be seen with increasing significance as the aluminium content increases.

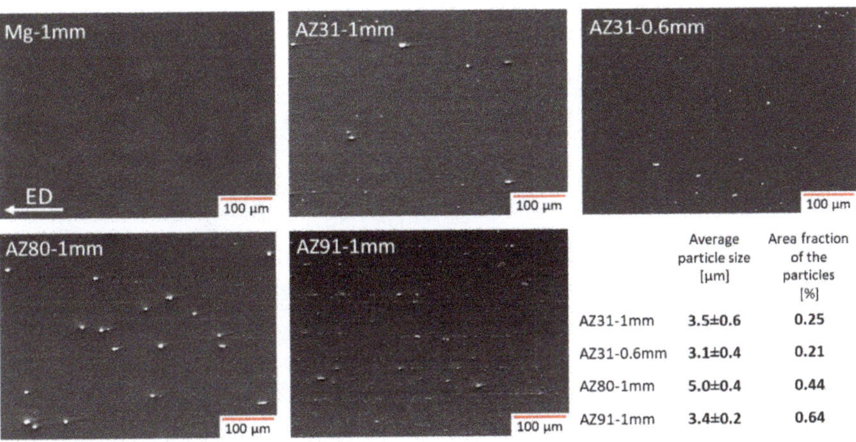

	Average particle size [µm]	Area fraction of the particles [%]
AZ31-1mm	3.5±0.6	0.25
AZ31-0.6mm	3.1±0.4	0.21
AZ80-1mm	5.0±0.4	0.44
AZ91-1mm	3.4±0.2	0.64

Figure 3. Scanning electron microscope (SEM)-Image with backscatter contrast (BSE)—contrast of all wires from the longitudianlsection in 500× magnification.

While the precipitates are aligned as particle stringers or single particles with an ellipsoid form elongated to the extrusion direction, this is consistent with precipitates, which underwent deformation during extrusion and as a result aligned linearly in the extrusion direction. A potential deformation induced precipitation of these particles in not consistent with such an alignment but could result in a more random distribution of particles. Considering the particle size, AZ91 has the largest amount of precipitates, but AZ80 has by far the largest average particle size with 5.0 ± 0.4 µm. There is no significant difference in average particle size between AZ31 (1 mm and 0.6 mm wire) and AZ91.

The applied solid solution heat treatment at 400 °C prior to extrusion would preferentially lead to the dissolution of $Mg_{17}Al_{12}$ whereas Mn containing precipitates (Al_8Mn_5 or Al_4Mn have been reported) are stable up to the solidus and cannot be dissolved [47]. However, the lower extrusion temperature at 325 °C also allows the formation of these Mn-containing precipitates. However, the fast cooling of the thin wires—air cooling occurs after extrusion—supports the suppression of the formation of newly formed precipitates during the processing, especially during cooling below a solvus temperature for the respective precipitates.

There are also no differences in the distribution of precipitation between the 1 mm and 0.6 mm AZ31 wire. Thus, in this case the deformation degree does not seem to have a strong influence on the distribution or stringer formation at these high degrees of deformation.

Mechanical Properties and Forming behavior: Stress-strain diagrams from tensile tests of the wires are shown in Figure 4. Table 3 also shows the mechanical properties and their standard deviation in addition to the grain sizes taken from the EBSD measurements (Figure 2). Within the accuracy of the measurements, the results for AZ31 do not vary much. Higher Al content is consistent with an increase of both, stress and strain properties.

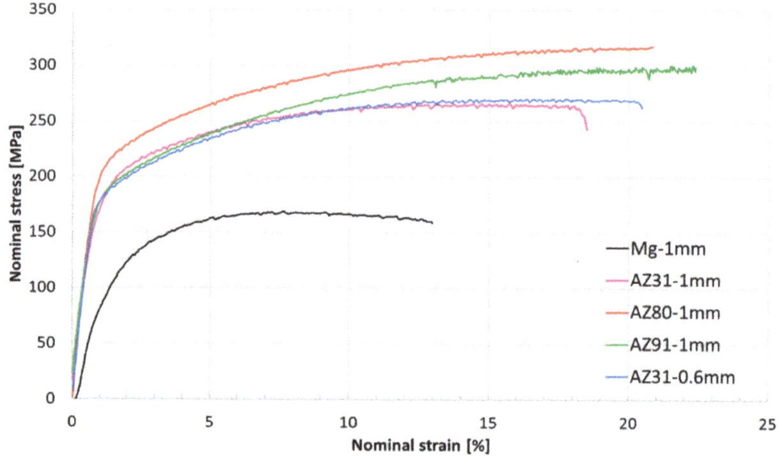

Figure 4. Stress-strain diagram in ED from the extruded 1 mm and 0.6 mm wires.

A continuous elasto-plastic yielding is followed by a typical strain hardening behavior with a continuous decrease of the slopes. The pure Mg wire is the only one that shows a following decrease of the stress levels after reaching a maximum stress level, obviously due to the occurrence of necking and the associated ductile behavior. Still, it reveals the lowest fracture strain with 11% and the lowest stress levels (yield stress (TYS) 90 MPa and maximum stress (UTS) 169 MPa) compared to the other wires of this study. For the Al-containing wires of AZ31, AZ80, and AZ91, the fracture strain increases with the Al content, from 16% for AZ31 to 21% for AZ91. None of these wires seems to exceed a uniform strain from where necking would become visible. However, while AZ31 seems to reach a limit of the stress level, strain hardening continues for AZ80 and especially for AZ91 until fracture.

The AZ80 wire shows the highest mechanical strength properties (TYS: 195 MPa; UTS 316 MPa) whereas AZ91 does not fully reach the same levels (TYS 180 MPa; UTS 302 MPa). Despite similar grain size, the strength properties of the AZ31 wires are even lower which is consistent with a lower strengthening effect due to the low Al content as an alloying element in solid solution [48]. It is not assumed that the Al-Mn-containing particles contribute to material strengthening to a visible extent [49]. Concurrent to their very similar microstructure properties, there are only very small variations in the mechanical behavior and the mechanical properties of the two AZ31 with different thickness, 1 mm and 0.6 mm, respectively. The variations of the comparably low average grain sizes will have a visible effect on the yield stress due to grain boundary strengthening (Hall-Petch relationship; see, e.g., Christian and Mahajan [50]), which corresponds well with the lower stress levels of AZ91 compared to AZ80, despite the quite similar alloy composition. Furthermore, the weaker texture of the wires with higher Al-content corresponds to the higher ability to strain hardening. It is hypothesized that the weaker texture, i.e., less significant alignment of basal planes parallel to the extrusion direction, enables higher ability to accommodate plastic strain due to enhanced basal slip. However, at the onset of plastic deformation, lower stress will be required to activate this slip mode; therefore, the corresponding stress levels will also be lower. A corresponding effect of the texture in compression along the extrusion direction is not accessible due to the small dimensions of the wires. It is assumed that weaker textures will reduce the asymmetric yielding behavior due to reduced activation ability of deformation twinning [24], which cannot be revealed experimentally in this work. It has been shown that this also has an impact towards increased fracture strains during compressive loading [34].

For an examination of the forming behavior of the wires with a combined effect of various strain, tensile and compressive strain, manual wrapping of the wires has been carried out with various inner core diameters. An exemplary cross-sectional grain map at mid-plane is shown in Figure 5a, resulting from an EBSD measurement from the AZ31 wire (1 mm thickness, inner core 2 mm in diameter). There is a considerable change visible, which is supposed to represent the major features of the microstructure and texture development. The left-hand side of the image, the outer surface that underwent tensile strain, does not show much variation from the original texture only with the highest intensity at the $<2\bar{1}\,\bar{1}0>$ pole, see section A, thus revealing no strong texture change compared to the original texture in Figure 2. This may be somewhat surprising if it is recalled that tensile strain typically leads to the formation of a prismatic plane alignment such that higher intensity is found for the $<10\bar{1}0>$ pole; see, e.g., Reference [51,52]. However, the strain level during wrapping may be too low to resolve such a change in the respective pole figure of the respective symmetric profile. Still, the grains in this range are in tendency elongated vertically which corresponds to the extrusion direction. Thus, the grain form was visibly influenced by the applied strain, which persists for a considerable 1/3 along the cross section. Concurrently towards the inner part of the wire, section B, basal planes appear to tilt further out of the extrusion direction, which is visible from the broader intensity distribution in the pole figure towards the $<0001>$ pole. On the right-hand side, close to the centerline of the bent wire in section C, this development continues, leaving the texture very weak. This does not persist towards the inner surface of the wire on the right-hand side, section D. The pole figure reveals a component with a full tilt of basal planes and the corresponding c-axis parallel to ED. This rotation is consistent with this fraction of the microstructure undergoing twinning. Note, that almost no boundaries consistent with the lattice rotation of ca. 86° of tensile twins are found throughout the measurement (not shown), which indicates full twinning of the respective re-oriented grains. Grains also appear to be flattened horizontally, i.e., perpendicular to the wire extrusion direction.

In Figure 5b, the same section is shown with highlighted fractions of orientations as selected from the corresponding (0001) pole figure. The blue and yellow fractions correspond to basal planes aligned parallel to the extrusion direction with c-axis vertical or horizontal to the measured mid-plane, respectively. Grains are oriented in correspondence to both fractions randomly on the outer surface side (left). The most inner part on the right-hand side is revealed in red which corresponds to a fraction of grains with c-axis parallel to ED. Assuming the associated grains as twinned grains it becomes visible that this affects only a small range of grains along the through-thickness measurement.

Results from wrapping tests in Figure 6 were achieved in accordance to standard ISO 7802:2013 [37]. For this purpose, the wires were manually wrapped around core sticks with defined diameters from 4 mm down to 0.6 mm at least 6 times. For each wire the figure shows one example of the minimum wrapping diameter before wire fracture (left picture), as well as one example of wrapping with wire fracture (right picture). The 1 mm thick Mg wire already showed failure when it was wrapped around a diameter of 3.5 mm and also showed cracks even at the 4 mm wrapping diameter, see Figure 6a.

The AZ80 wire achieves a wrap of 2 mm followed by the AZ31 wire of 1.2 mm diameter. Among the 1 mm thick wires, the AZ91 wire is the tightest to wrap (diameter 1 mm) and only fails at a wrapping diameter of 0.8 mm.

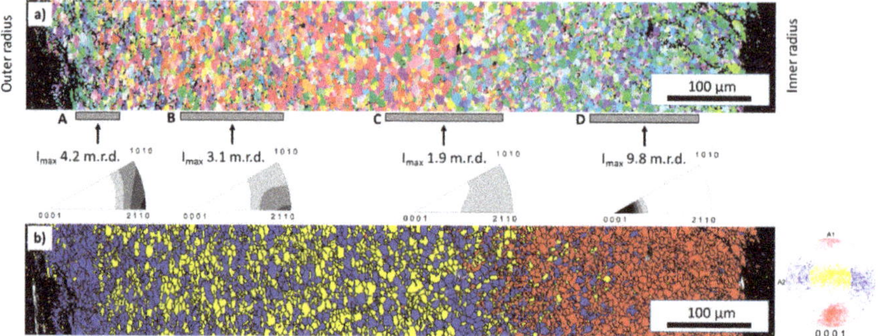

Figure 5. (**a**) Grain and orientation map of a cross section from a wrapped section of an AZ31 wire (thickness 1 mm) wrapped along a core with 2 mm diameter; outer radius is left, and inner radius in right. Sections are cropped for local inverse pole figure presentation along the extrusion direction (vertical): (A) outer surface, (B) tensile strain affected zone, (C) transition to compressive strain affected zone, and (D) inner surface; (**b**) same picture with highlighted grain orientations as labelled in the discrete (0001) pole figure.

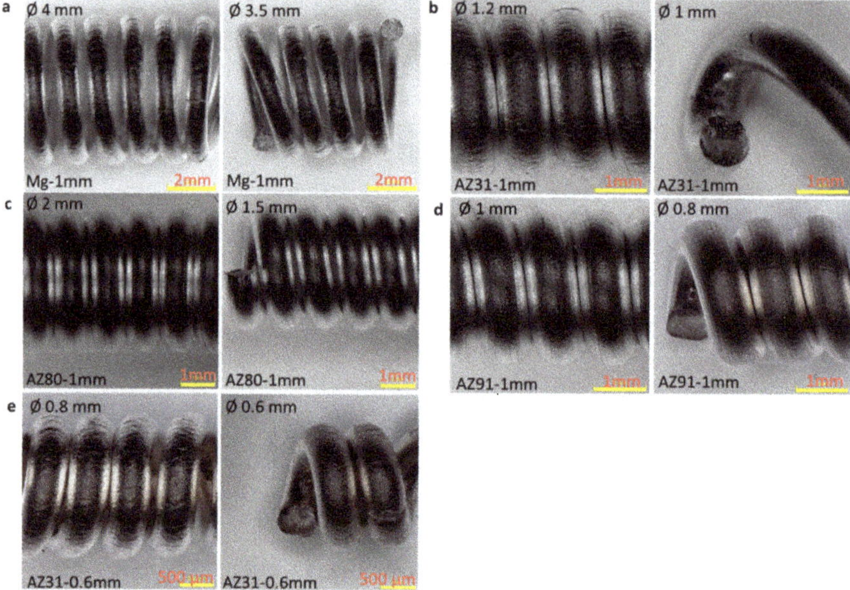

Figure 6. Macro pictures of the wrapped wires (left picture: one coil diameter setting before fracture, and right picture: coil diameter at material failure) for (**a**) Mg-1mm; (**b**) AZ31-1mm; (**c**) AZ80-1mm: (**d**) AZ91-1mm; and (**e**) AZ31-0.6mm; note the varying magnifications for improved surface visibility.

Considering the minimum diameter as a measure of the ductility for this multi-directional strain accommodation case, the results are quite comparable to those achieved from the tensile test. Only AZ80 shows a less ductile behavior compared to AZ31, which is contrary to the tensile test.

If related to the difference in the activation of deformation mechanisms and especially the potential role of twinning, this finding seems counterintuitive. Then, the obviously special distribution of precipitates in the alloys may act differently if local shear and compressive strain components occur. If the material still has large amounts of precipitation, as is the case with AZ80 (see Figure 3), stress peaks can arise, which can lead to a premature material failure. Beside this, it is confirmed that the higher Al-containing AZ91 wire shows excellent wrapping behavior compared to AZ31. The 0.6 mm thick AZ31 wire shows failure at a winding diameter of 0.6 mm and shows clear surface cracks even at 0.8 mm. Still, the behavior is consistent with the smaller wire thickness if compared to the 1 mm thick wire.

Surface characterization: Figure 7 shows the surface characteristics of the as-extruded wires in two magnifications, 50× and 500×. In addition, the surface roughness is shown as the average arithmetic height (Sa), as measured from the image with the higher magnification. There is an appreciable difference in the surface quality, associated with the alloy composition, as well as the resulting extrusion parameter variations (especially the speed). The pure Mg wire (1 mm) exhibits the highest roughness (0.46 μm) and shows an extremely uneven wavy surface compared to the Al-containing wires of AZ31, AZ80, and AZ91. The formation of the scratches in the extrusion direction becomes more pronounced with increasing Al content. This is also reflected in the roughness. Both AZ31 wires (0.6 and 1 mm thick) show the lowest roughness, despite the different extrusion speeds whereas the AZ91 wire shows the highest roughness (0.37 μm), except for pure Mg. The poor surface development of pure Mg has also been reported as a result of compressive strain in the case of forging [53].

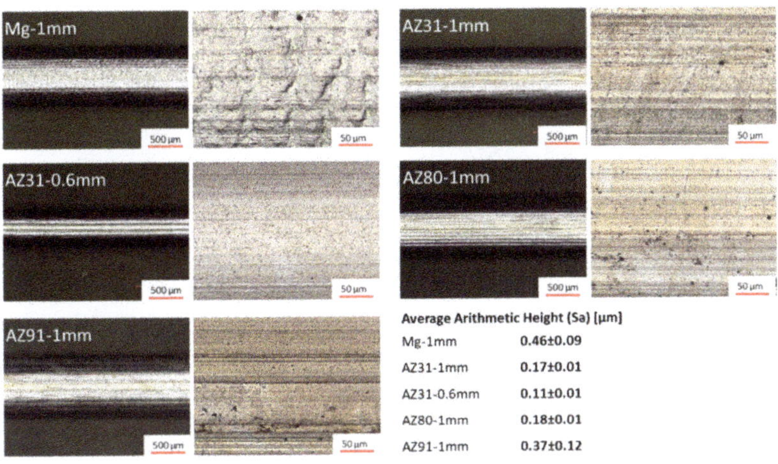

Figure 7. Surface characterization of the wires.

The number and depth of the grooves are detailed in Figure 8 for the wires, which shows a vertical surface profile measurement taken from the images at 500× magnification. In this figure, the depth of the grooves can be compared best in the form of the fluctuations of the curves. Again, AZ31 is very smooth in both cases. The AZ80 wire has a pronounced concave depression on the right side which exemplarily points towards a general flow instability during the extrusion process. The AZ91 shows grooves of up to 5 μm depth. The same effect seems less pronounced in case of pure Mg with its different surface morphology. While the surface of pure Mg is determined by an inhomogeneous deformation behavior, which leads to surface bulging, this behavior is reduced with increasing Al content. On the contrary, the tendency to groove formation is increased with the Al content.

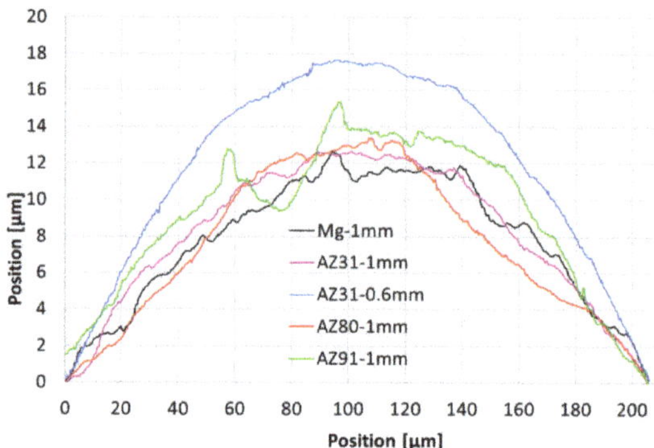

Figure 8. Surface contour measurements for the wires from illustrations in 500× magnification in Figure 7.

For the analysis of the influence of the wrapping test on the surface quality, roughness measurements were carried out on the wrapped wire surfaces. The results are presented in Figure 9 versus the wrapping diameter. The increasing wrapping diameter along the horizontal axis is in accordance to a less challenging bending test. The surface roughness (Sa) in μm of the original wires (initial roughness) is shown as a benchmark. The surface roughness increases for all wires with smaller wrapping diameters. The Mg wire has the highest initial roughness (0.46 ± 0.09 μm), as well by far (0.61 ± 0.09 μm) after wrapping towards fracture at a diameter as high as 3.5 mm. For the AZ31 (1 mm) wire Sa increases from 0.17 ± 0.01 μm to 0.54 ± 0.02 μm, which is somewhat similar to the 0.6 mm AZ31 wire where Sa increases from 0.11 ± 0.01 μm to 0.52 ± 0.06 μm at material failure. Note that, at the respective wrapping diameters, the thicker wire experiences higher strain at the surface; therefore, the respective roughness is always higher compared to the thinner wire.

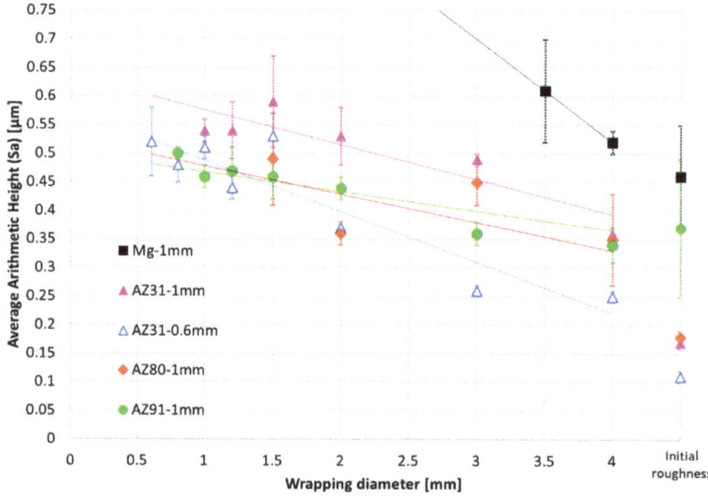

Figure 9. Averaged surface roughness (Sa) in μm vs. the wrapping diameter of the wrapped wires.

For the AZ80 wire, the roughness increase from 0.18 ± 0.01 μm to 0.49 ± 0.08 μm is somewhat unsteady and—if different in any way—possibly slightly lower than that of the AZ31 wire. Especially for the AZ91 wire, a remarkably low increase from the initial surface roughness of 0.37 ± 0.12 μm to 0.50 ± 0.01 μm is found. While the initial roughness is rather high (almost twice as high as that of the other Al-containing wires). Although the scattered results for AZ80 do not confirm this, the high Al content in AZ91 reveals a smoothening impact on the materials surface.

4. Conclusions

The direct extrusion of wires from magnesium alloys of the Mg-Al-Zn-series (AZ-alloys) is shown feasible, while the limitation towards thinner thickness of the wires has been related to the capacity limit of the used extrusion press. Rather low extrusion temperature and speed allowed the manufacturing of fine-grained wires with 1 mm in thickness. A thinner wire of AZ31 was also feasible to extrude, but alloys with higher Al-content reached over the capacity limit of the press at the extrusion parameters used in these studies (T = 325 °C; v = 0.1 mm/s).

Aluminium as an alloying element allowed maintaining fine-grained microstructures and texture weakening, where the distinct alignment of basal planes parallel to the extrusion direction is reduced. This is consistent with increasing stress properties but also with enhanced ductility in tensile tests and wrapping tests. While increasing the Al content the as-extruded surface shows grooves increasingly which differs from the general inhomogeneous deformation behavior of pure Mg. Wrapping of the wires increases the roughness of the wires considerably and obviously does so with lower wrapping diameter. This impact is however lowest at the alloy with the highest Al content, AZ91.

Author Contributions: Conceptualization and methodology, M.N., D.L., and J.B.; formal analysis, M.N. and J.B.; investigation, M.N., S.Y., and J.B.; supervision and validation: K.U.K. and D.L.; writing—original draft preparation, M.N. and J.B.; writing—review and editing, all authors; All authors have read and agreed to the published version of the manuscript.

Funding: This research received no external funding.

Acknowledgments: The authors appreciate the help of Daniel Strerath with the chemical analysis of the wires of this study and of Volker Heitmann with the mechanical testing setup, both at the Magnesium Innovation Centre.

Data Availability: The datasets generated for this study are available on request to the corresponding author.

References

1. Riekehr, S.; Ventzke, V.; Konovalovna, A.; Kashaev, N.; Enz, J. Microstructural characteristics of laser metal deposited magnesium alloy AZ31. *Mater. Sci. Forum* **2018**, *941*, 1004–1009. [CrossRef]
2. Witte, F. The history of biodegradable magnesium implants: A review. *Acta Biomater* **2010**, *6*, 1680–1692. [CrossRef] [PubMed]
3. Chen, W.Z.; Zhang, W.C.; Chao, H.Y.; Zhang, L.X.; Wang, E.D. Influence of large cold strain on the microstructural evolution for a magnesium alloy subjected to multi-pass cold drawing. *Mater. Sci. Eng. A* **2015**, *623*, 92–96. [CrossRef]
4. Maier, P.; Griebel, A.; Jahn, M.; Bechly, M.; Menze, R.; Bittner, B.; Schaffer, J. Corrosion Bending Fatigue of RESOLOY®and WE43 Magnesium Alloy Wires. In *Magnesium Technology 2019*; Springer: Cham, Switzerland, 2019; pp. 175–181. [CrossRef]
5. Maier, P.; Szakács, G.; Wala, M.; Hort, N. Mechanical and Corrosive Properties of Two Magnesium Wires: Mg4Gd and Mg6Ag. In *Magnesium Technology 2015*; Springer: Cham, Switzerland, 2015.
6. Sun, L.; Bai, J.; Xue, F.; Chu, C.; Meng, J. The work softening behavior of pure Mg wire during cold Drawing. *Materials* **2018**, *11*, 602. [CrossRef]
7. Bai, J.; Yin, L.; Lu, Y.; Gan, Y.; Xue, F.; Chu, C.; Yan, J.; Yan, K.; Wan, X.; Tang, Z. Preparation, microstructure and degradation performance of biomedical magnesium alloy fine wires. *Prog. Nat. Sci. Mater. Int.* **2014**, *24*, 523–530. [CrossRef]
8. Yan, K.; Sun, J.; Bai, J.; Liu, H.; Huang, X.; Jin, Z.; Wu, Y. Preparation of a high strength and high ductility Mg-6Zn alloy wire by combination of ECAP and hot drawing. *Mater. Sci. Eng. A* **2019**, *739*, 513–518. [CrossRef]
9. Seitz, J.-M.; Utermöhlen, D.; Wulf, E.; Klose, C.; Bach, F.-W. The manufacture of resorbable suture material from magnesium—Drawing and stranding of thin wires. *Adv. Eng. Mater.* **2011**, *13*, 1087–1095. [CrossRef]

10. Kustra, P.; Milenin, A.; Byrska-Wójcik, D.; Grydin, O.; Schaper, M. The process of ultra-fine wire drawing for magnesium alloy with the guaranteed restoration of ductility between passes. *J. Mater. Process. Technol.* **2017**, *247*, 234–242. [CrossRef]

11. Tesař, K.; Balík, K.; Sucharda, Z.; Jäger, A. Direct extrusion of thin Mg wires for biomedical applications. *Trans. Nonferrous Met. Soc. China* **2020**, *30*, 373–381. [CrossRef]

12. Jäger, A.; Habr, S.; Tesař, K. Twinning-detwinning assisted reversible plasticity in thin magnesium wires prepared by one-step direct extrusion. *Mater.Des.* **2016**, *110*, 895–902. [CrossRef]

13. Seitz, J.-M.; Wulf, E.; Freytag, P.; Bormann, D.; Bach, F.-W. The manufacture of resorbable suture material from magnesium. *Adv. Eng. Mater.* **2010**, *12*, 1099–1105. [CrossRef]

14. Johnston, S.; Shi, Z.; Dargusch, M.S.; Atrens, A. Influence of surface condition on the corrosion of ultra-high-purity Mg alloy wire. *Corros. Sci.* **2016**, *108*, 66–75. [CrossRef]

15. Zeng, Z.; Stanford, N.; Davies, C.H.J.; Nie, J.-F.; Birbilis, N. Magnesium extrusion alloys: A review of developments and prospects. *Int. Mater. Rev.* **2018**, *64*, 27–62. [CrossRef]

16. Liu, Q.; Zhou, X.; Zhou, H.; Fan, X.; Liu, K. The effect of extrusion conditions on the properties and textures of AZ31B alloy. *J. Magnes. Alloy.* **2017**, *5*, 202–209. [CrossRef]

17. Shahzad, M.; Wagner, L. Influence of extrusion parameters on microstructure and texture developments, and their effects on mechanical properties of the magnesium alloy AZ80. *Mater. Sci. Eng. A* **2009**, *506*, 141–147. [CrossRef]

18. Hu, M.-l.; Ji, Z.-s.; Chen, X.-y. Effect of extrusion ratio on microstructure and mechanical properties of AZ91D magnesium alloy recycled from scraps by hot extrusion. *Trans. Nonferrous Met. Soc. China* **2010**, *20*, 987–991. [CrossRef]

19. Liu, G.; Zhou, J.; Duszczyk, J. Prediction and verification of temperature evolution as a function of ram speed during the extrusion of AZ31 alloy into a rectangular section. *J. Mater. Process. Technol.* **2007**, *186*, 191–199. [CrossRef]

20. Atwell, D.L.; Barnett, M.R. Extrusion limits of magnesium alloys. *Metall. Mater. Trans. A* **2007**, *38*, 3032–3041. [CrossRef]

21. Nienaber, M.; Kainer, K.U.; Letzig, D.; Bohlen, J. Processing effects on the formability of extruded flat products of magnesium alloys. *Front. Mater.* **2019**, *6*. [CrossRef]

22. Dillamore, I.L.; Roberts, W.T. Preferred orientaion in wrought and annealed metals. *Metall. Rev.* **1965**, *10*, 271–380. [CrossRef]

23. Bohlen, J.; Yi, S.B.; Swiostek, J.; Letzig, D.; Brokmeier, H.G.; Kainer, K.U. Microstructure and texture development during hydrostatic extrusion of magnesium alloy AZ31. *Scr. Mater.* **2005**, *53*, 259–264. [CrossRef]

24. Bohlen, J.; Yi, S.; Letzig, D.; Kainer, K.U. Effect of rare earth elements on the microstructure and texture development in magnesium-manganese alloys during extrusion. *Mater. Sci. Eng. A* **2010**, *527*, 7092–7098. [CrossRef]

25. Bohlen, J.; Cano, G.; Drozdenko, D.; Dobron, P.; Kainer, K.; Gall, S.; Müller, S.; Letzig, D. Processing effects on the formability of magnesium alloy sheets. *Metals* **2018**, *8*, 147. [CrossRef]

26. Yi, S.; Brokmeier, H.-G.; Letzig, D. Microstructural evolution during the annealing of an extruded AZ31 magnesium alloy. *J. Alloy. Compd.* **2010**, *506*, 364–371. [CrossRef]

27. Chaudry, U.M.; Kim, T.H.; Park, S.D.; Kim, Y.S.; Hamad, K.; Kim, J.-G. Effects of calcium on the activity of slip systems in AZ31 magnesium alloy. *Mater. Sci. Eng. A* **2019**, *739*, 289–294. [CrossRef]

28. Ha, C.; Yi, S.; Bohlen, J.; Zhou, X.; Brokmeier, H.-G.; Schell, N.; Letzig, D.; Kainer, K.U. Deformation and Recrystallization Mechanisms and Their Influence on the Microstructure Development of Rare Earth Containing Magnesium Sheets. In *Magnesium Technology 2018*; Springer: Cham, Switzerland, 2018; pp. 209–216. [CrossRef]

29. Bohlen, J.; Wendt, J.; Nienaber, M.; Kainer, K.U.; Stutz, L.; Letzig, D. Calcium and zirconium as texture modifiers during rolling and annealing of magnesium-zinc alloys. *Mater. Charact.* **2015**, *101*, 144–152. [CrossRef]

30. Zeng, X.; Minárik, P.; Dobroň, P.; Letzig, D.; Kainer, K.U.; Yi, S. Role of deformation mechanisms and grain growth in microstructure evolution during recrystallization of Mg-Nd based alloys. *Scr. Mater.* **2019**, *166*, 53–57. [CrossRef]

31. Stanford, N.; Barnett, M.R. The origin of "rare earth" texture development in extruded Mg-based alloys and its effect on tensile ductility. *Mater. Sci. Eng. A* **2008**, *496*, 399–408. [CrossRef]

32. Dobroň, P.; Chmelík, F.; Yi, S.; Parfenenko, K.; Letzig, D.; Bohlen, J. Grain size effects on deformation twinning in an extruded magnesium alloy tested in compression. *Scr. Mater.* **2011**, *65*, 424–427. [CrossRef]

33. Victoria-Hernandez, J.; Yi, S.; Letzig, D.; Hernandez-Silva, D.; Bohlen, J. Microstructure and texture development in hydrostatically extruded Mg-Al-Zn alloys during tensile testing at intermediate temperatures. *Acta Mater.* **2013**, *61*, 2179–2193. [CrossRef]

34. Bohlen, J.; Chmelík, F.; Dobroň, P.; Letzig, D.; Lukáč, P.; Kainer, K.U. Acoustic emission during tensile testing of magnesium AZ alloys. *J. Alloy. Compd.* **2004**, *378*, 214–219. [CrossRef]

35. DIN. *EN ISO 25178-2:2012-09, Geometrische Produktspezifikation (GPS)—Oberflächenbeschaffenheit: Flächenhaft—Teil 2: Begriffe und Oberflächen-Kenngrößen (ISO 25178-2:2012)*; Beutch Verlag GmbH: Berlin, Germany, 2012. [CrossRef]

36. DIN. *EN ISO 25178-3:2012-11, Geometrische Produktspezifikation (GPS)—Oberflächenbeschaffenheit: Flächenhaft—Teil 3: Spezifikationsoperatoren (ISO 25178-3:2012)*; Beutch Verlag GmbH: Berlin, Germany, 2012.

37. DIN. *ISO 7802:2014-11, Metallische Werkstoffe—Draht—Wickelversuch (ISO 7802:2013)*; Beutch Verlag GmbH: Berlin, Germany, 2014.

38. Yu, H.; Hyuk Park, S.; Sun You, B.; Min Kim, Y.; Shun Yu, H.; Soo Park, S. Effects of extrusion speed on the microstructure and mechanical properties of ZK60 alloys with and without 1wt% cerium addition. *Mater. Sci. Eng. A* **2013**, *583*, 25–35. [CrossRef]

39. Park, S.S.; You, B.S.; Yoon, D.J. Effect of the extrusion conditions on the texture and mechanical properties of indirect-extruded Mg-3Al-1Zn alloy. *J. Mater. Process. Technol.* **2009**, *209*, 5940–5943. [CrossRef]

40. McQueen, H.J.; Ryan, N.D. Constitutive analysis in hot working. *Mater. Sci. Eng. A* **2002**, *322*, 43–63. [CrossRef]

41. Chaudry, U.M.; Kim, T.H.; Kim, Y.S.; Hamad, K.; Ko, Y.G.; Kim, J.-G. Dynamic recrystallization behavior of AZ31-0.5Ca magnesium alloy during warm rolling. *Mater. Sci. Eng. A* **2019**, *762*, 138085. [CrossRef]

42. Bohlen, J.; Meyer, S.; Wiese, B.; Luthringer-Feyerabend, B.J.C.; Willumeit-Romer, R.; Letzig, D. Alloying and processing effects on the microstructure, mechanical properties, and degradation behavior of extruded magnesium alloys containing calcium, cerium, or silver. *Materials* **2020**, *13*, 391. [CrossRef]

43. Hadorn, J.P.; Hantzsche, K.; Yi, S.; Bohlen, J.; Letzig, D.; Wollmershauser, J.A.; Agnew, S.R. Role of solute in the texture modification during hot deformation of Mg-Rare earth alloys. *Metall. Mater. Trans. A* **2011**, *43*, 1347–1362. [CrossRef]

44. Robson, J.D.; Henry, D.T.; Davis, B. Particle effects on recrystallization in magnesium-manganese alloys: Particle-stimulated nucleation. *Acta Mater.* **2009**, *57*, 2739–2747. [CrossRef]

45. Al-Samman, T. Modification of texture and microstructure of magnesium alloy extrusions by particle-stimulated recrystallization. *Mater. Sci. Eng. A* **2013**, *560*, 561–566. [CrossRef]

46. Cano-Castillo, G.; Victoria-Hernández, J.; Bohlen, J.; Letzig, D.; Kainer, K.U. Effect of Ca and Nd on the microstructural development during dynamic and static recrystallization of indirectly extruded Mg-Zn based alloys. *Mater. Sci. Eng. A* **2020**, 139527. [CrossRef]

47. Ohno, M.; Mirkovic, D.; Schmid-Fetzer, R. Liquidus and solidus temperatures of Mg-rich Mg–Al–Mn–Zn alloys. *Acta Mater.* **2006**, *54*, 3883–3891. [CrossRef]

48. Cáceres, C.H.; Rovera, D.M. Solid solution strenthening in concentrated Mg-Al alloys. *J. Light Met.* **2001**, *1*, 151–156. [CrossRef]

49. Laser, T.; Nürnberg, M.R.; Janz, A.; Hartig, C.; Letzig, D.; Schmid-Fetzer, R.; Bormann, R. The influence of manganese on the microstructure and mechanical properties of AZ31 gravity die cast alloys. *Acta Mater.* **2006**, *54*, 3033–3041. [CrossRef]

50. Christian, J.W.; Mahajant, S. Deformation twinning. *Prog. Mater. Sci.* **1995**, *39*, 1–157. [CrossRef]

51. Zhou, X.; Ha, C.; Yi, S.; Bohlen, J.; Schell, N.; Chi, Y.; Zheng, M.; Brokmeier, H.-G. Texture and lattice strain evolution during tensile loading of Mg-Zn alloys measured by synchrotron diffraction. *Metals* **2020**, *10*, 124. [CrossRef]

52. Dudamell, N.V.; Ulacia, I.; Gálvez, F.; Yi, S.; Bohlen, J.; Letzig, D.; Hurtado, I.; Pérez-Prado, M.T. Twinning and grain subdivision during dynamic deformation of a Mg AZ31 sheet alloy at room temperature. *Acta Mater.* **2011**, *59*, 6949–6962. [CrossRef]
53. Kurz, G.; Sillekens, W.H.; Swiostek, J.; Letzig, D. Alloy Development and Processing for the European Project MagForge. In Proceedings of the 15th Magnesium Automotive and end User Seminar, Aalen, Germany, 27–28 September 2007; p. 7.

Article

In Situ Characterization of the Effect of Twin-Microstructure Interactions on $\{10\bar{1}2\}$ Tension and $\{10\bar{1}1\}$ Contraction Twin Nucleation, Growth and Damage in Magnesium

William D. Russell [1,2], Nicholas R. Bratton [1,2], YubRaj Paudel [1,*], Robert D. Moser [3], Zackery B. McClelland [3], Christopher D. Barrett [1,2], Andrew L. Oppedal [1,2], Wilburn R. Whittington [1,2], Hongjoo Rhee [1,2], Shiraz Mujahid [1], Bhasker Paliwal [1], Sven C. Vogel [4] and Haitham El Kadiri [1,2,5,*]

[1] Center for Advanced Vehicular Systems, Mississippi State University, Mississippi State, MS 39762, USA; wrussell530@gmail.com (W.D.R.); brattonnicholas@gmail.com (N.R.B.); barrett@me.msstate.edu (C.D.B.); aoppedal@cavs.msstate.edu (A.L.O.); whittington@me.msstate.edu (W.R.W.); hrhee@me.msstate.edu (H.R.); shiraz@cavs.msstate.edu (S.M.); paliwb@rpi.edu (B.P.)

[2] Department of Mechanical Engineering, Mississippi State University, Mississippi State, MS 39762, USA

[3] U.S. Army Corps of Engineers, Engineer Research and Development Center, Vicksburg, MS 39180, USA; robert.d.moser@usace.army.mil (R.D.M.); zackery.b.mcclelland@usace.army.mil (Z.B.M.)

[4] Materials Science & Technology Division, Los Alamos National Laboratory, Los Alamos, NM 87545, USA; sven@lanl.gov

[5] School of Automotive Engineering, Université Internationale de Rabat, Rabat-Shore Rocade Rabat-Salé, Rabat 11103, Morocco

[*] Correspondence: yubraj@cavs.msstate.edu (Y.P.); elkadiri@me.msstate.edu (H.E.K.)

Received: 25 September 2020; Accepted: 16 October 2020; Published: 22 October 2020

Abstract: Through in situ electron backscatter diffraction (EBSD) experiments, this paper uncovers dominant damage mechanisms in traditional magnesium alloys exhibiting deformation twinning. The findings emphasize the level of deleterious strain incompatibility induced by twin interaction with other deformation modes and microstructural defects. A double fiber obtained by plane-strain extrusion as a starting texture of AM30 magnesium alloy offered the opportunity to track deformation by EBSD in neighboring grains where some undergo profuse $\{10\bar{1}2\}$ twinning and others do not. For a tensile loading applied along extrusion transverse (ET) direction, those experiencing profuse twinning reveal a major effect of grain boundaries on non-Schmid behavior affecting twin variant selection and growth. Similarly, a neighboring grain, with its $\langle c \rangle$-axis oriented nearly perpendicular to tensile loading, showed an abnormally early nucleation of $\{10\bar{1}1\}$ contraction twins (2% strain) while the same $\{10\bar{1}1\}$ twin mode triggering under $\langle c \rangle$-axis uniaxial compression have higher value of critical resolved shear stress exceeding the values for pyramidal $\langle c + a \rangle$ dislocations. The difference in nucleation behavior of contraction vs. compression $\{10\bar{1}1\}$ twins is attributed to the hydrostatic stresses that promote the required atomic shuffles at the core of twinning disconnections.

Keywords: EBSD; magnesium; deformation twinning; texture; damage initiation

---

1. Introduction

The reduction of greenhouse gases and the reduced dependence on hydrocarbon-based fuels are of key priority worldwide. The United States, Canada, China, and the Euro-zone have engaged in a plan to reduce the mass of CO_2 emitted by passenger vehicles a full 30% to 50% below current standards. Such tremendous improvements will require extensive vehicle mass reduction with lightweight materials allowing for a net fuel economy improvement in combustion engines and an increase in the range of electric vehicles, while maintaining/improving current standards for safety and crashworthiness. Two of the most attractive metals with potential to satisfy these conflicting demands, magnesium (Mg) and titanium (Ti) alloys, exhibit a hexagonal close-packed (HCP) crystal structure at room temperature [1]. While Ti is too expensive for most automotive applications, current structural applications of Mg alloys are limited to castings due to difficulties associated with forming wrought alloys [2].

The limitation in forming of Mg alloys is associated with the plastic anisotropy of a hexagonal lattice. The inability of Mg alloys to deform easily along the crystal $\langle c \rangle$-axis slip and satisfy the von Mises criterion of five independent deformation systems for polycrystalline plasticity stipulates plastic accommodation by deformation twins: $\{10\bar{1}2\}$ tension twins and $\{10\bar{1}1\}$ contraction twins. Studies have shown that profuse nucleation of $\{10\bar{1}2\}$ twins followed by twin interactions with other deformation modes and microstructural defects have led to asymmetry, anisotropy, and damage initiation in Mg alloys [3,4]. Likewise, $\{10\bar{1}1\}$ contraction twins are also associated with early damage nucleation in Mg alloys [5]. When a deformation twin nucleates at a grain boundary (GB), propagates across a grain, and reaches the opposite GB, the plastic shear brought about by this twin can be either accommodated by surface kinking, slip, or twinning at the GBs [6,7].

Accommodation effects by slip and kinking are of considerable importance to understand strain incompatibility induced by twinning. While kinking arises in the case where the twin emerges at the free surface, for example, single crystals, accommodation in polycrystals are mainly facilitated by means of a slip or another twin (interaction twins). If a GB has a low misorientation allowing the adjacent grain to have a high local resolved shear stress from the stress concentration ahead of shooting twin, twinning can be readily activated in the adjacent grain [8]. This is a particular case for sharp textures where twins could be observed traversing across multiple grains from one edge to another edge [9–11]. However, if the GB has a sufficiently high misorientation leading to a low local Schmid factor for twinning in the adjacent grain, which could arise in weak textures, slip is necessary or otherwise a crack may nucleate to drive fracture [12–14]. Crack nucleation at the termination of an incident twin meeting an obstacle twin has been actually reported as early as 1868 by Rose [15], who described "Rose channels", or voids, formed by intersecting twins in calcite. These were later described by Priestner [16] as microcracks in body-centered cubic (BCC) metals (Figure 1), and by Sleeswyk [17–19] to explain ductile–brittle fracture transition in BCC iron through a complex emissary dislocation proposed mechanism. Twin-accommodation effects at GBs and twin boundaries (TBs) bear critical implications for damage initiation in HCP structures [20]. Recently, Zhang et al. [4] showed via molecular dynamics (MD) simulations that the openings of cracks in BCC crystal molybdenum at a GB into which a deformation twin impinges (Figure 2).

In general, even for largely unconstrained single crystal, plastic deformation carried by twinning is partly accommodated by kinking and/or slip [21–23]. Holden [21] observed that non-basal slip triggers in most cases to relax any strain incompatibility in pure metals. However, in the case of stronger alloys, such non-basal slip systems tend to be much harder to activate than the basal slip counterparts, which would promote hot stress spots and local fracture initiation. This explains why twinning is not as deleterious to ductility in a face-centered cubic (FCC) as compared to HCP metals, considering that the critical resolved shear stress (CRSS) remains the same for all active slip dislocations (practically one slip mode). Although the role of a twin-accommodation slip in plasticity was invoked and emphasized since the 1950s and 1960s, current crystal plasticity, and thus, continuum mechanics models still largely ignore it. The difficulty of activating the slip might be exacerbated by intergranular particles and solutes, so cracks may readily open at the termination of twin interactions in alloys as suggested by Remy [24,25]. Finally, all these events are compounded with classical effects of slip on strain incompatibility and thus localization. GBs are lattice orientation discontinuities over which strain incompatibility arises. Local phenomena are more pronounced in triple junctions and quadruple points where deformation is more than two grains influencing the effective lattice rotation.

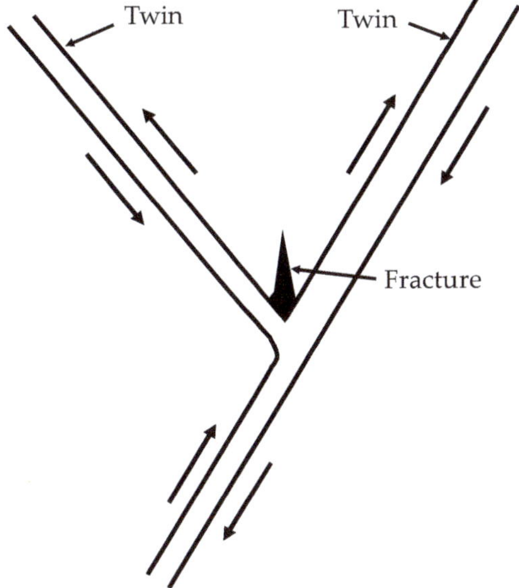

Figure 1. A schematic illustrating crack nucleation at the intersection of two twins, adapted from Priestner [16].

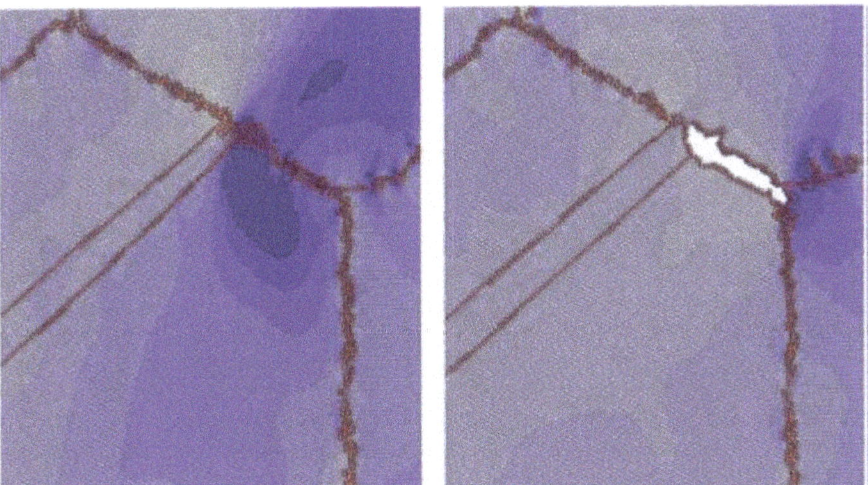

Figure 2. Twin–grain boundary (GB) interaction-induced regions of high stress (dark blue regions) leading to GB crack initiation in Mo according to molecular dynamics (MD) simulation by Zhang et al. [4] (with the permission of Elsevier, 2020). (Each pixelated dot represents an atom).

From the macroscopic standpoint, if one assumes that twinning is responsible for the limited ductility of magnesium, it is not yet clear why ductility is still unacceptable for fibers loaded under a loading orientation where $\{10\bar{1}2\}$ tension twinning activity is minimized. For instance, when a basal texture is pulled normal to the fiber axis, prismatic $\langle a \rangle$ and pyramidal $\langle c + a \rangle$ slip should be the dominant deformation mechanisms. Although an improvement in total elongation is obtained, compared to the case of profuse twinning having a sigmoidal stress-strain curve (20% vs. 12%, respectively), necking does not take place and fracture occurs in a brittle-like fashion. One may argue that locally, twinning occurs because of slip-induced lattice rotations. However, the amount of rotation remains largely incapable of bringing a favorable orientation of $\{10\bar{1}2\}$ twinning. There are reports of $\{10\bar{1}1\}$ [26,27] (also $\{10\bar{1}3\}$ and $\{10\bar{2}4\}$) formation in magnesium alloys that show deleterious effect in ductility [5,28]. Most of these reports suggest the higher values of CRSS for $\{10\bar{1}1\}$ such that they would only appear inside $\{10\bar{1}2\}$ twins during the saturation stage of stress after the rapid hardening correlated with profuse twinning [27].

In crystal plasticity simulations of pure magnesium, Oppedal et al. [29] used a CRSS value of 181 MPa for $\{10\bar{1}1\}$ twinning compared to 11 MPa for basal $\langle a \rangle$, 15 MPa for $\{10\bar{1}2\}$ twinning, 30 MPa for prismatic $\langle a \rangle$, and 50 MPa for the very hard second order pyramidal $\langle c + a \rangle$. In general, $\{10\bar{1}1\}$ twinning has a high Schmid factor in this loading orientation but because of its high CRSS, it is not expected to form at the relatively low saturation stress levels attained during tension normal to the basal fiber. Apart from the selection of deformation modes, studies show the competition within the twin variants of a tensile or a compression twin to accommodate deformation within a grain [30–32]. Researchers have associated twin variant selections to strain accommodation at the grain boundary, hydrostatic stresses, and twinning dissociation energies [33–35].

In an effort to bring experimental evidence and elucidate the dominant damage mechanisms involving twinning in HCP metals, we performed fractographical analysis as well as microstructural analysis on the interactions between twins with other deformation modes and microstructural defects. An in situ and interrupted electron backscatter diffraction characterization is performed on flats formed in a plane-strain constraint found in an extruded AM30 magnesium alloy "crush rail" showing a double-fiber texture. This texture allowed both profuse and no twinning in neighboring grains under the same uniaxial tension. This work closely studies the damage caused by various twin variant interactions as well as parsing their effect on texture evolution within each grain. Furthermore, double-fiber texture with grains, which have favorable Schmid values for twinning and are surrounded by grains prohibiting twinning, provided us an opportunity to study the effect of GB strain-accommodation on twin variant selections within a grain and damage initiation along the GBs.

2. Methods and Materials

2.1. Experimental Procedure

Magnesium alloy AM30 with a chemical composition (wt %) of 2.83 Al, 0.386 Mn, 0.0037 Zn, 0.03 Fe, <0.0015 Ni, <0.0005 Cu, balance Mg was the material of this study. Extrusion of the AM30 alloy was performed by Timminco. The process began with pre-extrusion to bring down the diameter of the cast billet from a 450 to 230 mm diameter cylinder through an extrusion press that ran up to a maximum pressure of 22 MPa and then pushed the 230 mm billet through the final die at 16.5 MPa. The entrance and the exit temperature were around 440 and 540 °C, respectively. The ram speed for the extrusion was about 1.3 mm s^{-1}, while the extrusion speed was about 45 mm s^{-1}. The extrusion ratio was 35.26:1. Lastly, the billet was extruded into the final shape, which corresponded to a hollow tube in "double-hat" like shape, known as a "crush rail" (Figure 3). This "crush rail" showed a texture that has a strong basal pole aligned with the in-plane direction of the sheet normal to the extrusion direction (extrusion traverse, or ET), and a weak basal texture component in the normal direction of the sheet (EN) (Figure 4). As such, under tension along ET (resp. EN), several grains belonging to the $ET\|(0001)$ fiber (resp. $EN\|(0001)$ fiber), would be subject to the formation of all six $\{10\bar{1}2\}$ twin variants contributing to their very large Schmid factor (0.499). While grains belonging to the $EN\|(0001)$ (resp. $ET\|(0001)$ fiber) fiber will experience favorable conditions for slip mechanisms to take place exclusively and perhaps $\{10\bar{1}1\}$ twins if enough stress is attained.

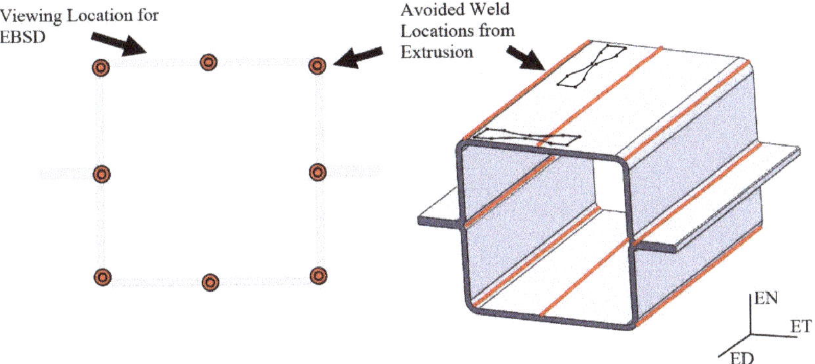

Figure 3. Magnesium AM30 crush rail is an extruded component used in an automotive shock absorption structure. This extrusion allows plane-strain conditions on the flat sections that promote a desirable double-fiber texture for studying twins as it promotes nucleation of multiple twin modes and variants in tension within neighboring grains.

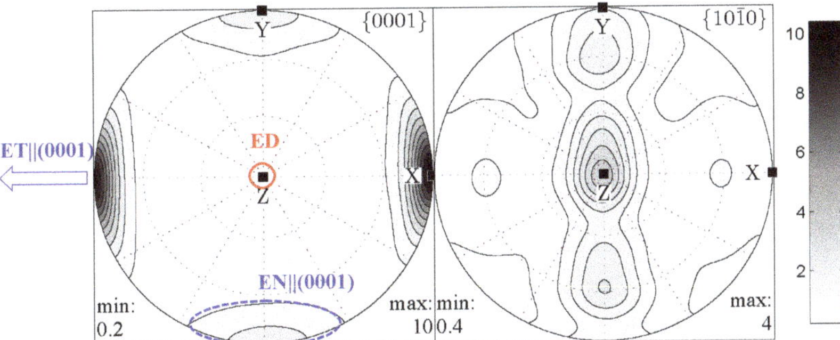

Figure 4. [0 0 0 1] and ⟨1 0 $\bar{1}$ 0⟩ pole figures obtained from neutron diffraction analysis showing $ET||\{0001\}$ plus $ED||\{0001\}$ double fiber texture of an AM30 magnesium alloy. The extrusion direction (ED) is along the z-axis with the extrusion transverse direction (ET) along the x-axis and the extrusion normal direction (EN) along the y-axis. The pole figures show high concentrations of [1 0 $\bar{1}$ 0] oriented grains parallel to the extrusion direction [36] (with the permission of Elsevier, 2020).

2.2. Fractography

Specimens machined in extrusion (ED) and extrusion transverse directions (ET) were planar ground using 4000 grit silicon carbide paper and then electropolished to a pristine finish. Through interrupted electron backscatter diffraction (EBSD) analysis on specific regions, the evolution of deformation and damage was studied. Figure 5 shows stress–strain curves obtained at $0.001\,\text{s}^{-1}$ strain rate when the specimens were taken to full fracture, which delineate traditional anisotropy of magnesium that is associated with basal and rod fiber textures. Under a tensile load in the ED direction, $\{10\bar{1}2\}$ twinning is nearly non-existent yielding to parabolic stress–strain behavior, while under ET all the grains belonging to the strong $ET||(0001)$ fiber experience profuse $\{10\bar{1}2\}$ twinning eventuating to a sigmoidal curve with very rapid hardening. After failure, fractographic analyses were conducted using a Zeiss Supra 40 field emission scanning electron microscope (FEG-SEM) to determine the overall causation of failure.

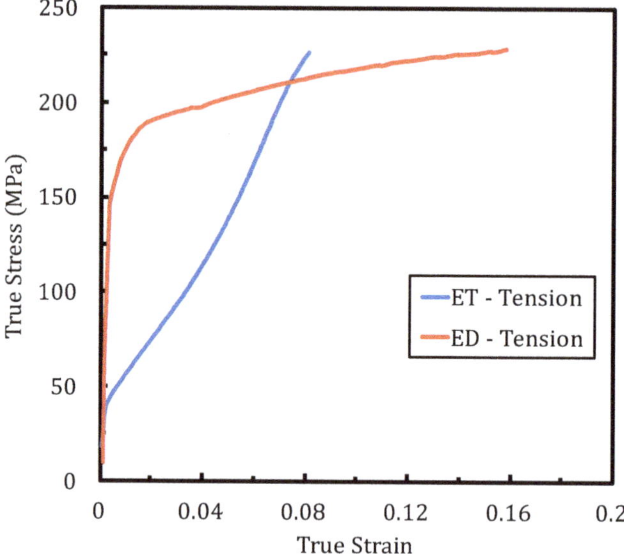

Figure 5. Stress-strain relationship of AM30 alloy in extrusion and extrusion transverse (ET) directions. A sigmoidal curve for tension along the ET direction indicates high strain hardening, which is characteristic of profuse tension twinning, while tension along the extrusion direction (ED) alludes to deformation due to slip with minimal twinning [36] (with the permission of Elsevier, 2020).

2.3. In Situ and Interrupted Electron Backscatter Diffraction (EBSD)

The interrupted EBSD characterization and collection were performed with a FEG-SEM equipped with orientation image microscopy (OIM, EDAX TSL) at a step size of 100 nm on large regions of the samples. These EBSD analyses were completed using the commercially available TSL OIM analysis software from EDAX. Neutron diffraction texture analysis was conducted with the HIPPO neutron time-of-flight diffractometer at LANSCE [37]. HIPPO consisted, at the time of the data collection, of 1320 ^{3}He detector tubes arranged on 30 panels on three rings, covering 14.2% of 4π steradian (sr) for a single rotation and 44.5% with four rotations [38]. The sample was glued on a sample holder, which was loaded on a robotic sample changer [39]. Beam collimation was 10 mm in diameter and count times were 10 min per rotation. The diffraction data were analyzed using the Rietveld method for the 120 diffraction histograms collected for four sample rotations of 0°, 45°, 67.5° and 90° around the vertical axis following procedures described in Wenk et al. [40]. The pole figure data were imported into MTEX [41] for further ODF (orientation distribution function) analysis. Intermittent monotonic tension tests were performed using an Instron 8856 universal test frame at a quasi-static strain rate (0.001 s^{-1}).

Most of the observations were conducted on sections normal to ET. The specimen was taken from the "crush rail" in such a way to avoid irregular texture created by welding in the extrusion process. Two contiguous areas of interest (AOI) as shown in the initial microstructure of Figure 6 were identified that are expected to show very different levels of twinning (Region AOI 1 and Region AOI 2). AOI 1 is composed of three primary grains all of which prescribe to the $[1\,1\,\bar{2}\,0]$ orientation with the $\langle c \rangle$-axis along the line of tension, and as such favorable for $\{1\,0\,\bar{1}\,2\}$ twinning (belonging to $ET||(0\,0\,0\,1)$ fiber). AOI 2 is also composed of the three primary grains showing a $[0\,0\,0\,2]$ basal texture meaning the $\langle c \rangle$-axis is perpendicular to the loading direction (belonging to $EN||(0\,0\,0\,1)$ fiber). This region should experience a dominance of slip deformation mechanisms, specifically basal $\langle a \rangle$, prismatic $\langle a \rangle$, and pyramidal $\langle c + a \rangle$ [36].

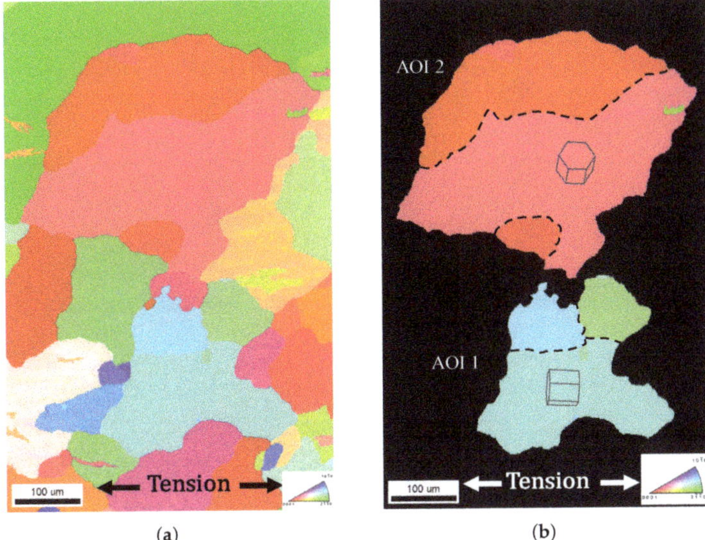

Figure 6. (**a**) Inverse pole figure (IPF) map generated through electron backscatter diffraction (EBSD) scans showing initial microstructure of Mg AM30 before deformation and (**b**) the highlighted partitions of the areas of interests (AOI 1 and AOI 2) with discrete unit crystal orientations that are the focus of study.

3. Results and Discussion

3.1. Fractographic Analysis of Twin-Interactions Induced Cracks

Coupled EBSD and FEG-SEM micrographic characterization at different strain levels revealed cracks nucleating at the intersection between two intersecting $\{10\bar{1}2\}$ twins, a twin and slip bands, and twin and grain boundaries. Figures 7–9 depict tension twin interactions with slip, GBs and other twins, respectively. In all these cases, the macroscopic SF of the twins implicated in damage was low and their growth was sluggish.

For the case of slip–twin interaction (Figure 7), one possible mechanism behind crack nucleation is the difficulty for the twin to facet or form a disconnection as a dislocation meets the interface [42–46]. This would lead to an accumulation of dislocations and potential slip bands that create strain incompatibility and this hots spots inside the twin domains. These phenomena could be exacerbated by the transmutation effect, which would increase the density of sessile dislocations [47].

Figure 7 shows the formation of micro-cracks due to interactions between tensile twins and basal slip bands. The slip–twin interaction creates a strain incompatibility at the twin interface, which leads to high stress concentration zones or hot spot regions. The occurrence and distribution of slip bands along with the glide of interstitial type dipole loops creates a vacancy, which in turn cause crack nucleation in slip–twin interfaces [48]. The interactions between basal dislocation and $\{10\bar{1}2\}$ twin could proceed in three different ways: (i) direct transmission that could be seen as mere cross slip of this dislocation into the basal plane of the twin, (ii) indirect transfer characterized by transmutation of the initial dislocation into a sessile [c]-containing dislocation on the prismatic plane leaving a disconnection along the boundary, and (iii) activation of non-basal slips [49]. The formation of sessile dislocation due to interaction between a slip–twin boundary, disconnections left in the interface, and pile ups of transmitted dislocation are consistent with the initiation of cracks [50].

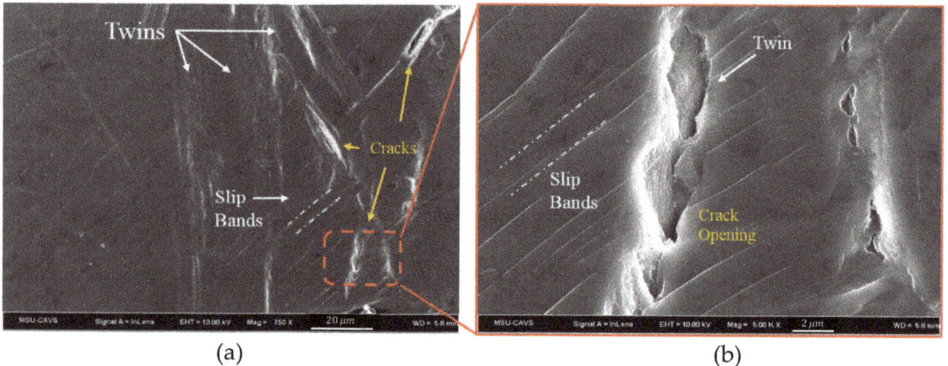

Figure 7. (**a**) Micrographs depicting slip–twin interactions induced crack nucleation and (**b**) zoomed image showing crack opening due to slip-twin interactions from the dashed red region in (**a**).

For the case of cracks nucleating at the twin-GB intersections revealed in Figure 8, one plausible explanation is the difficulty to accommodate strain at the GB due to the unfavorable orientation of the next grain to accommodate the local stress by easy slip. Propagating twins move through the grain until they reach the boundary at which the stress field carried ahead of the twin tip must be accommodated. In the case of a grain orientation favorable for twinning on any plane variant, which is a typical case of sharp textures, a new twin is formed in the next grain, giving rise to autocatalytic twinning. However, if the orientation requires activation of hard slip, i.e., hard neighboring grains, the twin may cease to propagate lengthwise beyond the GB. The twinning dislocations lying on the top of each other, which has been driving this propagation will pile-up at the GB, create a region of high stress concentration that would continue to exacerbate with further strain. As such, depending on the orientation of neighboring grains (i.e., GB misorientations) twin-GB regions of intersection are prone to hot spots and damage nucleation.

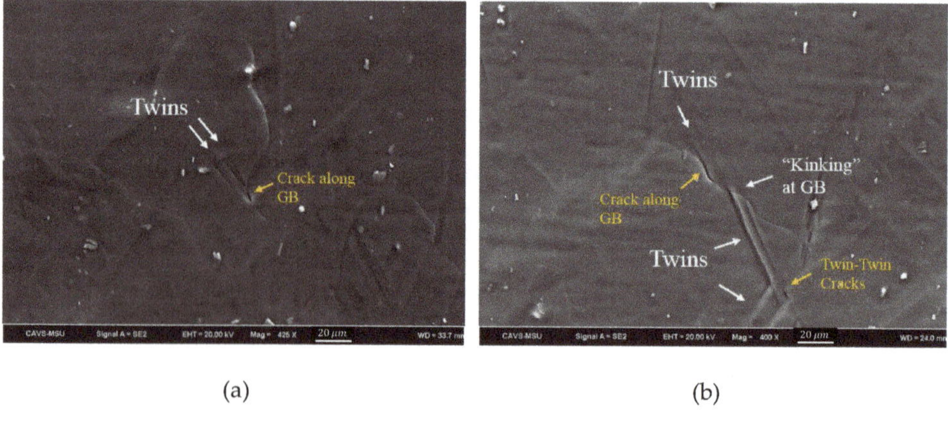

Figure 8. (**a**) Twin–grain boundary (GB) interactions leading to crack nucleation due to a lack of strain accommodation. (**b**) Proper accommodation of strain is found at the surface where GB emerges through "kinking". However, a crack is formed at the GB from interactions with another twin. Cracks are also noted from twin–twin interactions.

Figure 9 reveals cracks nucleating exactly at twin–twin intersections, which has been reported in early literature for BCC systems [24,25,51–53]. When an incident twin encounters an obstacle twin, the strain can be accommodated by either of the following mechanisms: (i) retwinning of the obstacle twin, (ii) slip in the incident or obstacle twin, or (iii) detwinning of the incident twin [17,51,54]. As described by Beyerlein and Tomé [55], grains with multivariant twinning are more likely to rapidly strain harden due to the fact of lower CRSS of twin propagation compared to nucleation. Intersecting twins have been actually associated with fractures in the literature since 1957 [56–58] or even earlier as noted in these papers. Further evidence of twins variants impeding the variants growth and building strain until a crack nucleation is observed in AOI 1 and discussed in the following subsection. Based on the local crystallography, two types of twin–twin interactions could occur depending on whether the two twin variants share the same $\langle 1\,1\,\overline{2}\,0 \rangle$ zone axis or not. In the first case, one twin does not transmit across another, whereas for the latter type, the twin could transmit through under proper conditions of loading. However, for most cases the former type of twin–twin interaction occurs with quilted-looking twin structures consisting of boundary dislocations [59]. This phenomenon might be exacerbated in the case of low SF twins, as their lack of thickening rate might promote strong strain incompatibilities at the triple junction and the likelihood of debonding between the two twins.

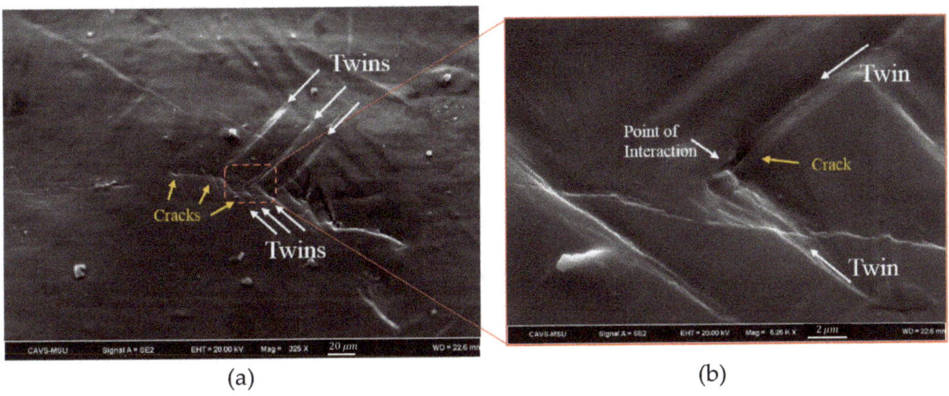

(a)   (b)

Figure 9. Micrographs of (**a**) $\{1\,0\,\overline{1}\,2\}$ twin–twin interactions and (**b**) their involvement in crack nucleation viewed through zoomed image of the dashed red region in (**a**).

Double twinning has been used to describe multifold twinning within an original twin as early as the 1960s [60]. Within Mg, double twinning enhances damage and commonly occurs with $\{1\,0\,\overline{1}\,1\}$ twins forming within a $\{1\,0\,\overline{1}\,2\}$ tensile twin [5,26,27,61–65]. However, in the case of this study, contraction twins are found to form at relatively low stresses and are associated with double twinning. In basal textured AM30 alloy, $\{1\,0\,\overline{1}\,1\}$-$\{1\,0\,\overline{1}\,2\}$ double twins were reported to be preferential sites for fracture initiation. Also, the sharp surface steps related to these double twins were associated with the formation of a crack [66,67]. This is evident in the massive cracks formed in Figures 10 and 11.

Notice the contraction twins are located in a parent grain and did not form through double twinning. Figure 11 shows the misorientation of the twin with the parent to be 56° and not the 38° of $(1\,2\,\overline{1}\,0)$ expected for double twinning [5,62]. In addition, the size and shape are that of a contraction twin, while a double twin would fill the silhouette of the original twin.

Contraction twin induced damage could be explained by the inability of a slip to penetrate the hard contraction twin boundary, creating large areas of cleavage and stepped terraces within the fracture surface commonly related to brittle failure (Figure 12). The tormented areas occasionally observed in the fracture surface are attributed to slip deformation around the twin and grain boundaries and are commonly related to a ductile failure.

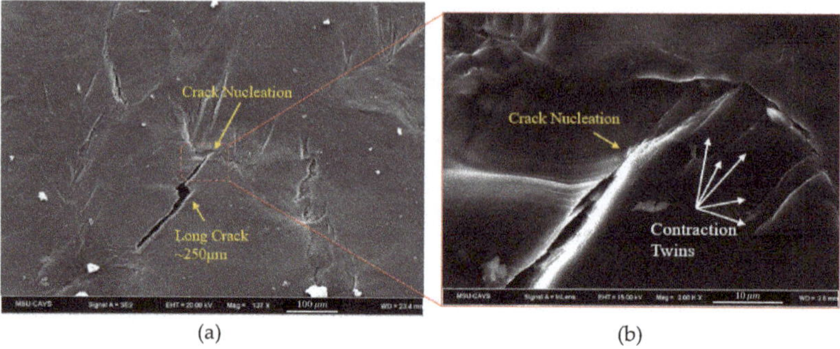

(a)   (b)

Figure 10. (**a**) Micrograph of evident contraction twins near long crack nucleation and (**b**) enlarged micrograph showing sharp surface steps made by these contraction twins.

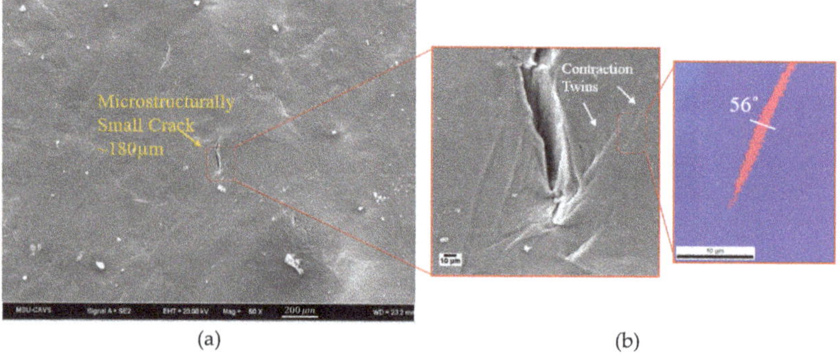

(a)   (b)

Figure 11. (**a**) Crack formation due to contraction twin interactions and (**b**) EBSD micrograph showing the contraction twins with a misorientation of 56°.

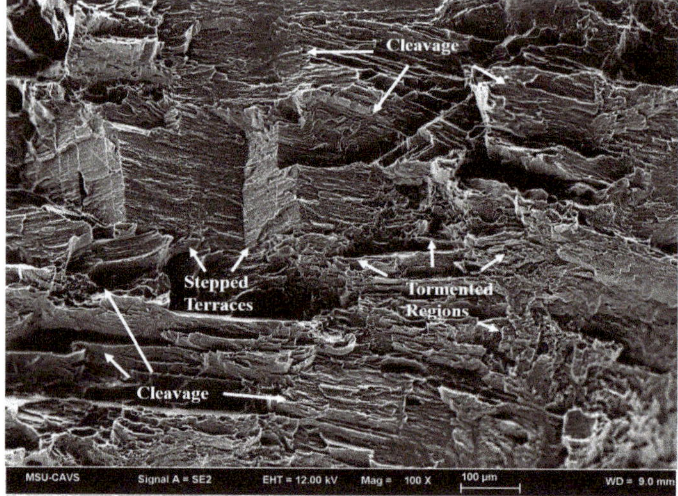

Figure 12. Micrograph shows evidence of dual phase fracture. Cleavage and stepped terraces formed from twin interactions caused brittle failure, whereas the tormented regions show ductile failure.

3.2. Interrupted EBSD Characterization in AOI 1

Figure 13 shows that AOI 1 readily experiences $\{10\bar{1}2\}$ tensile twinning caused by the extension of the $\langle c \rangle$-axis. Resolved Shear Stress (RSS) is commonly used to determine the likelihood of elevation of stress through slip or twins. The RSS is dependent on the Schmid factor (SF), $0 \leq |SF| < 0.5$, of slip or twin plane of interest and the applied stress (σ_{app}) of the material [29] (Table 1 describes the RSS of possible slip and twin systems). With all variants of $\{10\bar{1}2\}$ tensile twins having near maximum SF values, all variants are equally likely to be present in the EBSD micro textures.

Table 1. Values of resolved shear stress and Schmid factors determined by $[0\,0\,0\,1]$ loading with applied stress at yield of ET-Tension in Figure 2.

	Slip or Twin Plane	Schmid Factor	Resolved Shear Stress
	$(h\,k\,i\,l)\,[u\,v\,t\,w]$	SF	RSS (MPa)
Tension Twin Variants	$(\bar{1}012)\,[10\bar{1}1]$	0.499	19.96
	$(10\bar{1}2)\,[\bar{1}011]$	0.499	19.96
	$(0\bar{1}12)\,[01\bar{1}1]$	0.499	19.96
	$(01\bar{1}2)\,[0\bar{1}11]$	0.499	19.96
	$(1\bar{1}02)\,[\bar{1}101]$	0.499	19.96
	$(\bar{1}102)\,[1\bar{1}01]$	0.499	19.96
Slip Systems	Basal $\langle a \rangle$	0	-
	Pyramidal $\langle a \rangle$	0	-
	Pyramidal $\langle c + a \rangle$	0.401	16.04

When determining the twin variants, multiple factors are checked to confirm a twin variant. The simplest of these is boundary misorientation of twin and parent. In the case of all twins present, the misorientation is found to be $86° \pm 8°$. We also know that the $\langle c \rangle$-axis of the parent grain is in line with the loading direction allowing extension of the $\langle c \rangle$-axis, which promotes tensile twins. With evidence of confirmed tensile twins, we specified the Miller–Bravais indices of each twin based on a reference frame choice. The data from the EBSD analysis is used to create an orientation map, which highlights similar areas of lattice orientation based on the angle of the $\langle c \rangle$-axis. This map is shown in Figure 14 alongside the twin planes of the $\{10\bar{1}2\}$ system. The six twin variants found in the $\{10\bar{1}2\}$ system are made up of three distinct twin pairs and are highlighted as such in Figure 14. Each orientation of the present twins was used to calculate the SF of each twin plane. The average SF for each twin pair under a non-idealized $[10\bar{1}\ 30]$ loading is shown in Table 2.

Table 2. Schmid factors of each of the three distinct $\{10\bar{1}2\}$ pairs with $[10\bar{1}\ 30]$ loading.

Tensile Twin Pairs	$(0\bar{1}12)$	$(\bar{1}102)$	$(\bar{1}012)$
Schmid Factor (m)	0.219	0.479	0.245

Incremental ET tensile tests of the AM30 with interrupted EBSD scans show the nucleation and progression of twins in AOI 1 (Figure 13). At 0% strain, very small $\{10\bar{1}2\}$ tensile twins were observed in the top right grain of AOI 1. The twins likely nucleated after grinding and polishing and their orientations are found to have a Schmid factor of SF = 0.43. This further provides an opportunity to discuss on twin–twin interaction effects within a single grain. While the other two grains are engulfed by tension twin at 3%, the top right grain only has half of the region covered by twin. The pre-existing twin blocks the growth of the red twin across the other half of this grain. Ultimately, twin nucleates and grows from the grain boundary with the right-bottom grain. The stress–strain relationship is commonly categorized by three regimes. The first regime represents the elastic portion of the response while the second encompasses the plastic strain leading to an inflection point in this case, the transition from twin nucleation dominated to twin growth dominated strain accommodation and the final regime represents the continuation of plastic deformation until failure. Within Regime I of the strain relationship twins readily nucleate from relatively high misorientation GBs before 1% strain could have been reached. Although six tensile twin variants are present, a dominant variant of each distinct pair present is conspicuous. This was associated by the ability to nucleate at all possible sites while the recessive variant nucleates from other twin boundaries. The dominant variant of each pair is easily distinguished in the pole figures of Figure 15b.

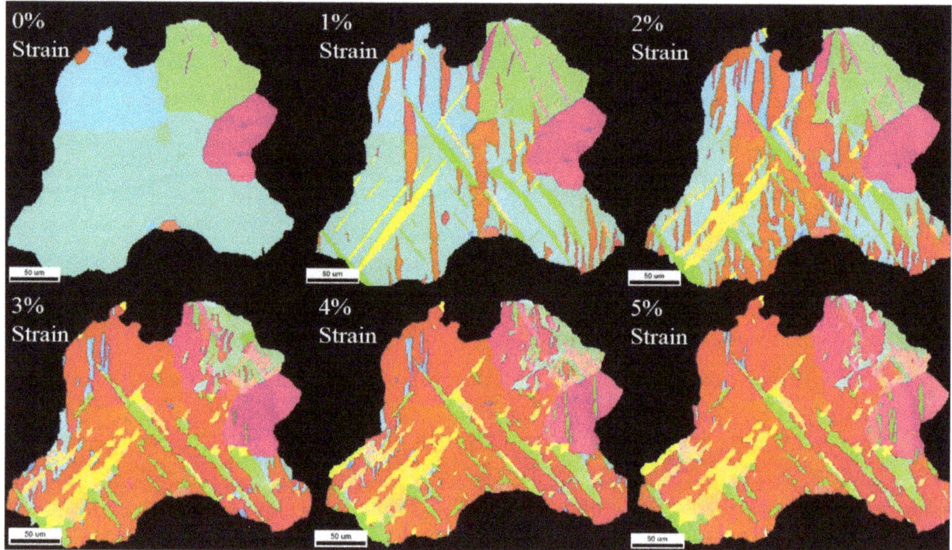

Figure 13. Inverse pole figure maps at various incremental strains ranging from 0% to 5%. The tensile load is applied horizontally with respect to the image. The six variants of the $\{10\bar{1}2\}$ tensile twins nucleate and propagate in the parent grains resulting in a fully reoriented grain by 5% strain.

To determine the effect of twin nucleation on growth, we measured the GB perimeter fraction at which each twin variant nucleated with respect to the overall GB length. Figure 14b shows a plot of this tendency, which outlines a noticeable advantage of the $(10\bar{1}2)$ variant. Coupled with twin–twin interactions between the $(01\bar{1}2)$ and $(\bar{1}102)$ variant pairs (green and yellow), stunting twin growth, enhanced nucleation at the GB allowed the $(10\bar{1}2)$ variant pair (red) to dominant overall twinning in AOI 1. By 5% strain, 97.3% of the parent grain has twinned. The twin variant area fractions are plotted in Figure 15.

The active tensile twins are composed of three distinct pairs. In Figure 14a, the tensile twin variants are highlighted in pairs with the respective colors of green yellow and red and further crystallographically determined in Figure 14c. This maps the twins by their orientation relative to each other instead of a global orientation given by the inverse pole figure. By doing so, all variants are easily distinguished at each strain level without confusing between new nuclei and twin thickening. Figure 14b plots each of the variants' nucleation occupancy in the GB at 1% strain. It is important to note that as soon as the strain reached 1%, nucleation nearly ceased and twin fraction dominantly increased by twin thickening. Therefore, the $(10\bar{1}2)$ variant occupies the majority of GB restricting growth of the $(01\bar{1}2)$ and $(\bar{1}012)$ twin variants. In fact, so few of the recessive variants occupy the GB, their growth is nearly negligible and mostly relies on forming from other twin boundaries.

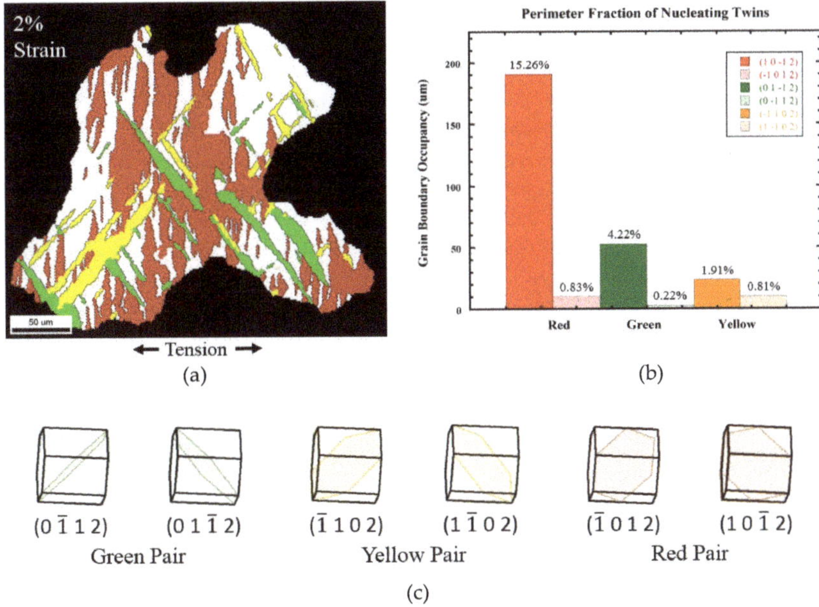

Figure 14. Twin variant selection in AOI 1. (a) The orientation map of the three evident twin pairs highlighted in green, yellow, and red. (b) By mapping the grain boundary occupancy of twins at 1% strain (just after nucleation), the perimeter fraction of twinning can be determined. The perimeter fraction shows a clear dominance of $(10\bar{1}2)$ twinning with evident dominate variants within the green and yellow pairs, i.e., $(01\bar{1}2)$ and $(\bar{1}102)$ respectively. (c) The proposed variant pair twin planes are shown within parent orientation with their respective colors.

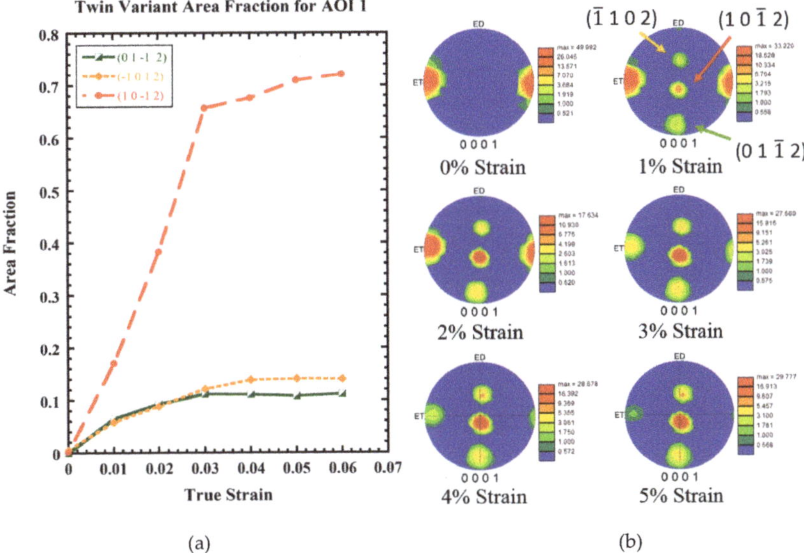

Figure 15. (a) Evolution of twin variant area fraction with strain for the three distinct pairs of the $\{10\bar{1}2\}$ system in AOI 1. (b) (0001) pole figures showing the nucleation of three dominate twin variants forming at 1% strain then growing in intensity as the parent orientation diminishes.

Even though the red $(10\bar{1}2)$ variant pair had a lower Schmid Factor compared to the yellow $(\bar{1}102)$ and green $(0\bar{1}12)$ variant pairs, they continued to engulf the parent grain while the yellow and green variants grew comparatively little after nucleation. This is thought to be an effect of grain boundary enhanced disconnection nucleation. In fact, the red $(10\bar{1}2)$ variant pair had a chance to thicken along the grain boundary segments of the grain, which is clearly correlated to the much higher thickness of these twin variants as they propagate along the grain boundary (Figure 14a). The other yellow and green variants were engulfed away from the grain boundary and thus were subject to substantial twin–twin interactions. Likewise, the pole figures in Figure 14b show that noticeable advantage of $(10\bar{1}2)$ twin variant can be the effect of a free surface on twin nucleation and growth. The pole for $(10\bar{1}2)$ variant is aligned towards the free surface compared to other twin variants, thus the free surface provides easy stress relief compared to the grain boundaries [68]. The EBSD micrographs provide a two-dimensional view of a three-dimensional twin with limited information on neighboring grain boundary effects. The understanding of the effect of neighboring grain boundaries on twin nucleation, propagation, and growth demands further research with three-dimensional observation of a twin through atomistics and modeling.

3.3. Characterization of Contraction Twinning in AOI 2

The second area of interest has a very different orientation compared to AOI 1. AOI 2 is composed of the three primary grains with a basal orientation (0001). This orientation belongs to the weak texture of the double-fiber. The basal orientation aligns the $\langle c \rangle$-axis perpendicular to the loading direction, which leads to contraction of the $\langle c \rangle$-axis.

Figure 16 shows a schematic allowing to compare the symmetric pair of in-plane compression (IPC) of ED leading to $\{10\bar{1}1\}$ compression twin and $[2\bar{1}\bar{1}0]$ tension of ET leading to $\{10\bar{1}1\}$ contraction twinning. IPC experiences a linear stress compressing the $\langle c \rangle$-axis resulting in a net loss of volume, while $[2\bar{1}\bar{1}0]$ tension experiences complex hydrostatic stress with maximum compression acting on the center plane and maximum extension occurring along the loading axis. This results in a net gain in volume, which would aid shuffles and thus nucleate $\{10\bar{1}1\}$ twinning.

Compression twins $\{\bar{1}011\}$ and shear bands, formed from compression twins, have been noted for failure in plane-strain compression [69]. Compression twins usually remain as thin and needle-like lamellae, a fact that was attributed to the low mobility of twinning disconnections due to the step height and complexity of atomic shuffling requirements [70]. The formation of compression twins is believed to occur at relatively high stress levels due to their very large CRSS [29]. Although compression twins and contraction twins are exactly of the same crystallographic nature, $\{\bar{1}011\}[10\bar{1}2]$, contraction twins in this study have been observed to nucleate at substantially lower CRSS, well below previously reported values in the literature when the $\langle c \rangle$-axis was under compression, not contraction.

The progression of damage in AOI 2 is shown in Figure 17 and, surprisingly, thin needle-like structures are nucleating at strains as low as 2%. At 0% strain, a tensile twin has formed likely due to handling of the specimen as the twin detwins with increasing strain. Since the focus in this region is to study contraction twinning and the small pre-existing tension twin has no effects on the nucleation or growth of the contraction twin, the effects of the tension twin is neglected. By 2% strain, contraction twins with SF of 0.209 and 0.448 have formed, while there is evidence of another about to nucleate. By 5% strain, multiple contraction twins are formed and large distortions can be observed within the parent grain. Like tensile twins, contraction twins are characterized based on their misorientation with the parent along the boundary (56°).

Like AOI 1, the SF of possible slip systems as well as the $\{\bar{1}011\}$ contraction twins are compared in Table 3. When comparing these values, contraction twins are not likely to contribute significantly to the deformation accommodation as prismatic and pyramidal slip modes.

Table 3. Values of resolved shear stress and Schmid factors determined by $[2\,\bar{1}\,\bar{1}\,0]$ loading with applied stress at 2% strain of ET-Tension in Figure 5.

	Slip or Twin Plane	Schmid Factor	Resolved Shear Stress
	$(h\,k\,i\,l)\;[u\,v\,t\,w]$	SF	RSS (MPa)
Comp. Twin Variants	$(\bar{1}011)\,[10\bar{1}2]$	−0.311	21.77
	$(10\bar{1}1)\,[\bar{1}012]$	−0.311	21.77
	$(0\bar{1}11)\,[01\bar{1}2]$	0	-
	$(01\bar{1}1)\,[0\bar{1}12]$	0	-
	$(1\bar{1}01)\,[\bar{1}102]$	−0.311	21.77
	$(\bar{1}101)\,[1\bar{1}02]$	−0.311	21.77
Slip Systems	Basal$\langle a\rangle$	0	-
	Prismatic $\langle a\rangle$	0.433	30.31
	Pyramidal $\langle a\rangle$	0.382	26.74
	Pyramidal $\langle c+a\rangle$	0.446	31.22

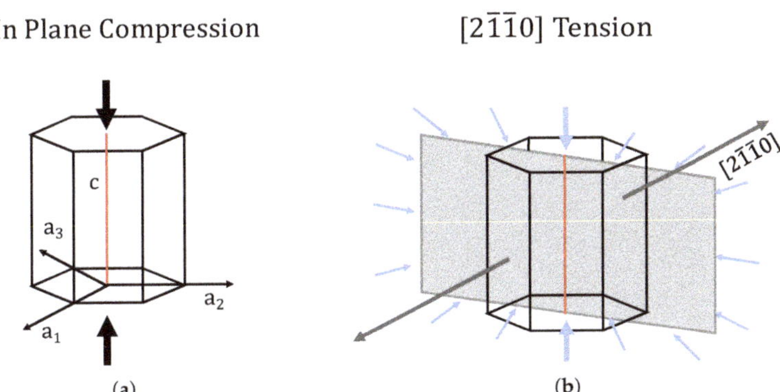

Figure 16. Comparison of the stress states experienced by (**a**) in-plane compression and (**b**) $[2\,\bar{1}\,\bar{1}\,0]$ tension. The representation of $[2\,\bar{1}\,\bar{1}\,0]$ tension is simplified to a two-dimensional plane stress representing the plane of maximum compression in an attempt to visualize the complexity of the stress state experienced by contraction of the $\langle c\rangle$-axis due to the Poisson effect.

In an effort to understand the reason for the $\{\bar{1}011\}$ twin to nucleate at lower stress levels in contraction than in compression stress states with respect to the $\langle c\rangle$-axis stress sign, we performed a more detailed characterization of their micro texture and microstructural conditions within AOI 2. Figure 18 displays the highlighted contraction twins are grouped by $\{10\bar{1}1\}$ distinct pairs. For this characterization, the $(\bar{1}101)$ pair is blue, $(0\bar{1}11)$ pair is green, and the $(\bar{1}011)$ pair is orange. The selection of the pairs were determined by relating the actual orientation of the twins (Pole Figure in Figure 18b) to the idealized orientation of $\{10\bar{1}1\}$ contraction twins (Pole Figure in Figure 18c). As expected, the actual orientations do not precisely relate to the idealized orientation but the general orientations match with reasonable accuracy.

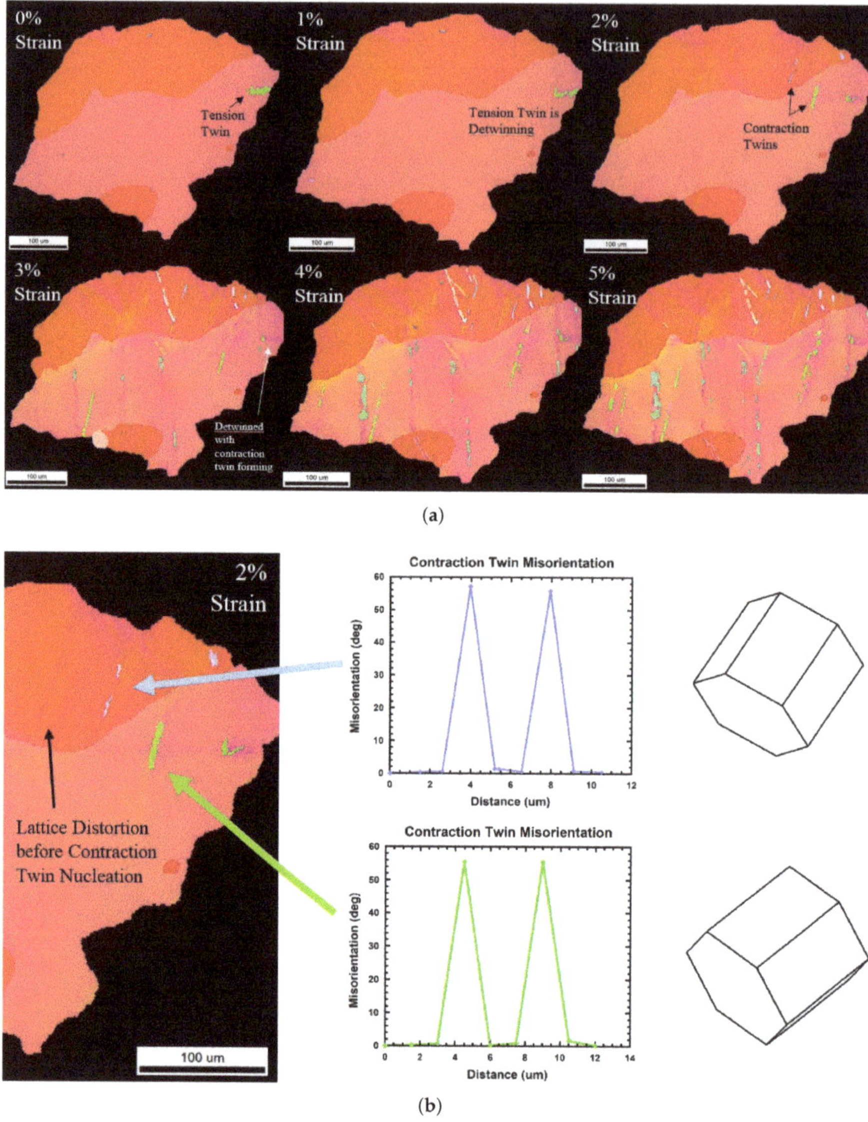

Figure 17. (**a**) Inverse pole maps revealing lattice reorientation due to profuse nucleation of $\{10\bar{1}1\}$ contraction twinning and slip as deformation proceeds from 0% to 5%. Although pyramidal slip $\langle c + a \rangle$ has a relatively higher SF, there is profuse nucleation of contraction twins. (**b**) The misorientation data of the twins indicates the presence of contraction twins as early as 2% strain with unit crystal orientation of both twins. The misorientation peak in the plot represents a twin boundary with parent grain.

However, by 4% strain, the contraction twins have begun to experience twinning in their lattice, a phenomenon known as double twinning or retwinning. This process has been characterized at length in early literature [3,26,71–74]. Double twinning occurs when a residual twin accommodates the increasing strain by retwinning its own lattice. This process is common to an extension twin in which a compression twin nucleates [27]. The misorientation of the parent grain is not 86°, as with most tensile twins, but instead 38° ± 7° of $(1\bar{2}10)$ [5]. In Figure 19, all of the boundaries meeting the prescribed misorientations is highlighted in black.

To verify the presence of double twinning, a twin of the $(0\bar{1}11)$ green pair was selected at 3% strain (Figure 19). This twin has completed its growth by 3% strain and shows evidence of double twinning by 4% strain, which creates an attractive location for further analysis. The misorientation data shows that the twin is indeed a contraction twin at 3% strain and then twins again by 4% strain making a misorientation with the parent grain of 36°. It looks like the contraction twin reorients the lattice to a very favorable orientation for tension twinning, which enjoys a very low critical resolved shear stress compared to the values of other non-basal deformation systems, prompting the double twinning effect. This event leads to an interface where perhaps neither of the tension or contraction glissile twinning disconnection can glide, and thus stopping the thickening of the twin lamella. This double twinning phenomenon clearly explains the thin-needle shape of contraction twins, as tension twinning and a modification of the atomic interface structure rapidly plague them.

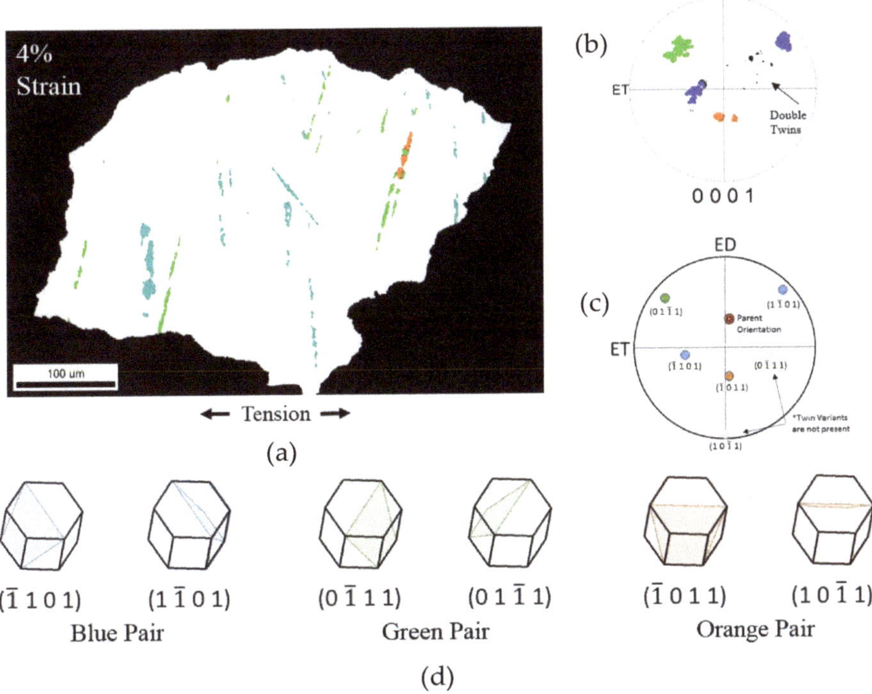

Figure 18. Characterization of $\{10\bar{1}1\}$ contraction twins at (**a**) 4% strain revealing continuous nucleation of new twin variants. The twins were individually selected to be partitioned from the parent grain. The texture of the partitioned twins is shown in part (**b**). Double twins are found among the partitioned twins. The idealized texture of $\{10\bar{1}1\}$ contraction twins is shown in part (**c**). Not all twin variants are present, but they are representatives of each twin pair. The contraction twin variants are characterized by their respective color pair highlighted in the parent grain (**a**), pole figures (**b**,**c**), and crystal lattice (**d**).

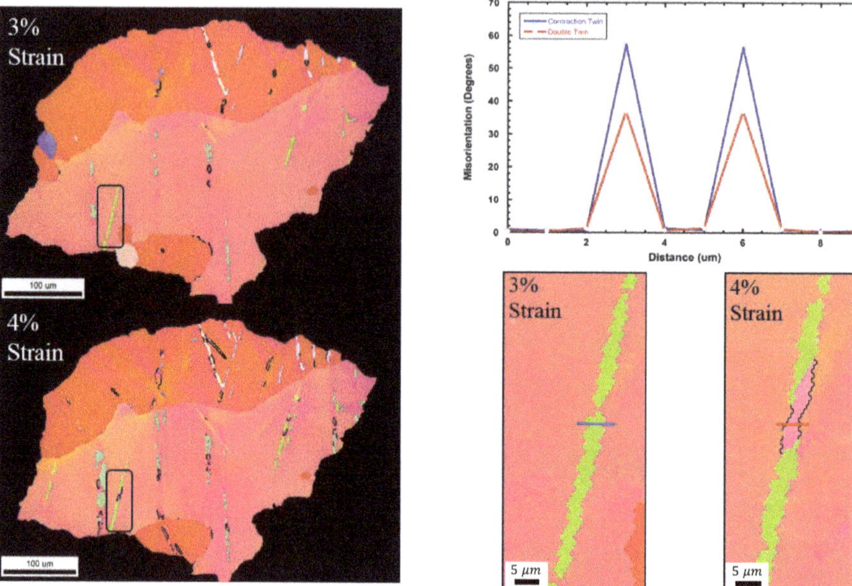

Figure 19. Characterization of double twinning at low stresses. The twin of interest is fully formed by 3% strain and does not show signs of double twinning. By 4% strain, the twin of interest begins to double twin at the middle. Misorientation analysis between the double twin and the contraction twin is compared to confirm the presence of contraction and double twins.

Nonetheless, the reason why $\{10\bar{1}1\}$ twinning is easier to nucleate in contraction than in compression will be difficult to explain or prove. This CRSS dependence on stress sign for twinning could, however, be addressed from the following perspective. It is known that twinning in HCP metals requires atomic movement by both shear and shuffle. El Kadiri et al. [70] advanced a theory that provides analytical solutions for the vector displacements for both shear and shuffle for any compound twin. Schmid effects are associated with accommodations through shear, being the most fundamental condition for dislocation and disconnection motion. However, shuffles can be viewed as corresponding to atomic movement due to diffusion. In general, the only component of stress that drives diffusion is the hydrostatic pressure or more precisely a gradient of hydrostatic pressure. As shuffle is a pure diffusion phenomenon, it could be readily sensitive to the local state of hydrostatic pressure in the disconnection core as its glide along the twin interface. That is, the inherent difference in the stress state between contraction and compression could have profound implications on the ease of shuffle and the mobility of disconnections. Previous atomistic investigations of deformation twinning in bcc metal showed the ease in twin nucleation with hydrostatic tension [75]. One can in fact hypothesize that a more complex stress state might be acting during $\langle c \rangle$-axis contraction compared to uniaxial $\langle c \rangle$-axis compression, aiding the atomic shuffle associated with twin formation.

4. Conclusions

We performed in situ electron backscatter diffraction and fracture analyses on a plane-strain extruded AM30 magnesium alloy with a double fiber texture exhibiting adjacent grains that are both favorably and unfavorably oriented to $\{10\bar{1}2\}$ and $\{10\bar{1}1\}$ twinning. The results show various interactions between twins, slip dislocations, grain boundaries and other twins, which lead to material damage and local cracks. Remarkable non-Schmid's behaviors were revealed for both $\{10\bar{1}2\}$ and

$\{1\,0\,\bar{1}\,1\}$ twinning modes, which are of important consideration in crystal plasticity simulations. Results from this work supports the following conclusions:

a   In situ electron backscatter diffraction (EBSD) and scanning electron microscopy (SEM) micrographs revealed the importance of the twin–slip, twin–GBs, and twin–twin interactions in the formation of large cracks inside the material. Twins nucleating under a very low macroscopic Schmid factor show a greater sensitivity to crack nucleation through these interactions.

b   During $\{1\,0\,\bar{1}\,2\}$ profuse twinning inside a favorably oriented grain, two variants of which were able to nucleate more substantially on grain boundaries grew much faster than all the other variants. This could be due to an enhanced nucleation of glissile disconnections at the intersection between grain boundary and twin boundary than between two twin boundaries.

c $\{1\,0\,\bar{1}\,1\}$ twins nucleate under $\langle c \rangle$-axis contraction at much lower CRSS values than those previously reported in the literature under $\langle c \rangle$-axis compression. This behavior was attributed to the greater hydrostatic pressure in the case of contraction compared to the case of uniaxial $\langle c \rangle$ compression, which may assist the complex shuffles required for this twinning mode.

Author Contributions: Conceptualization, W.D.R., N.R.B. and H.E.K.; investigation, W.D.R. and N.R.B.; methodology, W.D.R., N.R.B., R.D.M. and Z.B.M.; project administration, R.D.M. and Z.B.M.; resources, H.R. and S.C.V.; supervision, W.R.W., H.R. and H.E.K.; validation, Y.P., C.D.B., A.L.O., W.R.W. and H.E.K.; writing—original draft, W.D.R., Y.P. and A.L.O.; writing—review and editing, Y.P., C.D.B., A.L.O., S.M., B.P., S.C.V. and H.E.K. All authors have read and agreed to the published version of the manuscript.

Funding: The research described and the resulting data presented herein, unless otherwise noted, were funded under PE 0602784A, Project T53 "Military Engineering Applied Research", Topic 3 under Contract W56HZV-17-C-0095, managed by the US Army Engineer Research and Development Center. The work described in this document was conducted in the Center for Advanced Vehicular Systems at Mississippi State University. OPSEC permission was granted to publish this information.

Conflicts of Interest: The authors declare no conflict of interest. The funders had no role in the design of the study; in the collection, analyses, or interpretation of data; in the writing of the manuscript, or in the decision to publish the results.

References

1.   Pollock, T.M. Weight loss with magnesium alloys. *Science* **2010**, *328*, 986–987. [CrossRef] [PubMed]

2.   Mordike, B.L.; Ebert, T. Magnesium: Properties–applications–potential. *Mater. Sci. Eng. A* **2001**, *302*, 37–45. [CrossRef]

3.   Al-Samman, T.; Gottstein, G. Room temperature formability of a magnesium AZ31 alloy: Examining the role of texture on the deformation mechanisms. *Mater. Sci. Eng. A* **2008**, *488*, 406–414. [CrossRef]

4.   Zhang, Y.; Millett, P.C.; Tonks, M.; Biner, B. Deformation-twin-induced grain boundary failure. *Scr. Mater.* **2012**, *66*, 117–120. [CrossRef]

5.   Barnett, M.R. Twinning and the ductility of magnesium alloys: Part II."Contraction" twins. *Mater. Sci. Eng. A* **2007**, *464*, 8–16. [CrossRef]

6.   Pratt, P.L.; Pugh, S.F. Twin accommodation in zinc. *J. Inst. Met.* **1952**, *80*, 653–658.

7.   Christian, J.W.; Mahajan, S. Deformation twinning. *Prog. Mater. Sci.* **1995**, *39*, 1–157. [CrossRef]

8.   Kumar, M.A.; Beyerlein, I.J. Local microstructure and micromechanical stress evolution during deformation twinning in hexagonal polycrystals. *J. Mater. Res.* **2020**, *35*, 217–241. [CrossRef]

9.   Baird, J.C.; Li, B.; Parast, S.Y.; Horstemeyer, S.J.; Hector, L.G., Jr.; Wang, P.T.; Horstemeyer, M.F. Localized twin bands in sheet bending of a magnesium alloy. *Scr. Mater.* **2012**, *67*, 471–474. [CrossRef]

10. Paudel, Y.; Barrett, C.D.; Tschopp, M.A.; Inal, K.; El Kadiri, H. Beyond initial twin nucleation in hcp metals: Micromechanical formulation for determining twin spacing during deformation. *Acta Mater.* **2017**, *133*, 134–146. [CrossRef]

11. Paudel, Y.; Indeck, J.; Hazeli, K.; Priddy, M.W.; Inal, K.; Rhee, H.; Barrett, C.D.; Whittington, W.R.; Limmer, K.R.; El Kadiri, H. Characterization and modeling of $\{1\,0\,\bar{1}\,2\}$ twin banding in magnesium. *Acta Mater.* **2020**, *183*, 438–451. [CrossRef]

12. Gilman, J.J. Mechanism of ortho kink-band formation in compressed zinc monocrystals. *JOM* **1954**, *6*, 621–629. [CrossRef]

13. Rosenbaum, H.S. Non-basal slip and twin accommodation in zinc crystals. *Acta Metall.* **1961**, *9*, 742–748. [CrossRef]

14. Rosenbaum, H.S. Nonbasal Slip in h.c.p. Metals and its Relation to Mechanical Twinning. In *Deformation Twinning*; Reed-Hill, R.E., Hirth, J.P., Rogers, H.C., Eds.; Grodon and Breach Science Publishers: New York, NY, USA, 1964; pp. 43–76.

15. Rose, G. Über die im kalkspath vorkommenden hohlen canäle. *Aus Den Abh. Königlichen Akad. Wissenschalten* **1868**, *23*, 57–79.

16. Priestner, R. The Relationship Between Brittle Cleavage and Deformation Twinning in BCC Metals. In *Deformation Twinning*; Reed-Hill, R.E., Hirth, J.P., Rogers, H.C., Eds.; Grodon and Breach Science Publishers: New York, NY, USA, 1964; pp. 321–355.

17. Sleeswyk, A.W. Emissary dislocations: Theory and experiments on the propagation of deformation twins in α-iron. *Acta Metall.* **1962**, *10*, 705–725. [CrossRef]

18. Sleeswyk, A.W. Twinning and the origin of cleavage nuclei in α-iron. *Acta Metall.* **1962**, *10*, 803–812. [CrossRef]

19. Sleeswyk, A.W. $1/2\langle 1\,1\,1\rangle$ screw dislocations and the nucleation of $\{112\}\langle 1\,1\,1\rangle$ twins in the bcc lattice. *Philos. Mag.* **1963**, *8*, 1467–1486.

20. Barnett, M.R.; Stanford, N.; Cizek, P.; Beer, A.; Xuebin, Z.; Keshavarz, Z. Deformation mechanisms in Mg alloys and the challenge of extending room-temperature plasticity. *JOM* **2009**, *61*, 19–24. [CrossRef]

21. Holden, J. Plastic deformation features on cleavage surfaces of metal crystals. *Philos. Mag.* **1952**, *43*, 976–984. [CrossRef]

22. Zerilli, F.J.; Armstrong, R.W. Constitutive relations for the plastic deformation of metals. AIP conference proceedings. *Am. Inst. Phys.* **1994**, *309*, 989–992.

23. Zerilli, F.J.; Armstrong, R.W. Dislocation mechanics based analysis of material dynamics behavior: Enhanced ductility, deformation twinning, shock deformation, shear instability, dynamic recovery. *J. Phys. IV* **1997**, *7*, 637–642. [CrossRef]

24. Remy, L. Twin-slip interaction in fcc crystals. *Acta Metall.* **1977**, *25*, 711–714. [CrossRef]

25. Remy, L. The interaction between slip and twinning systems and the influence of twinning on the mechanical behavior of fcc metals and alloys. *Metall. Trans. A* **1981**, *12*, 387–408. [CrossRef]

26. Ma, Q.; El Kadiri, H.; Oppedal, A.L.; Baird, J.C.; Horstemeyer, M.F.; Cherkaoui, M. Twinning and double twinning upon compression of prismatic textures in an AM30 magnesium alloy. *Scr. Mater.* **2011**, *64*, 813–816. [CrossRef]

27. Ma, Q.; El Kadiri, H.; Oppedal, A.L.; Baird, J.C.; Li, B.; Horstemeyer, M.F.; Vogel, S.C. Twinning effects in a rod-textured AM30 magnesium alloy. *Int. J. Plast.* **2012**, *29*, 60–76. [CrossRef]

28. Barnett, M.R. Twinning and the ductility of magnesium alloys: Part I: "Tension" twins. *Mater. Sci. Eng. A* **2007**, *464*, 1–7. [CrossRef]

29. Oppedal, A.L.; El Kadiri, H.; Tomé, C.N.; Kaschner, G.C.; Vogel, S.C.; Baird, J.C.; Horstemeyer, M.F. Effect of dislocation transmutation on modeling hardening mechanisms by twinning in magnesium. *Int. J. Plast.* **2012**, *30*, 41–61. [CrossRef]

30. Martin, É.; Capolungo, L.; Jiang, L.; Jonas, J.J. Variant selection during secondary twinning in Mg–3% Al. *Acta Mater.* **2010**, *58*, 3970–3983. [CrossRef]

31. Jonas, J.J.; Mu, S.; Al-Samman, T.; Gottstein, G.; Jiang, L.; Martin, È. The role of strain accommodation during the variant selection of primary twins in magnesium. *Acta Mater.* **2011**, *59*, 2046–2056. [CrossRef]

32. Mu, S.; Jonas, J.J.; Gottstein, G. Variant selection of primary, secondary and tertiary twins in a deformed Mg alloy. *Acta Mater.* **2012**, *60*, 2043–2053. [CrossRef]

33. Paudel, Y. Micromechanics-Crystal Plasticity Links for Deformation Twinning. Ph.D. Thesis, Mississippi State University, Mississippi State, MS, USA, 2018.

34. Paudel, Y.; Barrett, C.D.; El Kadiri, H. Full-Field Crystal Plasticity Modeling of $\{1\,0\,\bar{1}\,2\}$ Twin Nucleation. In *Magnesium Technology 2020*; Jordon, J., Miller, V., Joshi, V., Neelameggham, N., Eds.; Springer: Cham, Switzerland, 2020; pp. 141–146.

35. Cheng, J.; Ghosh, S. Crystal plasticity finite element modeling of discrete twin evolution in polycrystalline magnesium. *J. Mech. Phys. Solids* **2017**, *99*, 512–538. [CrossRef]

36. El Kadiri, H.; Baird, J.C.; Kapil, J.; Oppedal, A.L.; Cherkaoui, M.; Vogel, S.C. Flow asymmetry and nucleation stresses of $\{1\,0\,\bar{1}\,2\}$ twinning and non-basal slip in magnesium. *Int. J. Plast.* **2013**, *44*, 111–120. [CrossRef]

37. Wenk, H.R.; Lutterotti, L.; Vogel, S. Texture analysis with the new HIPPO TOF diffractometer. *Nucl. Instrum. Meth. A* **2003**, *515*, 575–588. [CrossRef]

38. Takajo, S.; Vogel, S.C. Determination of pole figure coverage for texture measurements with neutron time-of-flight diffractometers. *J. Appl. Crystallogr.* **2018**, *51*, 895–900. [CrossRef]

39. Losko, A.S.; Vogel, S.C.; Reiche, H.M.; Nakotte, H. A six-axis robotic sample changer for high-throughput neutron powder diffraction and texture measurements. *J. Appl. Crystallogr.* **2014**, *47*, 2109–2112. [CrossRef]

40. Wenk, H.R.; Lutterotti, L.; Vogel, S.C. Rietveld texture analysis from TOF neutron diffraction data. *Powder Diffr.* **2010**, *25*, 283–296. [CrossRef]

41. Hielscher, R.; Schaeben, H. A novel pole figure inversion method: Specification of the MTEX algorithm. *J. Appl. Crystallogr.* **2008**, *41*, 1024–1037. [CrossRef]

42. Barrett, C.D.; El Kadiri, H.; Tschopp, M.A. Breakdown of the Schmid law in homogeneous and heterogeneous nucleation events of slip and twinning in magnesium. *J. Mech. Phys. Solids* **2012**, *60*, 2084–2099. [CrossRef]

43. Barrett, C.D.; El Kadiri, H. The roles of grain boundary dislocations and disclinations in the nucleation of $\{10\bar{1}2\}$ twinning. *Acta Mater.* **2014**, *63*, 1–15. [CrossRef]

44. El Kadiri, H.; Barrett, C.D.; Wang, J.; Tomé, C.N. Why are $\{10\bar{1}2\}$ twins profuse in magnesium? *Acta Mater.* **2015**, *85*, 354–361. [CrossRef]

45. Barrett, C.D.; El Kadiri, H. Impact of deformation faceting on $\{10\bar{1}2\}$, $\{10\bar{1}1\}$ and $\{10\bar{1}3\}$ embryonic twin nucleation in hexagonal close-packed metals. *Acta Mater.* **2014**, *70*, 137–161. [CrossRef]

46. Barrett, C.D.; Tschopp, M.A.; El Kadiri, H. Automated analysis of twins in hexagonal close-packed metals using molecular dynamics. *Scr. Mater.* **2012**, *66*, 666–669. [CrossRef]

47. El Kadiri, H.; Oppedal, A.L. A crystal plasticity theory for latent hardening by glide twinning through dislocation transmutation and twin accommodation effects. *J. Mech. Phys. Solids* **2010**, *58*, 613–624. [CrossRef]

48. Partridge, P.G. Effect of cyclic stresses on the microstructures of hexagonal close packed metals. *Czech. J. Phys. Sect. B* **1969**, *19*, 323–332. [CrossRef]

49. Molodov, K.D.; Al-Samman, T.; Molodov, D.A. Profuse slip transmission across twin boundaries in magnesium. *Acta Mater.* **2017**, *124*, 397–409. [CrossRef]

50. Wang, F.; Barrett, C.D.; McCabe, R.J.; El Kadiri, H.; Capolungo, L.; Agnew, S.R. Dislocation induced twin growth and formation of basal stacking faults in $\{10\bar{1}2\}$ twins in pure Mg. *Acta Mater.* **2019**, *165*, 471–485. [CrossRef]

51. Mahajan, S. Twin-slip and twin-twin interactions in Mo-35 at.% Re alloy. *Philos. Mag.* **1971**, *23*, 781–794. [CrossRef]

52. Mahajan, S.; Chin, G.Y. Twin-slip, twin-twin and slip-twin interactions in Co-8 wt.% Fe alloy single crystals. *Acta Metall.* **1973**, *21*, 173–179. [CrossRef]

53. Mahajan, S.; Chin, G.Y. Formation of deformation twins in fcc crystals. *Acta Metall.* **1973**, *21*, 1353–1363. [CrossRef]

54. Sleeswyk, A.W.; Verbraak, C.A. Incorporation of slip dislocations in mechanical twins–I. *Acta Metall.* **1961**, *9*, 917–927. [CrossRef]

55. Beyerlein, I.J.; Tomé, C.N. A dislocation-based constitutive law for pure Zr including temperature effects. *Int. J. Plast.* **2008**, *24*, 867–895. [CrossRef]

56. Reed-Hill, R.E. Twin intersections & Cahns continuity conditions. *Trans. Met. Soc. AIME* **1964**, *230*, 809.

57. Reed-Hill, R.E.; Robertson, W.D. The crystallographic characteristics of fracture in magnesium single crystals. *Acta Metall.* **1957**, *5*, 728–737. [CrossRef]

58. Chin, G.Y. The role of preferred orientation in plastic deformation. In *The Inhomogeneity of Plastic Deformation*; Reed-Hill, R.E., Ed.; American Society for Metals: Russell, OH, USA, 1973; Chapter 4, pp. 83–112.

59. Yu, Q.; Wang, J.; Jiang, Y.; McCabe, R.J.; Li, N.; Tomé, C.N. Twin–twin interactions in magnesium. *Acta Mater.* **2014**, *77*, 28–42. [CrossRef]

60. Crocker, A.G. Double twinning. *Philos. Mag.* **1962**, *7*, 1901–1924. [CrossRef]

61. Hartt, W.H.; Reed-Hill, R.E. Internal deformation and fracture of second-order $\{10\bar{1}1\}$-$\{10\bar{1}2\}$-twins in magnesium. *Trans. Met. Soc. AIME* **1968**, *242*, 1127–1133.

62. Wonsiewicz, B.C. Plasticity of Magnesium Crystals. Ph.D. Thesis, Massachusetts Institute of Technology, Cambridge, MA, USA, 1966.

63. Cizek, P.; Barnett, M.R. Characteristics of the contraction twins formed close to the fracture surface in Mg–3Al–1Zn alloy deformed in tension. *Scr. Mater.* **2008**, *59*, 959–962. [CrossRef]

64. Koike, J. Enhanced deformation mechanisms by anisotropic plasticity in polycrystalline Mg alloys at room temperature. *Metall. Mater. Trans. A* **2005**, *36*, 1689–1696. [CrossRef]

65. Koike, J.; Fujiyama, N.; Ando, D.; Sutou, Y. Roles of deformation twinning and dislocation slip in the fatigue failure mechanism of AZ31 Mg alloys. *Scr. Mater.* **2010**, *63*, 747–750. [CrossRef]

66. Ando, D.; Koike, J.; Sutou, Y. The role of deformation twinning in the fracture behavior and mechanism of basal textured magnesium alloys. *Mater. Sci. Eng. A* **2014**, *600*, 145–152. [CrossRef]

67. Ando, D.; Koike, J. Relationship between Deformation-Induced Surface Relief and Double Twinning in AZ31 Magnesium Alloy. *J. Jpn. Inst. Met.* **2007**, *71*, 684–687. [CrossRef]

68. Hazeli, K.; Cuadra, J.; Streller, F.; Barr, C.; Taheri, M.; Carpick, R.; Kontsos, A. Three-dimensional effects of twinning in magnesium alloys. *Scr. Mater.* **2015**, *100*, 9–12. [CrossRef]

69. Kelley, E.W.; Hosford, W.F., Jr. Plane-strain compression of magnesium and magnesium alloy crystals. *Trans. Met. Soc. AIME* **1968**, *242*, 5–13.

70. El Kadiri, H.; Barrett, C.D.; Tschopp, M.A. The candidacy of shuffle and shear during compound twinning in hexagonal close-packed structures. *Acta Mater.* **2013**, *61*, 7646–7659. [CrossRef]

71. Keshavarz, Z.; Barnett, M.R. EBSD analysis of deformation modes in Mg–3Al–1Zn. *Scr. Mater.* **2006**, *55*, 915–918. [CrossRef]

72. Xu, S.W.; Kamado, S.; Matsumoto, N.; Honma, T.; Kojima, Y. Recrystallization mechanism of as-cast AZ91 magnesium alloy during hot compressive deformation. *Mater. Sci. Eng. A* **2009**, *527*, 52–60. [CrossRef]

73. Barnett, M.R.; Keshavarz, Z.; Beer, A.G.; Ma, X. Non-Schmid behaviour during secondary twinning in a polycrystalline magnesium alloy. *Acta Mater.* **2008**, *56*, 5–15. [CrossRef]

74. Beyerlein, I.J.; Wang, J.; Barnett, M.R.; Tomé, C.N. Double twinning mechanisms in magnesium alloys via dissociation of lattice dislocations. *Proc. R. Soc. A Math. Phys. Eng. Sci.* **2012**, *468*, 1496–1520. [CrossRef]

75. Xu, D.S.; Chang, J.P.; Li, J.; Yang, R.; Li, D.; Yip, S. Dislocation slip or deformation twinning: Confining pressure makes a difference. *Mater. Sci. Eng. A* **2004**, *387-389*, 840–844. [CrossRef]

Publisher's Note: MDPI stays neutral with regard to jurisdictional claims in published maps and institutional affiliations.

Article

Microstructures and Mechanical Properties of Precipitation-Hardenable Magnesium–Silver–Calcium Alloy Sheets

Mingzhe Bian *, Xinsheng Huang and Yasumasa Chino

Multi-Material Research Institute, National Institute of Advanced Industrial Science and Technology (AIST), Nagoya, Aichi 463-8560, Japan; huang-xs@aist.go.jp (X.H.); y-chino@aist.go.jp (Y.C.)
* Correspondence: mingzhe.bian@aist.go.jp

Received: 17 November 2020; Accepted: 2 December 2020; Published: 4 December 2020

Abstract: Precipitation hardening provides one of the most common strengthening mechanisms for magnesium (Mg) alloys. Here, we report a new precipitation-hardenable Mg sheet alloy based on the magnesium–silver–calcium system. In a solution treated condition (T4), the strength of Mg–xAg–0.1Ca alloys is enhanced with increasing the Ag content from 1.5 wt.% to 12 wt.%. The Mg–12Ag–0.1Ca (wt.%) alloy sheet shows moderate tensile yield strengths of 193 MPa, 130 MPa, 117 MPa along the rolling direction (RD), 45° and transverse direction (TD) in the T4-treated condition. Subsequent artificial aging at 170 °C for 336 h (T6) increases the tensile yield strengths to 236 MPa, 163 MPa and 143 MPa along the RD, 45° and TD, respectively. This improvement in the tensile yield strength by the T6 treatment can be ascribed to the formation of AgMg$_4$ precipitates lying on the $\{11\bar{2}0\}_a$ and pyramidal planes. Our finding is expected to stimulate the development of precipitation-hardenable Mg–Ag-based wrought alloys with high strength.

Keywords: magnesium alloys; rolling; strength; segregation; precipitate

---

1. Introduction

The lightest structural metal magnesium (Mg) and its alloys have attracted significant interest in the past two decades due to their potential applications in the automotive sector [1–3]. However, current applications of wrought Mg alloys as a structural component are very limited, which can be ascribed to their inferior mechanical properties at room temperature (RT) and poor corrosion resistance [4]. Precipitation-hardening, also known as age-hardening, is one of the most effective ways to strengthen Mg alloys [5,6]. For example, an extraordinary high tensile yield strength of 473 MPa was obtained in a heavy rare-earth (RE) containing Mg–10Gd–5.7Y–1.6Zn–0.7Zr (wt.%) alloy processed by a conventional extrusion process followed by 200 °C aging for 64 h [7]. A RE-free Mg–6.2Zn–0.4Ag–0.2Ca–0.4Zr (wt.%) alloy subjected to twin-roll cast (TRC), hot-rolling and artificial aging at 160 °C for 24 h also yielded a high tensile yield strength of 316 MPa with a large fracture elongation of 17% [8]. Recent studies reported that Mg–Al–Ca–Mn(–Zn) dilute alloys could reach their peak hardness condition at 200 °C within 1 h that is significantly shorter than the time to reach the peak-hardness conditions of concentrated Mg–Gd, Mg–Zn and Mg–Sn based alloys [9–11]. Microstructure characterization by transmission electron microscopy (TEM), in conjunction with atom probe tomography, revealed that mono-atomic layer Guinier–Preston (G.P.) zones lying on the (0002) basal plane are formed in the peak-aged samples, and their dense distribution is responsible for the strength improvement in Mg–Al–Ca–Mn(–Zn) alloy sheets [12]. A more recent study demonstrated that bake-hardenability is even obtainable after 170 °C aging for only 20 min in a TRC Mg–1.3Al–0.8Zn–0.7Mn–0.5Ca (wt.%) alloy sheet, which opened up a possibility to develop bake-hardenable Mg alloys [13]. Those studies suggest that the development of precipitation-hardenable alloys is a promising approach to strengthen wrought Mg alloys.

In a recent study, we found that the co-addition of Ag (1.5 wt.%) and Ca (0.1 wt.%) to pure Mg could significantly enhance the stretch formability compared to the single addition of Ag or Ca [14]. A subsequent study achieved a high tensile yield strength of 182 MPa along the rolling direction (RD) by increasing the Ag content to 6 wt.%, while keeping a high index Erichsen value of 8.7 mm [15]. In a more recent study, we found that a compositionally optimized Mg–1.5Ag–2Ca (wt.%) alloy is non-flammable up to 1000 °C due to the formation of compact and dense CaO film on the surface. [16]. If precipitation-hardenability can be obtained by artificial aging, the Mg–Ag–Ca system is expected to attract more attention. Based on the binary Mg–Ag phase diagram, the maximum solid solubility of Ag is about 15 wt% in Mg at 472 °C, and it decreases to about 0 wt.% at 200 °C [17]. As shown in Figure 1, the equilibrium mole fraction of $AgMg_4$ phase is about 0.19 at 200 °C for the binary Mg–15Ag (wt.%) alloy, indicating there is a high possibility to obtain precipitation-hardenability from the Mg–Ag system. In fact, the Ag addition was reported to enhance the age-hardening response and thus increase mechanical properties of Mg–Nd alloys more than 50 years ago, which has facilitated the development of a commercial casting alloy QE22 (Mg–2.5Ag–2Nd–0.7Zr in wt.%) [18]. Subsequent studies showed that the Ag addition is also very effective in improving the age-hardening response of Mg–6Gd–0.6Zr and Mg–6Y–1Zn–0.6Zr (all in wt.%) casting alloys [19,20]. TEM observations revealed that the remarkable improvement in age-hardening could be ascribed to a dense and uniform distribution of nano-scale basal precipitates. On the other hand, adding a trace amount of Ag (0.4 wt.%) to a binary Mg–6.2Zn alloy was reported to enhance the age hardening response significantly due to a substantial refinement of $MgZn_2$ precipitates [21]. However, a closer look at those precipitation-hardenable Ag-containing alloys developed until now, we could find that Ag was mainly added to refine the size of precipitates formed in Mg–RE or Mg–Zn based alloys. To the authors' knowledge, there is no prior research on the development of precipitation-hardenable Mg–Ag based wrought alloys, which motivated us to explore the feasibility of developing precipitation-hardenable alloy based on the Mg–Ag–Ca system.

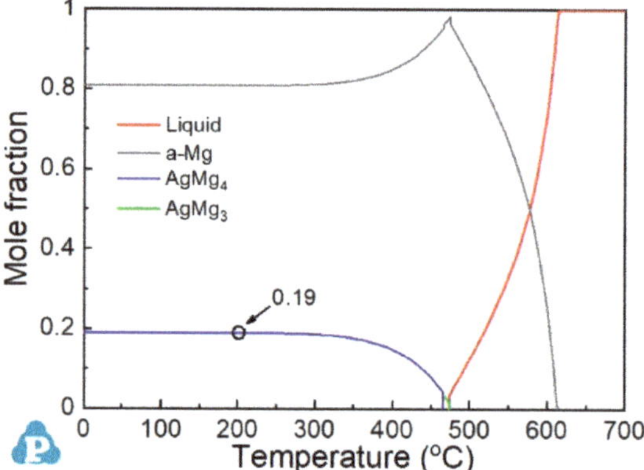

Figure 1. Calculated equilibrium mole fraction of phases as a function of temperature for a binary Mg–15Ag (wt.%) alloy using the PANDAT software.

In this study, Mg–xAg–0.1Ca alloys with a composition range from 1.5 wt.% to 12 wt.% were prepared and their age-hardening response at 170 °C was evaluated. Tensile properties of Mg–1.5Ag–0.1Ca, Mg–6Ag–0.1Ca and Mg–12Ag–0.1Ca alloy sheets were evaluated at RT, and a systematic microstructure characterization was conducted to clarify the strengthening mechanism in precipitation-hardenable Mg–Ag–Ca alloys.

2. Materials and Methods

Mg–xwt.%Ag–0.1wt.%Ca (x = 1.5, 6 and 12) ingots were produced using pure Mg (>99.9%), pure Ag (>99.9%) and Mg–4.92 wt%Ca master alloy by an induction furnace (IR) under an Argon (Ar) atmosphere. The chemical composition of alloys used in the present study was analyzed by inductively coupled plasma-optical emission spectroscopy (ICP-OES), and the analyzed results were given in Table 1. The as-cast alloys were initially extruded as sheets of 5 mm in thickness at 380 °C. The extrusion ratio and the ram speed were 6 and 5 mm/min, respectively. The extruded sheets were homogenized at 400 °C for 18 h to avoid hot-cracking during rolling, particularly the concentrated Mg–12Ag–0.1Ca alloy. The homogenized sheets were rolled from 5 mm to 1 mm in thickness with ~ 21% thickness reduction per pass by 7 passes. After each pass, the rolled sheets were immediately quenched into cold water. The water quenched sheets were then re-heated to 350 °C prior to subsequent rolling, and rollers were heated to 90 °C during rolling. The rolled sheets were solution-treated at 450 °C for 1 h (T4) and then quenched into cold water. Some of them were subsequently aged in a silicone oil bath at 170 °C.

Table 1. Chemical composition of Mg–xAg–0.1Ca (x = 1.5, 6 and 12) alloys analyzed by ICP-OES.

Alloy	Ag (wt.%)	Ca (wt.%)
Mg–1.5Ag–0.1Ca	1.37	0.10
Mg–6Ag–0.1Ca	5.68	0.11
Mg–12Ag–0.1Ca	11.0	0.10

Dog-bone shaped tensile samples having a parallel length of 12 mm, a width of 4 mm and a thickness of 1 mm were machined from the T4-treated sheets along the RD (0°), 45° and TD (90°). Figure 2 shows a schematic drawing of tensile test samples used in the present study. A screw driven Instron 5565 tensile testing machine was used to evaluate RT tensile properties with a constant testing speed of 2 mm/min. The age hardening response was evaluated by a HM-200 micro Vickers hardness tester (Mitutoyo Corporation, Kawasaki, Japan) under a load of 200 g with a holing time of 10 s. Ten points were measured in each condition. The maximum and minimum hardness values were removed, and the remaining 8-point values were averaged. X-ray diffraction (XRD) patterns were obtained from the mid-layers of sheets using Rigaku RINT Ultima III operating at 40 kV and 40 mA (Rigaku Corporation, Akishima, Japan). Samples for secondary scanning electron microscope (SEM) and electron backscatter diffraction (EBSD) observations were prepared using silicon carbide (SiC) papers, 60 nm alumina suspension, and Ar ion beam using an ELIONIX EIS-200ER ion beam shower system (ELIONIX Inc., Hachioji, Japan). SEM observation was performed at 15 kV using a JEOL JSM-IT500 equipped with a JEOL EX-74600U4L2Q EDS detector (JEOL Ltd., Akishima, Japan). EBSD measurements were performed at 20 kV using the same SEM equipped with TSL OIM 7.0 data collection software (EDAX Inc., Mahwah, NJ, USA). Samples for TEM observation were mechanically ground to about 150 µm in thickness using SiC papers (500, 1000, 2400, and 4000 grit) and subsequently punched to disks of 3 mm in diameter. The disks were twin-jet electropolished at about −50 °C with a solution of 15.9 g lithium chloride, 33.6 g magnesium perchlorate, 1500 mL methanol and 300 mL 2-butoxy-ethanol, and finally ion-milled using a Gatan Precision Ion Polishing System (PIPS) (Gatan Inc., Pleasanton, CA, USA). TEM observations were carried out on JEM-2010 and C_s-corrected JEM-ARM200F TEMs operating at 200 kV (JEOL Ltd., Akishima, Japan). The thermodynamic simulation was carried out with the PANDAT software 2020 (CompuTherm LLC, Middleton, WI, USA) [22].

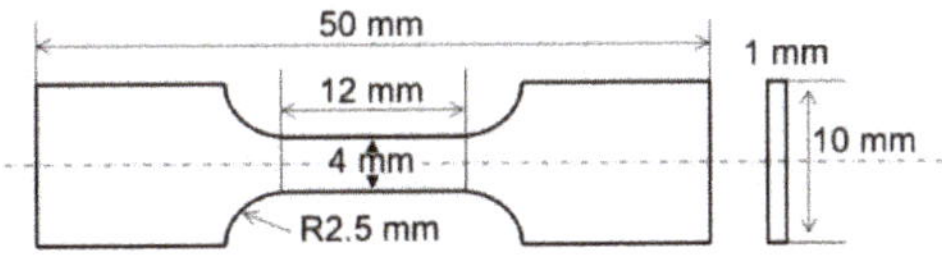

Figure 2. Schematic diagram showing the dimensions of tensile test sample used in this study.

3. Results

Figure 3 shows the age-hardening response of T4-treated Mg–xAg–0.1Ca alloy sheets (x = 1.5, 6 and 12) during isothermal aging at 170 °C. The 1.5Ag containing alloy sheet shows a low Vickers hardness value of 41.4 HV in the initial condition (T4), and it exhibits a negligible age-hardening. Increasing the Ag content not only leads to a higher hardness value in the T4-treated condition but also enhances the age-hardening response during aging. The 6Ag containing alloy sheet and the 12Ag containing alloy sheet show hardness values of 51.1 HV and 67.0 HV, respectively. After aging for 336 h, the hardness value of the former is increased to 57.4 HV, and that of the latter is increased to 80.0 HV. Table 2 summarizes initial hardness, peak-hardness, time to reach peak hardness and hardness increment of Mg–xAg–0.1Ca alloy sheets.

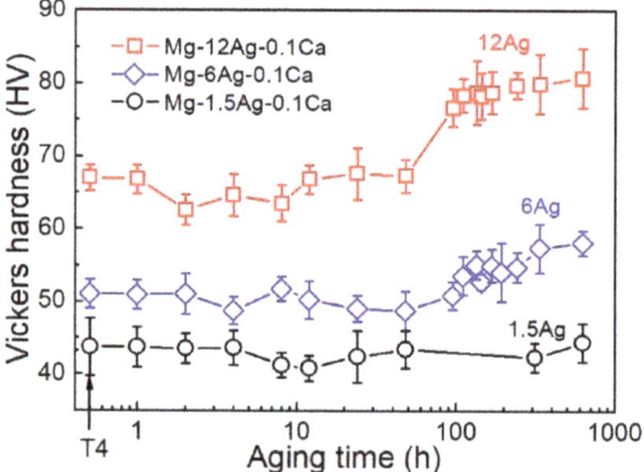

Figure 3. Age-hardening response of 1.5Ag, 6Ag and 12Ag containing alloy sheets at 170 °C.

Table 2. Initial hardness (T4 condition), peak-hardness, time to reach peak hardness and hardness increment of 1.5Ag, 6Ag and 12Ag containing alloy sheets.

Alloy	Initial Hardness (HV)	Peak Hardness (HV)	Time to Reach Peak Hardness (h)	Hardness Increment (HV)
Mg–1.5Ag–0.1Ca	41.4	43.7	0.5	2.3
Mg–6Ag–0.1Ca	51.1	57.4	336	6.3
Mg–12Ag–0.1Ca	67.0	80.0	336	13.0

Figure 4 shows tensile properties of the T4- and T6-treated 6Ag and 12Ag containing alloy sheets at RT. For the purpose of comparison, tensile curves obtained from the T4-treated 1.5Ag containing alloy sheet are given as Figure 4a. The tensile yield strength (TYS) of the T4-treated 1.5Ag containing alloy sheet is measured to be only 85 MPa, 57 MPa and 47 MPa along the RD, 45° and TD, respectively. Increasing the Ag content to 6 wt.% increases the TYS to 112 MPa, 70 MPa and 61 MPa along the RD, 45° and TD, Figure 4b. When the Ag content is further increased to 12 wt.%, the TYS is significantly increased to 193 MPa, 130 MPa and 117 MPa, Figure 4c. Subsequent artificial aging (T6) at 170 °C for 336 h further increases the TYS of 12Ag containing alloy sheet to 236 MPa, 163 MPa and 143 MPa along the RD, 45° and TD, respectively. The uniform elongation (UE) and fracture elongation (FE) of T4-treated 6Ag and 12Ag containing alloy sheets are all higher than 20%. However, both UE and FE are substantially reduced to about 10% after the T6 treatment. The tensile properties of 1.5Ag, 6Ag and 12Ag containing alloy sheets such as TYS, ultimate tensile strength (UTS), UE and FE are summarized in Table 3.

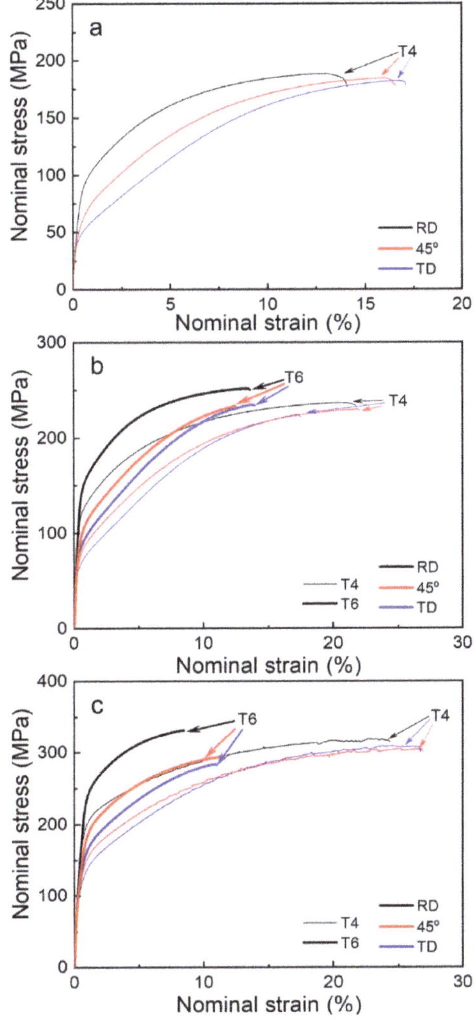

Figure 4. Tensile curves obtained from (**a**) T4-treated Mg–1.5Ag–0.1Ca alloy sheet, and T4- and T6-treated (**b**) Mg–6Ag–0.1Ca and (**c**) Mg–12Ag–0.1Ca alloy sheets along RD, 45° and TD at RT.

Figure 5 show EBSD inverse pole figure (IPF) maps and corresponding (0002) and (10$\bar{1}$0) PFs obtained from the T4-treated 1.5Ag, 6Ag and 12Ag containing alloy sheets. The 1.5Ag containing alloy sheet shows a coarse-grained microstructure of ~277 μm, and abnormal grain growth occurs. Increasing the Ag content to 6 wt.% refines the microstructure and the average grain size is decreased by about one half (~124 μm). In addition, the microstructure homogeneity is enhanced. An increased addition of the Ag content to 12 wt.% causes further refinement in the microstructure. It is to be noted that a homogeneous microstructure consisting of equiaxed grains with an average grain size of ~74 μm is developed in the 12Ag containing alloy. All alloy sheets show a TD-split texture, in which the (0002) basal poles are tilted toward the TD.

Table 3. RT tensile properties of 1.5Ag, 6Ag and 12Ag containing alloy sheets stretched along the RD, 45° and TD.

Alloy	Condition	Direction	TYS (MPa)	UTS (MPa)	UE (%)	FE (%)
Mg–1.5Ag–0.1Ca	T4	RD	85	189	12	13
		45°	57	185	15	16
		TD	47	182	16	16
Mg–6Ag–0.1Ca	T4	RD	112	237	20	21
		45°	70	230	21	22
		TD	61	224	17	17
	T6	RD	149	252	13	13
		45°	89	233	12	12
		TD	81	235	13	14
Mg–12Ag–0.1Ca	T4	RD	193	319	22	24
		45°	130	306	24	26
		TD	117	310	23	26
	T6	RD	236	331	8	8
		45°	163	294	11	11
		TD	143	284	10	10

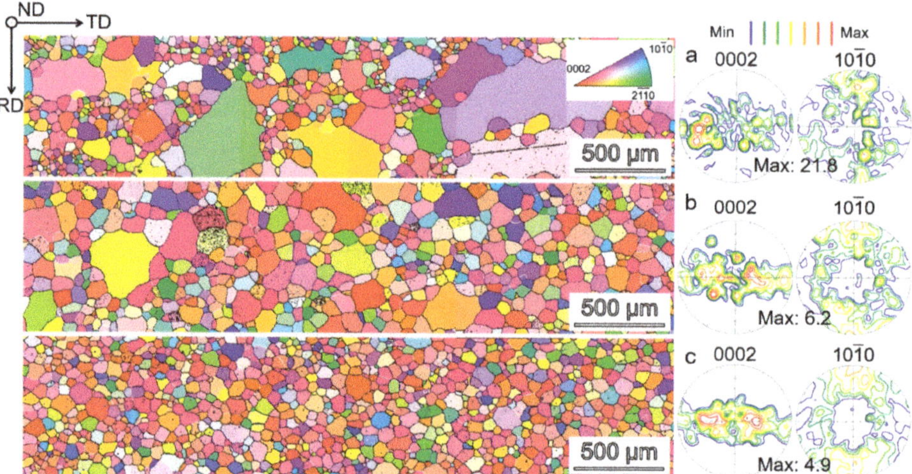

Figure 5. EBSD inverse pole figure (IPF) maps and corresponding (0002) and (10$\bar{1}$0) PFs showing microstructures and textures of T4-treated (**a**) Mg–1.5Ag–0.1Ca, (**b**) Mg–6Ag–0.1Ca, and (**c**) Mg–12Ag–0.1Ca alloy sheets.

To understand the reasons why the grain size is decreased with increasing the Ag content, the microstructures of T4-treated samples were observed by SEM. As can be seen from Figure 6a,b, the microstructures of the 1.5Ag and 6Ag containing alloy sheets consist of only α-Mg, indicating that solute atoms are fully dissolved into the Mg matrix after annealing at 450 °C for 1 h. In contrast, the 12Ag containing alloy sheet contains coarse second phase particles that are located both within the grains and along the grain boundaries, Figure 6c. The backscattered electron (BSE) image and corresponding Ag and Ca EDS maps indicate that these particles are mainly enriched with Ag, Figure 6d–f. Point analysis on a particle (marked with a triangle in Figure 6d shows that an atomic ratio between Mg and Ag is close to 4, Figure 6g. Based on our previous study, these particles are believed to be AgMg$_4$ phase [15]. The microstructure of the T4-treated Mg–12Ag–0.1Ca alloy sheet was further observed by TEM. Figure 7a shows the high-angle annular dark-field scanning transmission electron microscopy

(HAADF-STEM) image, in which the contrast is proportional to the square of the atomic number [23,24]. The grain boundary is therefore believed to be enriched with Ag (Z = 47) and or Ca (Z = 20) (Mg, Z = 12) because the grain boundary looks much brighter than adjacent grains. The corresponding Ag and Ca EDS maps reveal that Ag and Ca are co-segregated along the grain boundary, Figure 7b,c.

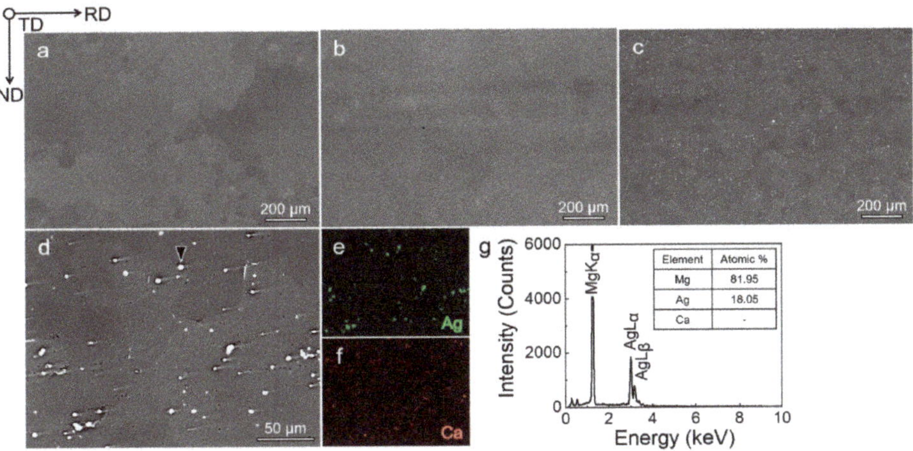

Figure 6. Backscattered electron (BSE) images showing microstructures of T4-treated (**a**) Mg–1.5Ag–0.1Ca, (**b**) Mg–6Ag–0.1Ca, and (**c**) Mg–12Ag–0.1Ca alloy sheets. (**d**) BSE image, and corresponding (**e**) Ag and (**f**) Ca elemental maps obtained from the Mg–12Ag–0.1Ca alloy sheet. (**g**) EDS spectrum obtained from a second phase particle (marked with a triangle) in (**d**).

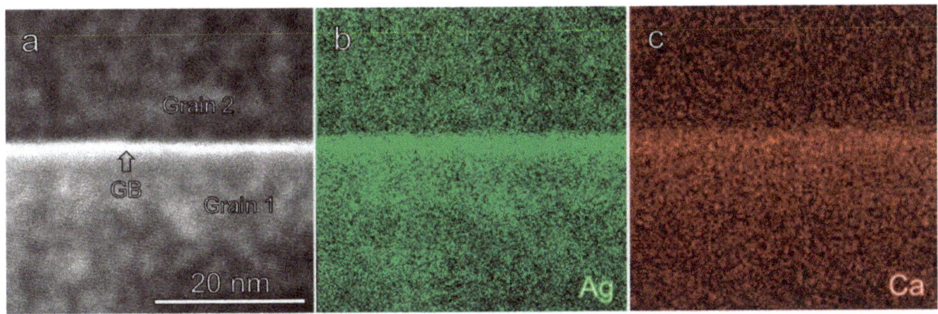

Figure 7. (**a**) High-angle annular dark-field-STEM image and (**b**) Ag and (**c**) Ca elemental maps showing the co-segregation of Ag and Ca atoms along a grain boundary of T4-treated Mg–12Ag–0.1Ca alloy sheet.

To clarify the reasons for the enhanced TYS after aging, the microstructure of the T6-treated 12Ag containing alloy sheet was observed by TEM. Figure 8a,b show the bright field (BF) TEM images taken with the incident beam along the $[0001]_\alpha$ and $[10\bar{1}0]_\alpha$ zone axes, respectively. The dispersion of precipitates is homogeneous, and the morphologies of the precipitates seem to consist of rod and polygonal shapes. The rod-type precipitates with a length of 0.5–2.5 µm are on the $\{11\bar{2}0\}_\alpha$ plane and their growing direction is parallel to the $<10\bar{1}0>_\alpha$ or $[0001]_\alpha$ directions. It is to be noted that some rod type precipitates lay on the pyramidal plane, as indicated by triangles in Figure 8b.

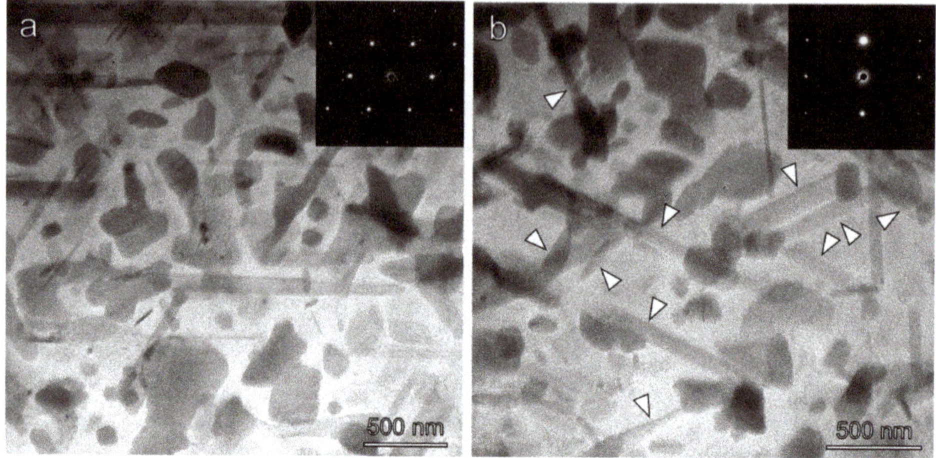

Figure 8. Bright field-TEM images showing the microstructure of T6-treated Mg–12Ag–0.1Ca alloy sheet along (**a**) $[0001]_a$ zone axis and (**b**) $[10\bar{1}0]_a$ zone axis.

Figure 9a shows the XRD pattern obtained from the T6-treated Mg–12Ag–0.1Ca alloy sheet. For the purpose of comparison, the XRD pattern obtained from the T4-treated sheet is given in Figure 9b. As can be seen that the T4-treated sample mainly consists of α-Mg phase. After the T6-treatment, new peaks are clearly observable, and they are confirmed to be generated mainly by MgAg$_4$ phase. Therefore, the improvement in the tensile yield strength is associated with the formation of AgMg$_4$ precipitates by the T6 treatment.

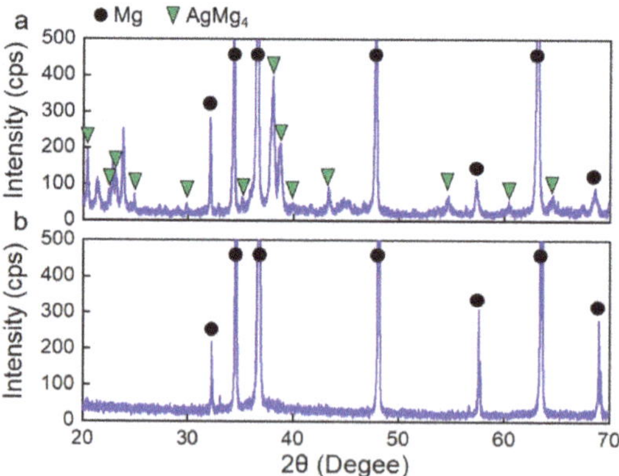

Figure 9. (**a**) X-ray diffraction (XRD) pattern obtained from T6-treated 12Ag containing alloy sheet. For the purpose of comparison, XRD pattern obtained from T4-treated 12Ag containing alloy sheet is given in (**b**).

To know whether there is an orientation relationship (OR) between AgMg$_4$ phase and α-Mg matrix, further microstructure analysis was carried out on the T6-treated 12Ag containing alloy sheet. Figure 10a shows the HAADF-STEM image with the zone axis of $[0001]_a$, and Figure 10b shows the high resolution (HR) HAADF-STEM image that is enlarged from the rectangular region in Figure 10a. Figure 10c,d are the fast Fourier transform (FFT) patterns generated from the left-hand region and right-hand region of Figure 10b, respectively. Analysis of these patterns reveals that left-hand region is the α-Mg matrix with the zone axis of $[0001]_\alpha$ and the right-hand region is the AgMg$_4$ phase with the zone axis of $[0001]_{AgMg4}$. By further analyzing the FFT patterns, the OR between the α-Mg matrix and AgMg$_4$ is confirmed to be $(0001)_\alpha \parallel (0001)_{AgMg4}$, $[\bar{2}110]_\alpha \parallel [10\bar{1}0]_{AgMg4}$.

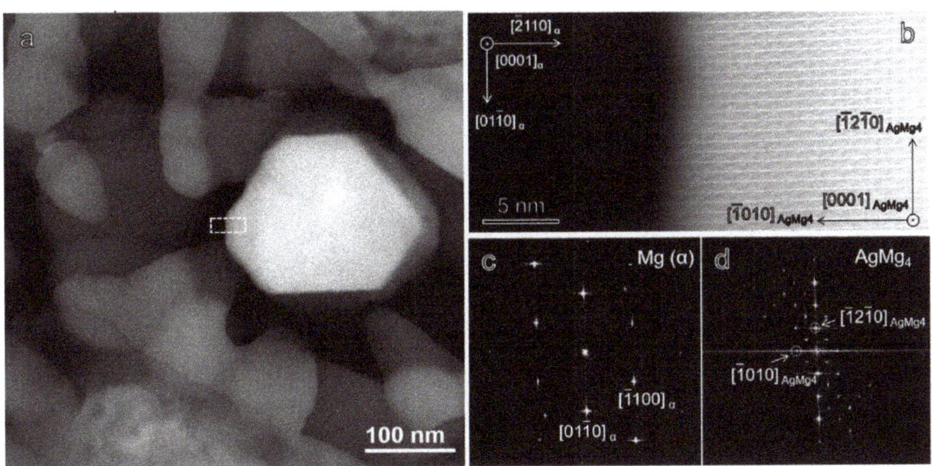

Figure 10. (**a**) HAADF-STEM image recorded along the $[0001]_\alpha$ direction. (**b**) Enlargement of the marked region in (a). FFT patterns generated from (**c**) α-Mg matrix and (**d**) AgMg$_4$ phase showing their orientation relationship.

4. Discussion

In the present study, we have successfully developed a new type of precipitation-hardenable Mg–Ag–Ca sheet alloys. In the T4-treated condition, the 1.5Ag containing alloy sheet shows a large average grain size of ~277 μm. In addition, abnormal grain growth occurs, as shown in Figure 5a. From our previous study, this alloy sheet subjected to 350 °C annealing for 1.5 h exhibits a homogeneous microstructure that consists of equiaxed grains with an average grain size of 20 μm [14], indicating abnormal grain growth is likely to occur at higher temperatures. In the case of the pure Mg and Mg alloy AZ31 (Mg–3Al–1Zn–0.3Mn in wt.%) [25,26], it seems that abnormal grain growth occurs at intermediate temperatures rather than higher temperatures, which is different from the trend observed in the present study. For example, abnormal grain growth occurs at an annealing temperature of 220 °C, while normal grain grow occurs at a higher temperature of 350 °C [25]. It was reported that recrystallized grains with the $<11\bar{2}0> \parallel$ RD orientation grow preferentially at the expense of deformed matrix grains close to the $<10\bar{1}0> \parallel$ RD orientation and neighboring recrystallized small grains, thereby evolving to abnormally large sizes in pure Mg and AZ31 alloy [27,28]. Figure 11 shows EBSD IPF maps and corresponding IPFs from abnormally coarse grains (>100 μm) and normal-size grains ($\leq$100 μm) in the T4-treated 1.5Ag containing alloy sheet. As can be seen, the normal-size grains have mainly the $<11\bar{2}0> \parallel$ RD orientation while the abnormally coarse grains do not have such orientation. These results suggest that the mechanism responsible for abnormal grain growth in Mg–Ag–Ca alloys is likely to be different from that in pure Mg and AZ31 alloy. It is to be noted that a homogeneous microstructure is developed in the 12Ag containing alloy sheet, Figure 5c. Ag and Ca atoms are demonstrated to be enriched in grain boundaries of the T4-treated condition, as shown in Figure 7. It is believed that solute segregation to grain boundaries occurs in the 1.5Ag containing alloy as well. However, the degree of solute segregation should be much lower than that in the 12Ag containing alloy. It is thus hypothesized that the segregated solute atoms effectively reduce grain boundary mobility via solute drag effects, thereby leading to a homogeneous microstructure. This hypothesis is supported by the homogeneous microstructure developed in the 1.5Ag containing alloy sheet after 350 °C annealing for 1.5h. Based on the PANDAT calculation, the solid solubility of Ag in Mg is substantially decreased from 10 wt.% to 1.42 wt.% with decreasing the temperature from 450 °C to 350 °C, as shown in Figure 12. Considering less Ag atoms will be dissolved into α-Mg matrix and more Ag atoms segregate to grain boundaries at 350 °C, the degree of solute segregation is expected to be much stronger than that at 450 °C. As such grain boundaries are difficult to break away from the solute drag atmosphere and remain pinned in the 1.5Ag containing alloy after 350 °C annealing.

Figure 11. EBSD IPF maps and corresponding inverse pole figures showing textures of (**a**) abnormally coarse grains and (**b**) normal-size grains in the T4-treated Mg–1.5Ag–0.1Ca alloy sheet.

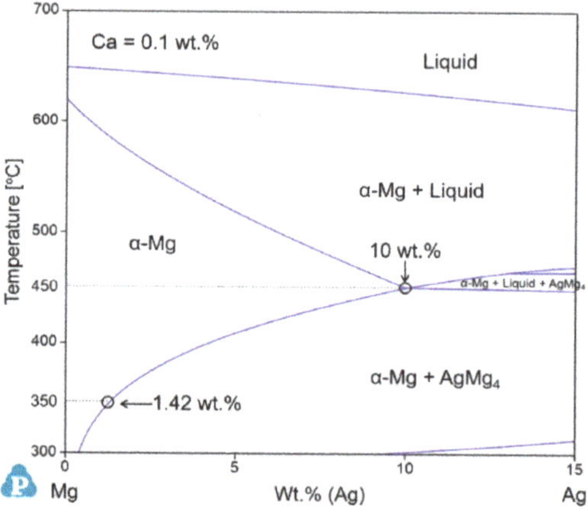

Figure 12. Calculated Mg-Ag phase diagram using the PANDAT software, in which the Ca content is fixed as 0.1 wt.%.

In the T4-treated condition, the TYS strength is substantially enhanced by increasing the Ag content from 1.5 wt.% to 12 wt.%. The 12Ag containing alloy sheet shows a homogeneous microstructure with an average grain size of ~74 μm. Thus, the increased TYS can be mainly ascribed to the refinement in the microstructure and solid solution strengthening effect. Nonetheless, the average grain size of the Mg–12Ag–0.1Ca alloy is still much coarser than the those developed in precipitation-hardenable Mg-Zn–Ca–Zr and Mg–Al–Ca–Mn(–Zn) alloys (<10 μm) [8,11,12,29]. This is because fine second phase particles that can effectively retard the growth of recrystallized grains are not present in the 12Ag containing alloy. Addition of small amounts of Al and Mn or Zn and Zr to in the Mg–12Ag–0.1Ca alloy might be promising to reduce the average grain size in the T4-condition by forming a densely distributed nano-scale Al_8Mn_5 or Zn_2Zr_3 particles [30,31]. On the other hand, all alloys exhibit a higher

TYS along the RD than along the TD due to the formation of the TD-split texture. Such a texture is beneficial for (0001)<11$\bar{2}$0> basal slip and {10$\bar{1}$2} tensile twin to accommodate the plastic deformation along the TD than the RD as their Schmid factor values are higher along the TD than the RD [32–36].

The TYS of the 12Ag containing alloy sheet is 193 MPa, 130 MPa and 117 MPa along the RD, 45° and TD, respectively. The T6 treatment further increases the TYS to 236 MPa, 163 MPa and 143 MPa. The strength enhancement by the T6-treatment is rather low compared to other alloys such as Mg–Gd [37], Mg–Zn [38] and Mg–Sn [39] based wrought alloys. Microstructure characterization reveals that non-basal AgMg$_4$ precipitates are formed in the T6-treated sample and their coarse microstructure is the main reason for the inferior age hardenability. There are several ways to refine the size of precipitates by increasing the number density of nucleation sites in early stage of aging: (i) deformation prior to aging treatment [40,41]; (ii) double aging consist of low temperature aging followed by high temperature aging [42–44]; and (iii) addition of trace amounts of elements [21,45,46]. Further optimization of the alloy composition incorporated with the modification of thermomechanical process is therefore expected to improve the age-hardening response and accelerate aging kinetics of Mg-Ag-Ca alloys.

5. Conclusions

Precipitation-hardenable wrought Mg alloy has been successfully developed based on the Mg–Ag–Ca system. In a T4-treated condition, the TYS of Mg–1.5Ag–0.1Ca alloy sheet is only 85 MPa, 57 MPa and 47 MPa along the RD, 45° and TD, respectively. With increase in the Ag content to 12 wt.%, the TYS is increased to 193 MPa, 130 MPa and 117 MPa along the RD, 45° and TD, which can be mainly ascribed to a refined microstructure and solid solution strengthening effect. Artificial aging at 170 °C for 336 h (T6) further increases the TYS of Mg–12Ag–0.1Ca alloy sheet to 236 MPa, 163 MPa and 143 MPa along the RD, 45° and TD. AgMg$_4$ precipitates lying on the {11$\bar{2}$0}$_a$ and pyramidal planes are responsible for the strength improvement.

Author Contributions: Conceptualization, M.B.; methodology, M.B.; validation, X.H. and Y.C.; investigation, M.B. and X.H.; writing—original draft preparation, M.B.; writing—review and editing, X.H. and Y.C.; project administration, Y.C.; funding acquisition, M.B. and X.H. All authors have read and agreed to the published version of the manuscript.

Funding: This research was funded by JSPS KAKENHI, grant numbers JP20K15067 and JP18K04787.

Conflicts of Interest: The authors declare no conflict of interest. The funders had no role in the design of the study; in the collection, analyses, or interpretation of data; in the writing of the manuscript, or in the decision to publish the results.

References

1. Hirsch, J.; Al-Samman, T. Superior light metals by texture engineering: Optimized aluminum and magnesium alloys for automotive applications. *Acta Mater.* **2013**, *61*, 818–843. [CrossRef]
2. Joost, W.J.; Krajewski, P.E. Towards magnesium alloys for high-volume automotive applications. *Scr. Mater.* **2017**, *128*, 107–112. [CrossRef]
3. Letzig, D.; Bohlen, J.; Kurz, G.; Victoria-Hernandez, J.; Hoppe, R.; Yi, S. Development of magnesium sheets. In *Magnesium Technology 2018*; The Minerals, Metals and Materials Series; Springer International Publishing: Berlin/Heidelberg, Germany, 2018; Volume F7, pp. 355–360.
4. Kim, N.J. Magnesium sheet alloys: Viable alternatives to steels? *Mater. Sci. Technol.* **2014**, *30*, 1925–1928. [CrossRef]
5. Nie, J.-F. Precipitation and hardening in magnesium alloys. *Metall. Mater. Trans. A* **2012**, *43*, 3891–3939. [CrossRef]
6. Hono, K.; Mendis, C.L.; Sasaki, T.T.; Oh-ishi, K. Towards the development of heat-treatable high-strength wrought Mg alloys. *Scr. Mater.* **2010**, *63*, 710–715. [CrossRef]
7. Homma, T.; Kunito, N.; Kamado, S. Fabrication of extraordinary high-strength magnesium alloy by hot extrusion. *Scr. Mater.* **2009**, *61*, 644–647. [CrossRef]

8.   Mendis, C.L.; Bae, J.H.; Kim, N.J.; Hono, K. Microstructures and tensile properties of a twin roll cast and heat-treated Mg–2.4Zn–0.1Ag–0.1Ca–0.1Zr alloy. *Scr. Mater.* **2011**, *64*, 335–338. [CrossRef]

9.   Nakata, T.; Mezaki, T.; Ajima, R.; Xu, C.; Oh-Ishi, K.; Shimizu, K.; Hanaki, S.; Sasaki, T.T.; Hono, K.; Kamado, S. High-speed extrusion of heat-treatable Mg-Al-Ca-Mn dilute alloy. *Scr. Mater.* **2015**, *101*, 28–31. [CrossRef]

10.   Nakata, T.; Xu, C.; Ajima, R.; Shimizu, K.; Hanaki, S.; Sasaki, T.T.; Ma, L.; Hono, K.; Kamado, S. Strong and ductile age-hardening Mg-Al-Ca-Mn alloy that can be extruded as fast as aluminum alloys. *Acta Mater.* **2017**, *130*, 261–270. [CrossRef]

11.   Bian, M.Z.; Sasaki, T.T.; Suh, B.C.; Nakata, T.; Kamado, S.; Hono, K. A heat-treatable Mg–Al–Ca–Mn–Zn sheet alloy with good room temperature formability. *Scr. Mater.* **2017**, *138*, 151–155. [CrossRef]

12.   Li, Z.H.; Sasaki, T.T.; Bian, M.Z.; Nakata, T.; Yoshida, Y.; Kawabe, N.; Kamado, S.; Hono, K. Role of Zn on the room temperature formability and strength in Mg–Al–Ca–Mn sheet alloys. *J. Alloys Compd.* **2020**, *847*, 156347. [CrossRef]

13.   Bian, M.Z.; Sasaki, T.T.; Nakata, T.; Yoshida, Y.; Kawabe, N.; Kamado, S.; Hono, K. Bake-hardenable Mg–Al–Zn–Mn–Ca sheet alloy processed by twin-roll casting. *Acta Mater.* **2018**, *158*, 278–288. [CrossRef]

14.   Bian, M.; Huang, X.; Chino, Y. A room temperature formable magnesium–silver–calcium sheet alloy with high ductility. *Mater. Sci. Eng. A* **2020**, *774*, 138923. [CrossRef]

15.   Bian, M.; Huang, X.; Chino, Y. A combined experimental and numerical study on room temperature formable magnesium–silver–calcium alloys. *J. Alloys Compd.* **2020**, 155017. [CrossRef]

16.   Bian, M.; Huang, X.; Chino, Y. Improving flame resistance and mechanical properties of magnesium–silver–calcium sheet alloys by optimization of calcium content. *J. Alloys Compd.* **2020**, *837*, 155551. [CrossRef]

17.   Nayeb-Hashemi, A.A.; Clark, J.B. *Phase Diagrams of Binary Magnesium Alloys*; ASM International: Metals Park, OH, USA, 1988; ISBN 9780871703286.

18.   Payne, R.J.M.; Bailey, N. Improvement of the age hardening properties of magnesium-rare-earth alloys by addition of silver. *J. Inst. Met.* **1960**, *88*, 417–427.

19.   Gao, X.; Nie, J.F. Enhanced precipitation-hardening in Mg–Gd alloys containing Ag and Zn. *Scr. Mater.* **2008**, *58*, 619–622. [CrossRef]

20.   Zhu, Y.M.; Morton, A.J.; Nie, J.F. Improvement in the age-hardening response of Mg-Y-Zn alloys by Ag additions. *Scr. Mater.* **2008**, *58*, 525–528. [CrossRef]

21.   Mendis, C.L.; Oh-ishi, K.; Hono, K. Enhanced age hardening in a Mg–2.4 at.% Zn alloy by trace additions of Ag and Ca. *Scr. Mater.* **2007**, *57*, 485–488. [CrossRef]

22.   Cao, W.; Chen, S.L.; Zhang, F.; Wu, K.; Yang, Y.; Chang, Y.A.; Schmid-Fetzer, R.; Oates, W.A. PANDAT software with PanEngine, PanOptimizer and PanPrecipitation for multi-component phase diagram calculation and materials property simulation. *Calphad* **2009**, *33*, 328–342. [CrossRef]

23.   Pennycook, S.J.; Jesson, D.E. High-resolution Z-contrast imaging of crystals. *Ultramicroscopy* **1991**, *37*, 14–38. [CrossRef]

24.   Pennycook, S.J.; Jesson, D.E. Atomic resolution Z-contrast imaging of interfaces. *Acta Metall. Mater.* **1992**, *40*, S149–S159. [CrossRef]

25.   Pei, R.; Korte-Kerzel, S.; Al-Samman, T. Normal and abnormal grain growth in magnesium: Experimental observations and simulations. *J. Mater. Sci. Technol.* **2020**, *50*, 257–270. [CrossRef]

26.   Bhattacharyya, J.J.; Agnew, S.R.; Muralidharan, G. Texture enhancement during grain growth of magnesium alloy AZ31B. *Acta Mater.* **2015**, *86*, 80–94. [CrossRef]

27.   Gottstein, G.; Alsamman, T. Texture development in pure Mg and Mg alloy AZ31. *Mater. Sci. Forum* **2005**, *195–197*, 623–632. [CrossRef]

28.   Steiner, M.A.; Bhattacharyya, J.J.; Agnew, S.R. The origin and enhancement of {0001}⟨11-20⟩ texture during heat treatment of rolled AZ31B magnesium alloys. *Acta Mater.* **2015**, *95*, 443–455. [CrossRef]

29.   Li, Z.H.; Sasaki, T.T.; Shiroyama, T.; Miura, A.; Uchida, K.; Hono, K. Simultaneous achievement of high thermal conductivity, high strength and formability in Mg-Zn-Ca-Zr sheet alloy. *Mater. Res. Lett.* **2020**, *8*, 335–340. [CrossRef]

30.   Trang, T.T.T.; Zhang, J.H.; Kim, J.H.; Zargaran, A.; Hwang, J.H.; Suh, B.-C.; Kim, N.J. Designing a magnesium alloy with high strength and high formability. *Nat. Commun.* **2018**, *9*, 2522. [CrossRef]

31.   Gao, X.; Muddle, B.C.; Nie, J.F. Transmission electron microscopy of Zr-Zn precipitate rods in magnesium alloys containing Zr and Zn. *Philos. Mag. Lett.* **2009**, *89*, 33–43. [CrossRef]

32. Bohlen, J.; Cano, G.; Drozdenko, D.; Dobron, P.; Kainer, K.; Gall, S.; Müller, S.; Letzig, D. Processing Effects on the Formability of Magnesium Alloy Sheets. *Metals* **2018**, *8*, 147. [CrossRef]
33. Yan, H.; Xu, S.W.; Chen, R.S.; Kamado, S.; Honma, T.; Han, E.H. Activation of {10-12} twinning and slip in high ductile Mg-2.0Zn-0.8Gd rolled sheet with non-basal texture during tensile deformation at room temperature. *J. Alloys Compd.* **2013**, *566*, 98–107. [CrossRef]
34. Chino, Y.; Sassa, K.; Mabuchi, M. Texture and stretch formability of a rolled Mg–Zn alloy containing dilute content of Y. *Mater. Sci. Eng. A* **2009**, *513–514*, 394–400. [CrossRef]
35. Chino, Y.; Huang, X.; Suzuki, K.; Sassa, K.; Mabuchi, M. Influence of Zn concentration on stretch formability at room temperature of Mg–Zn–Ce alloy. *Mater. Sci. Eng. A* **2010**, *528*, 566–572. [CrossRef]
36. Bohlen, J.; Nürnberg, M.R.; Senn, J.W.; Letzig, D.; Agnew, S.R. The texture and anisotropy of magnesium–zinc–rare earth alloy sheets. *Acta Mater.* **2007**, *55*, 2101–2112. [CrossRef]
37. He, S.M.; Zeng, X.Q.; Peng, L.M.; Gao, X.; Nie, J.F.; Ding, W.J. Microstructure and strengthening mechanism of high strength Mg–10Gd–2Y–0.5Zr alloy. *J. Alloys Compd.* **2007**, *427*, 316–323. [CrossRef]
38. Mendis, C.L.; Oh-ishi, K.; Kawamura, Y.; Honma, T.; Kamado, S.; Hono, K. Precipitation-hardenable Mg-2.4Zn-0.1Ag-0.1Ca-0.16Zr (at.%) wrought magnesium alloy. *Acta Mater.* **2009**, *57*, 749–760. [CrossRef]
39. Sasaki, T.T.; Elsayed, F.R.; Nakata, T.; Ohkubo, T.; Kamado, S.; Hono, K. Strong and ductile heat-treatable Mg–Sn–Zn–Al wrought alloys. *Acta Mater.* **2015**, *99*, 176–186. [CrossRef]
40. The effect of cold work on precipitation in alloy WE54. In Proceedings of the Magnesium Alloys and Their Applications, Wolfsburg, Germany, 28–30 April 1998; pp. 329–334.
41. Shi, G.L.; Zhang, D.F.; Zhang, H.J.; Zhao, X.B.; Qi, F.G.; Zhang, K. Influence of pre-deformation on age-hardening response and mechanical properties of extruded Mg-6%Zn-1%Mn alloy. *Trans. Nonferrous Met. Soc. China* **2013**, *23*, 586–592. [CrossRef]
42. Oh-ishi, K.; Hono, K.; Shin, K.S. Effect of pre-aging and Al addition on age-hardening and microstructure in Mg-6 wt% Zn alloys. *Mater. Sci. Eng. A* **2008**, *496*, 425–433. [CrossRef]
43. Sasaki, T.T.; Oh-ishi, K.; Ohkubo, T.; Hono, K. Effect of double aging and microalloying on the age hardening behavior of a Mg-Sn–Zn alloy. *Mater. Sci. Eng. A* **2011**, *530*, 1–8. [CrossRef]
44. Mendis, C.L.; Oh-ishi, K.; Ohkubo, T.; Shin, K.S.; Hono, K. Microstructures and mechanical properties of extruded and heat treated Mg–6Zn–1Si–0.5Mn alloys. *Mater. Sci. Eng. A* **2012**, *553*, 1–9. [CrossRef]
45. Mendis, C.L.; Bettles, C.J.; Gibson, M.A.; Hutchinson, C.R. An enhanced age hardening response in Mg–Sn based alloys containing Zn. *Mater. Sci. Eng. A* **2006**, *435–436*, 163–171. [CrossRef]
46. Sasaki, T.T.; Oh-ishi, K.; Ohkubo, T.; Hono, K. Enhanced age hardening response by the addition of Zn in Mg–Sn alloys. *Scr. Mater.* **2006**, *55*, 251–254. [CrossRef]

Publisher's Note: MDPI stays neutral with regard to jurisdictional claims in published maps and institutional affiliations.

Article

Interaction of Migrating Twin Boundaries with Obstacles in Magnesium

Andriy Ostapovets [1],*, Konstantin Kushnir [1], Kristián Máthis [2] and Filip Šiška [1]

[1] CEITEC-IPM, Institute of Physics of Materials, Czech Academy of Sciences, Žižkova 22, 61600 Brno, Czech Republic; kushnir@ipm.cz (K.K.); siska@ipm.cz (F.Š.)
[2] Department of Physics of Materials, Faculty of Mathematics and Physics, Charles University, Ke Karlovu 3, 12116 Prague, Czech Republic; mathis@met.mff.cuni.cz
* Correspondence: ostapov@ipm.cz; Tel.: +420-5-3229-0429

Abstract: Interaction of migrating $\{10\bar{1}2\}$ twin boundary with obstacles was analyzed by atomistic and finite elements computer simulations of magnesium. Two types of obstacles were considered: one is a non-shearable obstacle and another one is the void inside bulk material. It is shown that both types of obstacles inhibit twin growth and increased stress is necessary to engulf the obstacle in both cases. However, the increase of critical resolved shear stress is higher for the passage of the twin boundary through raw of voids than for interaction with non-shearable obstacles.

Keywords: magnesium; twinning; modeling; void; rigid inclusion

Citation: Ostapovets, A.; Kushnir, K.; Máthis, K.; Šiška, F. Interaction of Migrating Twin Boundaries with Obstacles in Magnesium. *Metals* **2021**, *11*, 154. https://doi.org/10.3390/met11010154

Received: 25 November 2020
Accepted: 13 January 2021
Published: 15 January 2021

Publisher's Note: MDPI stays neutral with regard to jurisdictional claims in published maps and institutional affiliations.

1. Introduction

Magnesium and its alloys are prospective light-weight materials [1]. However, they have poor formability in comparison with aluminium alloys, which leads to limitation in the application of magnesium alloys in automotive and aerospace industries. The pure formability of magnesium is attributed to its hexagonal close packed structure, which does not provide a sufficient number of slip systems. Plastic deformation of magnesium is therefore characterized by significant anisotropy. Basal slip systems allow easy dislocation glide. However, non-basal slip systems are hard, and deformation twinning is often activated instead of non-basal slip [2,3]. Abundant twinning can lead to the formation of structure inhomogeneity during plastic deformation of these materials. Such inhomogeneity can be a serious restriction for engineering applications.

Magnesium formability can be improved by decreasing its plastic anisotropy, which can be overcome by either decreasing the critical resolved shear stresses (CRSS) for non-basal slip systems or by increasing the basal slip CRSS. The former can be reached by alloying of magnesium, for instance, with rare-earth elements such as Y or Gd [4–6]. These elements decrease stacking fault energies on non-basal crystallographic planes. The latter approach based on hardening of the basal slip can be achieved by introducing precipitates inside the alloy microstructure. These precipitates cause the effect called precipitation hardening of material [7,8] due to their interactions with dislocations. The study of dislocation interaction with obstacles has a long history. The nature of dislocation–precipitate interactions is well described by the Orowan mechanism [9]. In contrast, twin-precipitate interactions are more complex and less understood. This has attracted significant attention of material research at the present time [10–16]. For instance, the effect of precipitation on yield elongation in an extruded Mg–4.5 wt% Zn alloy was investigated in [10]. It was shown that precipitation led to the disappearance of the yield plateau on the stress-strain curve. Aging led to an increase in yield stress of about 60 MPa. Twins in the as-extruded sample were thicker and fewer in number than those in the aged sample. Applicability of the Orowan precipitate hardening equation to twin propagation was studied in [11] by two-dimensional dislocation dynamics simulations of a twin tip impinging upon a line of obstacles. It was shown that reasonable prediction of propagation stress can occur. However, this approach has limits, e.g., the twin boundary itself is not present in dislocation dynamics simulations. More precise consideration of the twin boundary-precipitate interaction is possible by atomistic simulations. It was demonstrated in [12] by molecular dynamics simulation that spherical precipitates have the strongest hardening effect on the twinning, while the rod-like precipitates have the weakest. Nonetheless, atomistic simulations are applicable, as a rule, to small volumes and high deformation rates only. Other approaches are also useful for analysis of twin-precipitate interactions, especially in the interplay with the dislocation slip. Recently, finite element method analysis allowed the conclusion that precipitates affect twin thickening by changing the slip CRSS values and acting as obstacles for twin propagation [13]. The value of pyramidal CRSS is the most critical one because the activity of the pyramidal slip significantly increases in the vicinity of the twin with an aspect ratio > 0.1. Visco-plastic self-consistent modeling was used to show that the twin systems are hardened more by the precipitation than by the basal planes [14].

Another possible type of obstacle for twin boundary migration is voids [15–17]. Nanoscale twinning is an energy dissipating mechanism for ductile fracture in fine-grained Mg alloys [15]. On the other hand, the void can be formed by particle cracking during deformation of fine-grained magnesium alloys [16]. Recently, interaction of an incoherent basal-prismatic twinning interface with the void was studied by atomistic simulations [17]. It was shown that the location of the void affects the start position of basal-prismatic boundary migration. The void serves as an obstacle during migration, and parts of the boundary are pinned by the void surface.

Suppressing of the twinning by twin interactions with obstacles can also be considered as a way to decrease plastic anisotropy and to improve the mechanical properties of the material.

The present study is focused on the numerical analysis of the interaction between the growing tensile twin and obstacles in magnesium. The study is based on a multiscale approach as the twin growth is simulated using an atomistic model while the overall stress state is evaluated using a finite element method.

2. Model

2.1. Atomistic Simulations

The atomistic simulations were performed by using LAMMPS (Sandia National Laboratories, Livermore, CA, USA, lammps.sandia.gov) [18] for simulations and OVITO (Darmstadt University of Technology, Darmstadt, Germany, www.ovito.org) [19] for visualization. The magnesium potential, published in Liu et al. [20], was used in the calculations. The scheme of the used simulation block is shown in Figure 1. The rectangular simulation blocks have their x-axis close to the $[\bar{1}011]$ crystal direction, y-axis along the $[1\bar{2}10]$ direction, and z-axis close to normal to the $(10\bar{1}2)$ plane. The x and z axes are inclined from corresponding directions by angle α. The blocks were prepared by the following procedure. The twin boundary was created inside the simulation block by the merging of crystals rotated by the misorientation angle ~86°. Then, the lattice was rotated by the angle α. The value of α was selected to satisfy the continuity of each $(10\bar{1}2)$ plane through periodic boundary conditions in the x direction. Each $(10\bar{1}2)$ plane is continued into the $(10\bar{1}2)$ plane separated by two interplanar distances from it. Due to periodicity, steps on the twin boundary were produced at the block edges. These steps are relaxed into twinning disconnections [21,22] during energy minimization of the block. Thereby, the studied configuration contains the twin boundary with disconnections periodically placed within it.

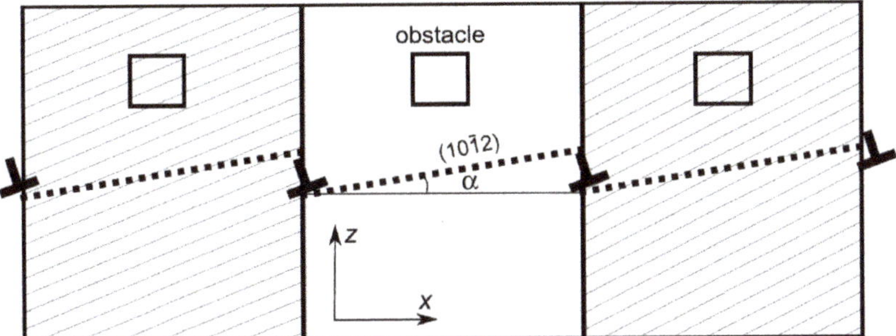

Figure 1. Scheme of the simulation block. The dashed line corresponds to the twin boundary. Shaded areas are periodic images of the central block. Twinning disconnections produce steps on the twin boundary due to periodic boundary conditions.

Two different distances between disconnections (23 nm or 45 nm) were used in our simulations, and different block sizes were used in order to study the influence of the disconnection density on the twin boundary behavior. All blocks have a size of 1.9 nm in the *y*-direction and 50 nm in the *z*-direction. However, the size in the *x*-direction was variable. The block sizes were selected as multiples of the above-mentioned distances between disconnections (23 nm or 45 nm). The larger blocks were produced by merging of several small blocks. Consequently, the larger block contained several disconnections inside and disconnection density was preserved.

The obstacle for twin boundary migration was also inserted into the simulation block. Two types of obstacles were considered. The first one is a void with a squared shape of its xz-section and with infinite size in the *y*-direction. The size of void was 2 × 2 nm in the xz plane. Another type of obstacle was obtained by freezing atoms inside the selected volume. The size of the obstacle was the same as the size of the void. Such obstacle can be considered a model of non-shearable inclusion. The distance between obstacles was equal to periodicity in the x-direction. The size of obstacle was selected to be small enough in comparison to the simulation block in order to leave space for twin boundary migration between obstacles.

Shear strain (ε_{xz}) was applied to the block in order to model the interaction of the migrating twin boundary with rigid inclusion. The strain was applied in small steps, $\Delta\varepsilon_{xz} = 0.0005$. Energy minimization by the conjugate gradient method was performed after each step. The stress value of the block was monitored after energy minimization.

2.2. FE Model

FEM analysis was used to analyze the stress distribution in the vicinity of the precipitate or cavity. These stresses can describe possible preferable locations of twin nucleation and also probable twin growth.

The FE model was made within the code Z-set (Mines ParisTech, Paris, France, www.zset-software.com) and its layout is provided in Figure 2. It is a 2D planar system that has a rectangular inclusion with an aspect ratio (width/length—*t*/*l*) 1/10 in the middle. The model represents an idealized case of an obstacle perpendicular to the (10$\bar{1}$2) extension twin plane which is inclined to the global *X*-axis. The loading and boundary conditions are applied as follows. The external tensile stress is applied in the X-direction on the vertical sides which represents tension in the crystallographic c-direction. The horizontal sides are kept straight and parallel. The length of each side is ten times the length (*l*) of the obstacle to avoid the influence of boundary conditions on the stresses in the obstacle vicinity. The analyzed stress is taken within the matrix along the normal axis of the inclusion (red line). The stress component that is analyzed is the shear stress along the (10$\bar{1}$2) plane.

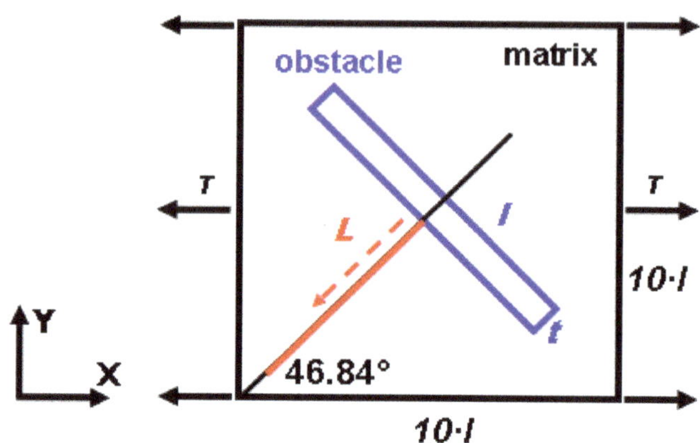

Figure 2. The FE model geometry, with loading conditions and location for stress analysis (red line). The size of the obstacle is not to scale.

A matrix is simulated as anisotropic elastic material with elastic constants of magnesium [23] which are summarized in Table 1. The purely elastic behavior is used to model the situation where twinning is the first inelastic deformation mode occurring prior to slip. The precipitate is isotropic elastic with Young's modulus $E = 80$ GPa and Poisson's ratio 0.35, which are the typical values used for magnesium alloys [24]. The void is simulated by a material with a very low Young's modulus, $E = 0.008$ GPa. Such an approximation was chosen to keep the FE mesh identical in both cases. The "void" elastic constants are four orders of magnitude smaller than the matrix ones which makes it a valid approximation of empty space.

Table 1. Elastic constants of Magnesium.

Tensor Component	Value [GPa]
C_{1111}	59.7
C_{2222}	59.7
C_{3333}	61.7
C_{1212}	16.8
C_{2323}	16.4
C_{3131}	16.4
C_{1122}	26.2
C_{2233}	20.8
C_{3311}	20.8

3. Results

3.1. Atomistic Simulations

Figure 3 shows strain-stress curves for the migration of the twin boundary in the block without obstacles and for the interaction of the twin boundary with the void. The stress provided in Figure 3 is total shear stress (σ_{xz}) in the simulation block. It can be seen that the presence of the obstacle potentially inhibits twin growth. The blue curve shows continuous growth of the stress up to a value of 190 MPa. The movement of disconnections begins at this stress level. Then, the twin boundary begins to migrate continuously under practically constant stress. It is worth noting that calculations are done at 0K and no temperature effect is considered. This is the reason why the observed stress is high.

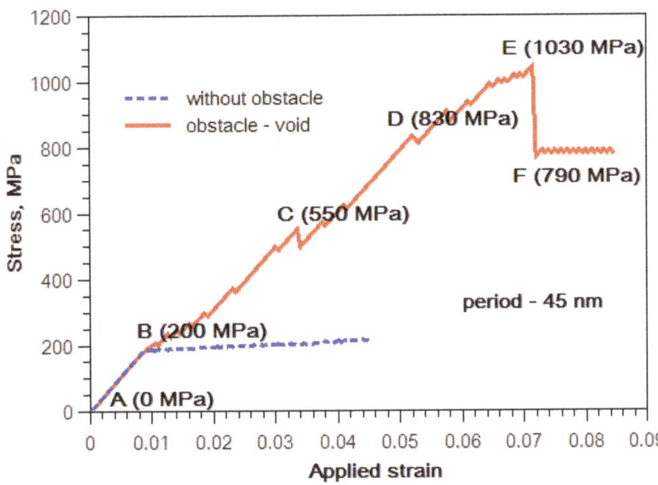

Figure 3. Strain-stress curves for the case of twin boundary migration in the block without the obstacle (blue line) and in the block containing the void (red line).

The stress increases to much higher values than 190 MPa if the void is present (red curve in Figure 3). The disconnections also begin to move at values close to 200 MPa (point B) in this case (Figure 4B). However, the presence of the obstacle leads to a further increase in the stress with the decreasing distance between the twin boundary and the cavity. After reaching the void, the twin boundary stays blocked on it (Figure 4C). The existing twinning disconnections are absorbed by the cavity surface at this moment. New disconnections begin to nucleate at the stress level 550 MPa (point C). They are also absorbed in the cavity surfaces. The twin boundary moves along the cavity during this process. It reaches the upper surface of the cavity at point D, which corresponds to the stress level 830 MPa (Figure 4D). Then, the boundary continues its migration but its stays anchored at the void. The boundary is connected to the void by segments of basal-prismatic interfaces (Figure 4E). These facets grow by 'pile-up' of disconnections gliding on parallel planes, similar to the case described in [25]. The point E in the stress-strain curve corresponds to the moment when basal-prismatic interfaces begin to move by nucleation of disconnection dipoles on them. This process is analogous to the process described in [22]. It happens at a considerably high stress level of 1030 MPa. A sharp decrease in the stress takes place in this moment. However, the stress does not relax completely and subsequent boundary migration takes place at the stress level of about 790 MPa. It is due to stress accumulated inside the twin, which cannot be immediately relaxed by migration of the existing twin boundary. The basal-prismatic interfaces coalesce and the twin boundary is straightened again (Figure 4F).

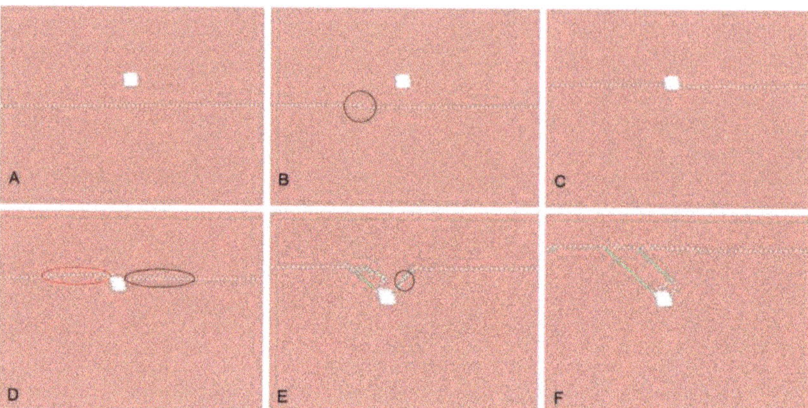

Figure 4. Interaction of the migrating twin boundary with the void: (**A**) initial configuration; (**B**) migration of the pre-existing disconnection along the boundary. The disconnection is marked by a black circle; (**C**) the boundary is blocked in the void; (**D**) nucleation of disconnection dipoles in the boundary and subsequent absorption of disconnections by the void surface; The dipoles are marked by ellipses. The red ellipse marks the already developed dipole and the black ellipse marks the location of dipole nucleation in progress. (**E**) formation and migration of basal-prismatic interfaces. Disconnection on the basal-prismatic interface is marked by a circle; (**F**) detachment of the boundary from the void with the formation of two basal stacking faults. The projection of the figure is in the [$1\bar{2}10$] direction. The atoms in the boundary are colored in white and atoms in stacking faults are colored in green.

Figure 4E,F show configurations with migrating basal-prismatic facets and after their coalescence. The green atoms indicate neighboring, which corresponds to the fcc structure. Consequently, two basal stacking faults are nucleated during detachment of the boundary from the obstacle. These stacking faults are trailed by the migrating boundary and grow together with the twin.

Dependences of the critical shear stress, which is necessary to push the boundary through the row of obstacles on the density of the obstacle (i.e., block size in the x direction), are shown in Figure 5. This stress corresponds to the moment when basal-prismatic interfaces begin to move (such as at point E in Figure 3). It can be seen that this stress is dependent on the type of obstacle as well as on the initial density of disconnections. It is interesting that voids serve as a stronger barrier for twin boundary migration than non-shearable obstacles.

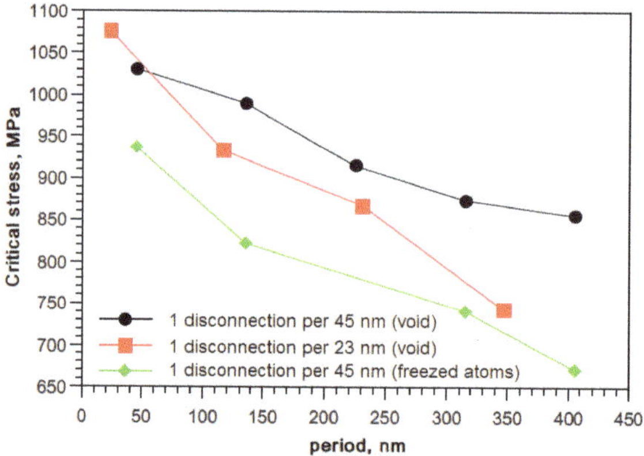

Figure 5. Dependence of critical stresses on the distance (period) between obstacles for different densities disconnections and kinds of obstacle (cavity or non-sheared obstacles).

3.2. FE Model

The relative shear stress distribution in the twin direction in the vicinity of the obstacle (along the red line in Figure 2) is shown in Figure 6. The stress is normalized by the applied external stress (τ_{app}), and the distance (L) from the precipitate/void is normalized by its thickness (t). It is clear that the precipitate causes a slight increase in the applied stress in its close vicinity and the stress linearly decreases to the distant applied stress level. The presence of the void shows a different picture. The shear stress is zero at the boundary due to the void's free surface. The stress increases with the distance from the void's surface. It reaches the value for the precipitate at the distance of 2.5 of the void's thickness and further increases to values of 1.3 of the applied stress.

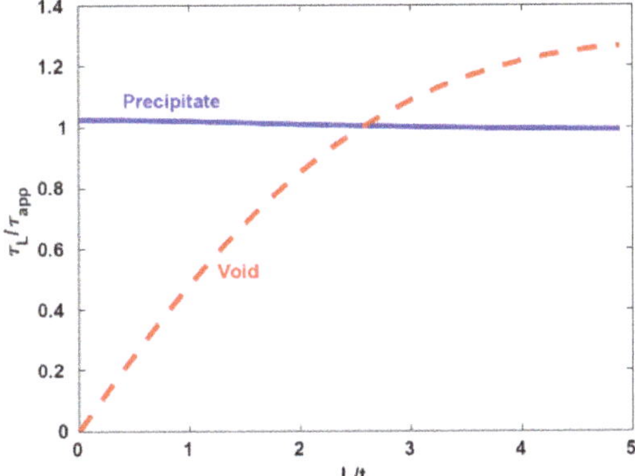

Figure 6. Stress distribution inside the matrix in the direction of the precipitate/void normal axis. Shear stress is normalized by externally applied stress τ_{app} and distance L is normalized by precipitate/void thickness t.

4. Discussion

The stress level obtained from atomistic simulations in this study is quite high in comparison to the experimental CRSS of the twin. It is caused by the fact that existing interatomic potentials were not fitted primarily for these values. However, it is possible to expect that qualitative trends as well as atomistic mechanisms predicted by the model are correct. On the other hand, it worth noting that the experimental increase in the yield stress for the aged AZ91 alloy is at the level of about 100 MPa [26], i.e., they are comparable with our values in order of magnitude. This increase is partially caused by twin interactions with precipitates.

Our simulation revealed that the void is a stronger barrier for twin boundary migration than non-shearable obstacles. Recently, the interaction of the basal-prismatic twin boundary with the void was studied by computer simulations by Xu et al. [17]. It was demonstrated in [17] that boundary migration is hindered by the void. In contrast to our simulation, the boundary was not completely pushed through the void. However, the presence of the void led to a decrease in yield stress in comparison to boundary migration with the absence of the void. This trend is opposite to our observations shown in Figure 3, where the presence of the void led to a slight increase in yield stress. This situation is probably caused by the difference in the initial boundary configuration. In the present paper, we studied the migration of the coherent ($10\bar{1}2$) twin boundary, which already contained disconnections and was already ready to migrate. The elastic field of the void slowed down the motion of the pre-existing disconnections. However, yielding takes place by nucleation of new boundary defects and subsequent detachment of basal-prismatic segments from raw immobile misfit dislocation [17]. The stress field around the void helps in the nucleation of such defects in the basal-prismatic boundary as claimed by the authors of [17].

The FE model results correspond with the atomistic simulations. The shear stress decrease in the vicinity of the void's surface implies that higher applied stress is needed to move the twinning boundary. Atomistic simulations show about a fivefold increase in the applied stress for initiation of boundary motion near the void (see Figure 3). Similarly, the estimated local shear stress is about fivefold lower in the close vicinity of the void's surface in the FE model (approximately $0.25t$), which suggests the necessity of an increase in the applied stress for twin initiation. In general, the void can be seen as a limiting case of a very soft precipitate, so the presented simulation shows the effect of precipitate stiffness on the local stress distribution. Soft precipitates inhibit twin boundary motion in their vicinity and increase the probability of twin initiation between precipitates due to the stress increase at this location.

5. Conclusions

Basal-prismatic facets can be formed by the glide of disconnections on parallel $\{10\bar{1}2\}$ planes. The mechanism of their formation agrees with the existing models of twinning such as the theory of admissible interfacial defects. The basal-prismatic facet on the straight twin boundary is equivalent to a wall of twinning disconnections.

The interaction of twin boundaries with obstacles is governed by the mobility of disconnections and basal-prismatic interfaces.

The increase in critical resolved shear stress is higher for the interaction of the twin boundary with cavities than for the interaction with non-shearable obstacles.

Stacking faults can be formed by twin–twin interactions. These stacking faults can be recognized as partial stacking faults reported previously in the literature.

The continuum theory also predicts inhibition of twin boundary movement due to the significant decrease in the shear stress void's surface.

Author Contributions: Conceptualization, A.O. and K.M.; investigation, A.O., K.K., F.Š.; writing—original draft preparation, A.O. All authors have read and agreed to the published version of the manuscript.

Funding: This research was funded by CZECH SCIENCE FOUNDATION, grant number 18-07140S, and by the MINISTRY OF EDUCATION, YOUTH AND SPORTS OF THE CZECH REPUBLIC, grant number CEITEC 2020 (LQ1601).

Institutional Review Board Statement: Not applicable.

Informed Consent Statement: Not applicable.

Data Availability Statement: Data available on request due to restrictions. The data presented in this study are available on request from the corresponding author. The data are not publicly available due to authors' employer restrictions.

Conflicts of Interest: The authors declare no conflict of interest.

References

1. Hirsch, J.; Al-Samman, T. Superior light metals by texture engineering: Optimized aluminum and magnesium alloys for automotive applications. *Acta Mater.* **2013**, *61*, 818–843. [CrossRef]
2. Yoo, M.H. Slip, twinning, and fracture in hexagonal close-packed metals. *Met. Trans.* **1981**, *12*, 409–418. [CrossRef]
3. Yoo, M.H.; Agnew, S.R.; Morris, J.R.; Ho, K.M. Non-basal slip systems in HCP metals and alloys: Source mechanism. *Mater. Sci. Eng.* **2001**, *A319–A321*, 87–92. [CrossRef]
4. Sandlobes, S.; Friak, M.; Neugebauer, J.; Raabe, D. Basal and non-basal dislocation slip in Mg-Y. *Mater. Sci. Eng. A* **2013**, *A576*, 61–68. [CrossRef]
5. Zhang, J.; Liu, S.; Wu, R.; Hou, L.; Zhang, M. Recent developments in high-strength Mg-RE-based alloys: Focusing on Mg-Gd and Mg-Y systems. *J. Magnes. Alloy* **2018**, *6*, 277–291. [CrossRef]
6. Andritsos, E.I.; Paxton, A.T. Effects of calcium on planar fault energies in ternary magnesium alloys. *Phys. Rev. Mater.* **2019**, *3*, 013607. [CrossRef]
7. Jayara, J.; Mendis, C.L.; Ohkubo, T.; Ohishi, K.; Hono, K. Enhanced precipitation hardening of Mg-Ca alloy by Al addition. *Scr. Mater.* **2010**, *63*, 831–834. [CrossRef]
8. Tehranchi, A.; Yin, B.; Curtin, W.A. Solute strengthening of basal slip in Mg alloys. *Acta Mater.* **2018**, *151*, 56–66. [CrossRef]

9. Stanford, N.; Geng, J.; Chun, Y.B.; Davies, C.H.J.; Nie, J.F.; Barnett, M.R. Effect of plate-shape particle distributions on the deformation behavior of magnesium alloy AZ91 in tension and compression. *Acta Mater.* **2012**, *60*, 218–228. [CrossRef]

10. Wang, J.; Ferdowsi, M.R.G.; Kada, S.R.; Hutchinson, C.R.; Barnett, M.R. Influence of precipitation on yield elongation in Mg-Zn alloys. *Scr. Mater.* **2019**, *160*, 5–8. [CrossRef]

11. Barnett, M.R.; Wang, H.; Guo, T. An Orowan precipitate strengthening equation for mechanical twinning in Mg. *Int. J. Plast.* **2019**, *112*, 108–122. [CrossRef]

12. Fan, H.; Zhu, Y.; El-Awady, J.A.; Raabe, D. Precipitation hardening effects on extension twinning in magnesium alloys. *Int. J. Plast.* **2018**, *106*, 186–202. [CrossRef]

13. Siska, F.; Stratil, L.; Cizek, J.; Guo, T.; Barnett, M.R. Numerical analysis of twin-precipitate interactions in magnesium alloys. *Acta Mater.* **2021**, *202*, 80–87. [CrossRef]

14. Stanford, M.; Barnett, M.R. Effect of particles on the formation of deformation twins in a magnesium-based alloy. *Mater. Sci. Eng. A* **2009**, *516*, 226–234. [CrossRef]

15. Wang, Y.N.; Xie, C.; Fang, Q.H.; Liu, X.; Zhang, M.H.; Liu, Y.W.; Li, L.X. Toughening effect of the nanoscale twinning induced by particle/matrix interfacial fracture on fine-grained Mg alloys. *Int. J. Solids Struct.* **2016**, *102–103*, 230–237. [CrossRef]

16. Xie, C.; Wang, Y.N.; Fang, Q.H.; Ma, T.F.; Zhang, A.B.; Peng, W.F.; Shu, X.D. Effects of cooperative grain boundary sliding and migration on the particle cracking of fine-grained magnesium alloys. *J. Alloys Compd.* **2017**, *704*, 641–648. [CrossRef]

17. Xu, C.; Yuan, L.; Shan, D.; Guo, B. $\{10\bar{1}2\}$ twin boundaries migration accompanied by void in magnesium. *Comp. Mater. Sci.* **2020**, *184*, 109857. [CrossRef]

18. Plimpton, S. Fast parallel algorithms for short- range molecular dynamics. *J. Comp. Phys.* **1995**, *117*, 1–19. [CrossRef]

19. Stukowski, A. Visualization and analysis of atomistic simulation data with OVITO—The Open Visualization Tool. *Modell. Simul. Mater. Sci. Eng.* **2009**, *18*, 015012. [CrossRef]

20. Liu, X.Y.; Adams, J.B.; Ercolessi, F.; Moriarty, J.A. EAM potential for magnesium from quantum mechanical forces. *Modell. Simul. Mater. Sci. Eng.* **1996**, *4*, 293–303. [CrossRef]

21. Serra, A.; Pond, R.C.; Bacon, D.J. Computer simulation of the structure and mobility of twinning disclocations in hcp metals. *Acta Metall. Mater.* **1991**, *39*, 1469–1480. [CrossRef]

22. Ostapovets, A.; Gröger, R. Twinning disconnections and basal—Prismatic twin boundary in magnesium. *Model. Simul. Mater. Sci. Eng.* **2014**, *22*, 025015. [CrossRef]

23. Long, T.R.; Smith, C.S. Single-crystal elastic constants of magnesium and magnesium alloys. *Acta Metall.* **1957**, *5*, 200–207. [CrossRef]

24. Robson, J.D. The effect of internal stresses due to precipitates on twin growth in magnesium. *Acta Mater.* **2016**, *121*, 277–287. [CrossRef]

25. Ostapovets, A.; Serra, A. Characterization of the matrix–twin interface of a $(10\bar{1}2)$ twin during growth. *Philos. Mag.* **2014**, *94*, 2827–2839. [CrossRef]

26. Stanford, N.; Taylor, A.S.; Cizek, P.; Siska, F.; Ramajayam, M.; Barnett, M.R. $(10\bar{1}2)$ twinning in magnesium-based lamellar microstructures. *Scr. Mater.* **2012**, *67*, 704–707. [CrossRef]

metals

Article

Texture Selection Mechanisms during Recrystallization and Grain Growth of a Magnesium-Erbium-Zinc Alloy

Fatim-Zahra Mouhib *, Fengyang Sheng, Ramandeep Mandia, Risheng Pei, Sandra Korte-Kerzel and Talal Al-Samman

Institute for Physical Metallurgy and Materials Physics, RWTH Aachen, 52056 Aachen, Germany; fengyang.sheng@rwth-aachen.de (F.S.); rmandia@asu.edu (R.M.); pei@imm.rwth-aachen.de (R.P.); korte-kerzel@imm.rwth-aachen.de (S.K.-K.); alsamman@imm.rwth-aachen.de (T.A.-S.)
* Correspondence: mouhib@imm.rwth-aachen.de

Abstract: Binary and ternary Mg-1%Er/Mg-1%Er-1%Zn alloys were rolled and subsequently subjected to various heat treatments to study texture selection during recrystallization and following grain growth. The results revealed favorable texture alterations in both alloys and the formation of a unique ±40° transvers direction (TD) recrystallization texture in the ternary alloy. While the binary alloy underwent a continuous alteration of its texture and grain size throughout recrystallization and grain growth, the ternary alloy showed a rapid rolling (RD) to transvers direction (TD) texture transition occurring during early stages of recrystallization. Targeted electron back scatter diffraction (EBSD) analysis of the recrystallized fraction unraveled a selective growth behavior of recrystallization nuclei with TD tilted orientations that is likely attributed to solute drag effect on the mobility of specific grain boundaries. Mg-1%Er-1%Zn additionally exhibited a stunning microstructural stability during grain growth annealing. This was attributed to a fine dispersion of dense nanosized particles in the matrix that impeded grain growth by Zener drag. The mechanical properties of both alloys were determined by uniaxial tensile tests combined with EBSD assisted slip trace analysis at 5% tensile strain to investigate non-basal slip behavior. Owing to synergic alloying effects on solid solution strengthening and slip activation, as well as precipitation hardening, the ternary Mg-1%Er-1%Zn alloy demonstrated a remarkable enhancement in the yield strength, strain hardening capability, and failure ductility, compared with the Mg-1%Er alloy.

Keywords: magnesium-rare earth alloy; recrystallization; selective grain growth; texture

Citation: Mouhib, F.-Z.; Sheng, F.; Mandia, R.; Pei, R.; Korte-Kerzel, S.; Al-Samman, T. Texture Selection Mechanisms during Recrystallization and Grain Growth of a Magnesium-Erbium-Zinc Alloy. *Metals* **2021**, *11*, 171. https://doi.org/10.3390/met11010171

Received: 29 December 2020
Accepted: 15 January 2021
Published: 19 January 2021

Publisher's Note: MDPI stays neutral with regard to jurisdictional claims in published maps and institutional affiliations.

1. Introduction

As of late, there is an increasing trend for using magnesium alloys in the automotive and aerospace industries owing to their excellent specific strength properties and light-weighting potential. However, a broad technical application of wrought magnesium alloy products still struggles with the difficulty to form at temperatures below 150 °C caused by the formation of a sharp basal texture and limited activation of deformation modes outside of the basal plane [1,2]. Great research efforts have been explored to not only alter the sharp sheet texture but also to control its strength and evolution [2–4]. Magnesium rare earth (RE) alloys have proven very useful in enhancing the cold formability and the work hardening capability of conventional magnesium alloys. This has been attributed to the development of a unique sheet annealing texture characterized by an off-basal pole spread away from the normal direction of the sheet plane [4–7]. Numerous recrystallization and growth mechanisms were explored in an attempt to unravel the underlying mechanisms for the favorable texture development in magnesium rare earth alloys. These included particle stimulated nucleation, deformation twin nucleation, and shear band nucleation mechanisms, which are known to promote random orientations or off-basal orientations of recrystallization nuclei [8]. However, since these mechanisms can also occur in magnesium alloys containing no RE elements [9–11], the research on the origin of the specific texture selection in Mg-RE alloys is still ongoing.

Recent studies in that context looked into a connection between RE solutes and oriented nucleation/growth during recrystallization, given that specific RE-textures have been observed in dilute alloys in which the RE elements were mostly in solid solution. It was also shown that the addition of zinc to Mg-RE alloys has proven to further weaken the texture and augment its off-basal nature [12]. Hence, more studies are needed to investigate the interaction of multiple solute species, particularly ones with largely different atomic radii [12,13]. Additionally, it is essential to examine how different RE and non-RE solute-boundary interactions impact the growth selection during texture evolution in Mg-RE and Mg-RE-Zn alloys.

This study aims at understanding the formation of unique sheet textures and the ensuing selection mechanisms in Er-containing magnesium alloys with and without the presence of Zn. Reports on this alloy system with respect to deformation and recrystallization textures are scarce in the literature. It is therefore important to illustrate the great texture modification potential the present alloys have to offer. In particular, the formation of a double-peak split texture in the transverse direction, similar to what is common for hexagonal metals with c/a < 1.633, is a unique feature that is hardly seen in Mg alloys subjected to rolling and annealing treatments. The discussion addresses the texture development in both binary and ternary versions of the alloy during recrystallization and grain growth under the influence of solutes and second phase precipitates.

2. Materials and Methods

Binary and ternary Mg-1%Er-(1%Zn) (wt %) alloys were melted in an induction furnace under an Ar/CO_2 gas atmosphere and subsequently casted and homogenized at 420 °C for 960 min. The resulting microstructure is shown in Figure 1. Table 1 shows the chemical composition measured by ICP/OES (IME RWTH University, Aachen, Germany). Sheets of the dimension $60 \times 40 \times 4$ mm^3 were hot rolled at 400 °C with multiple rolling passes (eleven) and a final thickness reduction of 80%.

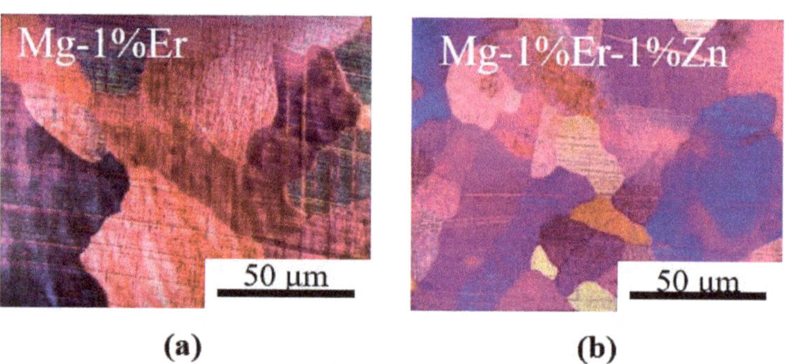

Figure 1. Micrographs of Mg-1%Er (**a**) and Mg-1%Er-1%Zn (**b**) after casting.

Table 1. Chemical composition in atomic and weight percentage of Mg-1%Er and Mg-1%Er-1%Zn.

Alloys	Mg (wt %/at %)		Er (wt %/at %)		Zn (wt %/at %)	
Mg-Er	99.04	99.86	0.96	0.14	/	
Mg-Er-Zn	98.01	99.46	0.93	0.14	1.06	0.40

Annealing treatments at 350 °C and 400 °C were conducted in a Heraeus RL200E air furnace (Heraeus Group, Hanau, Germany) for annealing times ranging from 5 to 1440 min and in a sand bath furnace for annealing times smaller than 5 min.

All samples were mechanically ground to the specimen mid-plane and polished with a diamond suspension up to 0.25 µm. For optical light microscopy (Leica microsystems, Wetzlar, Germany), the samples were electro-polished for 40 min in a 5:3 ethanol H_3PO_3 solution at 2 V and etched in an acetic picral solution. Grain size measurements from light microscopy images were performed by means of the linear intercept method. Texture measurements were conducted using a Bruker D8 advance diffractometer (Bruker, Billerica, MA, USA) operating at 30 V and 25 mA. Six incomplete pole figures [$\{10\bar{1}0\}$, $\{0002\}$, $\{10\bar{1}1\}$, $\{10\bar{1}2\}$, $\{11\bar{1}0\}$, $\{10\bar{1}3\}$] were measured, from which full pole figures and orientation distribution functions (ODFs) were calculated with the texture analysis toolbox MTEX (MTEX 5.3, Chemnitz, Germany) [14]. For selected samples intended for electron backscatter diffraction (EBSD) analysis under the electron beam an alternative electro-polishing procedure was employed using Lectro-Pol 5 in a Struers AC-2 solution at a voltage of 25 V and $-20\,°C$ for 120 s. Microstructure analysis via EBSD was performed using a LEO-1530 scanning electron microscope (Carl Zeiss Microscopy GmbH, Jena, Germany) operating at 20 kV, equipped with a HKL-Nordlys II EBSD detector. The detection step size varied according to the grain size in a range from 0.5 to 1.5 µm. EBSD and X-Ray diffraction (XRD) raw data were analyzed using the MTEX toolbox [14].

For the characterization of precipitates panorama back scatter electron (BSE)imaging (50 images per specimen) was conducted under a magnification of $10,000\times$ with a focused ion beam Helios 600i (Thermo Fisher Scientific, Waltham, MA, USA) equipped with a field emission electron gun operating at 5 kV.

Recrystallization (RX) kinetics for both alloys were obtained by Vickers micro-hardness measurements (HMV, Shimadzu, Nakagyo-ku, Kyōto, Japan) (load of 1 N) of specimens subjected to interrupted annealing at 350 °C for durations up to 1000 s. For each annealing condition, 15 indentations were used to determine the average hardness HV_t at time step t. The recrystallized fraction was calculated by the common Equation (1), where $HV_{initial}$ and HV_{final} denote the initial and final hardness values, respectively [15].

$$X = \frac{HV_{initial} - HV_t}{HV_{initial} - HV_{final}} \tag{1}$$

The mechanical properties were determined by uniaxial tensile tests at room temperature conducted along the rolling direction on fully recrystallized samples of both alloys. The tensile samples were carefully chosen to have a similar grain size in order to exclude grain size effects on the yield strength response of the material. Dog-bone-shaped samples with gauge dimensions of $5 \times 1.5 \times 0.8\ mm^3$ were cut out of the rolling sheets and subsequently annealed at 400 °C for 15 and 60 min for the binary and ternary alloys, respectively. Tensile tests were conducted using a ZWICK tension-compression testing machine. Specimens were strained under a constant strain rate of 2×10^{-4} to failure. Each tensile test was performed three times to ensure reproducibility and the average flow curve was considered for the analysis. In order to determine the operating slip systems during uniaxial tension, interrupted tensile tests were performed up to 5% strain on samples annealed at 400 °C for 60 min of both alloys. Subsequently, the slip systems were determined by comparison of slip lines from secondary electron (SE) images obtained at 20 kV and the grain orientations determined by EBSD of at least 120 grains each.

3. Results

3.1. Microstructure Evolution

Figure 2 shows the optical microstructures of rolled and annealed samples of the binary and ternary Mg-1%Er-(1%Zn) alloys. The annealing temperature was 400 °C and the annealing times were 15 and 360 min, respectively. The corresponding XRD textures, presented in terms of (0002) pole figures, as well as the evolution of the average grain size as a function of the annealing time are given in Figure 3. As it can be seen, both alloys exhibited a common deformation microstructure characterized by deformed grains and numerous deformation twins. The rolling texture of the binary alloy showed a typical basal component with a pole spread towards the rolling direction (RD) and moderate intensity. On the other hand, the ternary alloy exhibited a weaker and much softer rolling texture characterized by two off-basal components at ±20° RD that bare a significant pole spread in the transverse direction TD.

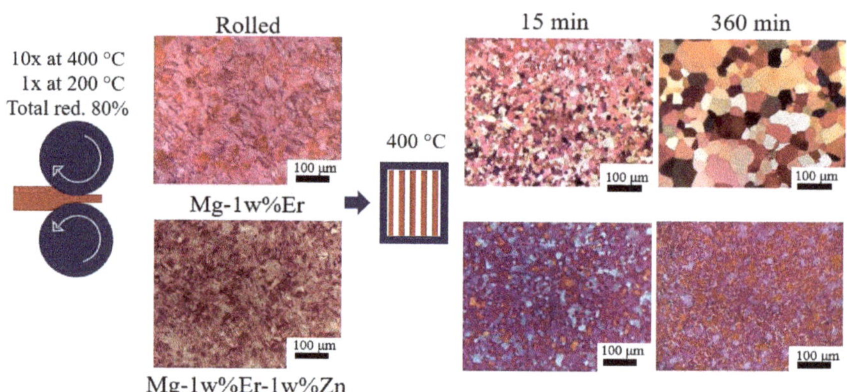

Figure 2. Micrographs of Mg-1%Er and Mg-1%Er-1%Zn in the as-rolled state and upon annealing at 400 °C for 15 and 360 min. The images are taken from the rolling plane (mid-thickness).

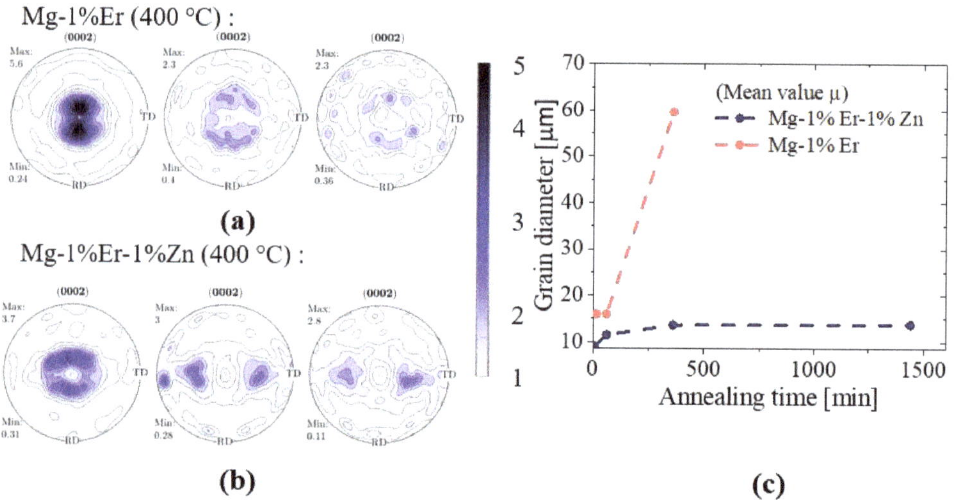

Figure 3. Texture evolution of binary and ternary (**a,b**) Mg-1%Er-(1%Zn) in the as rolled state and after annealing at 400 °C for 15 and 360 min, as well as the corresponding evolution of the mean grain diameter over time (**c**).

During a subsequent heat treatment at 400 °C, the ternary alloy seemed to develop a fine-grained annealing microstructure with a stable grain size (14 μm) that was not liable to additional grain growth at increased annealing durations (Figure 2). The corresponding annealing texture witnessed a unique transition demonstrated by new off-basal components appearing at ±40° TD. This type of annealing texture after rolling is only common for hcp metals like titanium [16], that have a lower c/a ratio than the ideal ratio of 1.633 for close-packed structures, and can thus easily activate prismatic and pyramidal slip. As to the binary alloy, it exhibited an initially stable grain size (16 μm) up to an annealing time of 60 min but underwent significant grain growth (60 μm) after 360 min of annealing

(Figure 3c). The texture of the binary Mg-1%Er alloy during early annealing conserved the dominant pole split in RD seen in the deformation texture but was also significantly weakened as a result of recrystallization (5.6 MRD → 2.3 MRD). With prolonged annealing up to 360 min, the texture spread around the normal direction ND seemed to increase during grain growth and the basal component in the center of the pole figure has vanished.

Given the interesting nature of the texture transition seen upon annealing of the Mg-1%Zn-1%Er alloy, additional annealing experiments were carried out at a lower temperature of 350 °C. The annealing durations were kept short—i.e., up to 15 min—in order to capture the effect of early recrystallization stages on the texture transition. The results are shown in Figure 4, which demonstrates how the ±20° RD deformation texture transforms into a ±40° TD annealing texture accompanied by full recrystallization of the specimen. Figure 5 shows the distribution of the basal poles (or c-axes) with respect to the rolling and transvers directions for the ternary (bottom) and binary alloy (top) along with their inverse pole figure (IPF) maps at 400 °C/360 min. Based on the annealing durations, the data clearly reveals that the texture established during recrystallization in the ternary alloy is preserved during subsequent grain growth, as the orientation distribution relative to the TD and RD remains virtually unchanged (Figure 5c). In case of the binary alloy, the total area fraction of grains with a large TD pole spread increases significantly with the annealing duration, whereas the area distribution of the RD spread remains almost constant (Figure 5b).

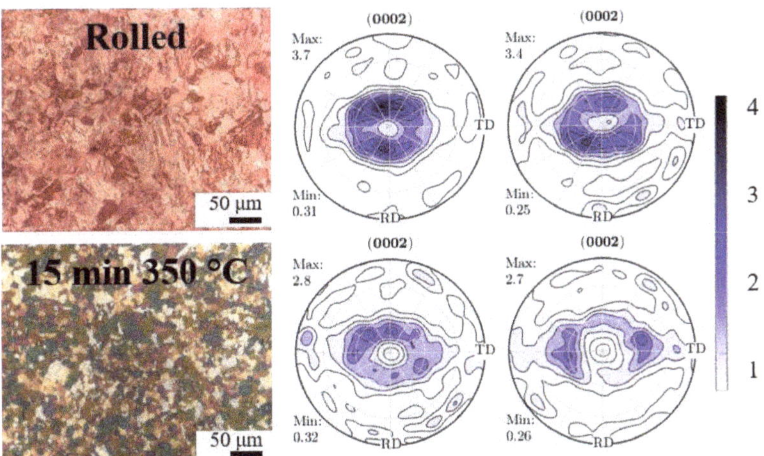

Figure 4. Texture evolution during recrystallization of Mg-1%Er-1%Zn annealed at 350 °C for 0, 5, 7, and 15 min.

With respect to second phase precipitates in the microstructure, the ternary alloy was found to exhibit a fine distribution of nano-scale precipitates and also larger ones that were sometimes clustered together, as shown in Figure 6. Energy dispersive X-ray spectroscopy (EDS) point analysis and elemental mapping were employed to investigate the chemical composition of these large precipitates. The results revealed a particle composition with a magnesium to zinc to erbium ratio of approximately 5:5:1. A similar zinc to erbium ratio was previously reported for several hexagonal and icosahedral quasi-crystalline $(Mg, Zn)_x Er$ compounds [17,18].

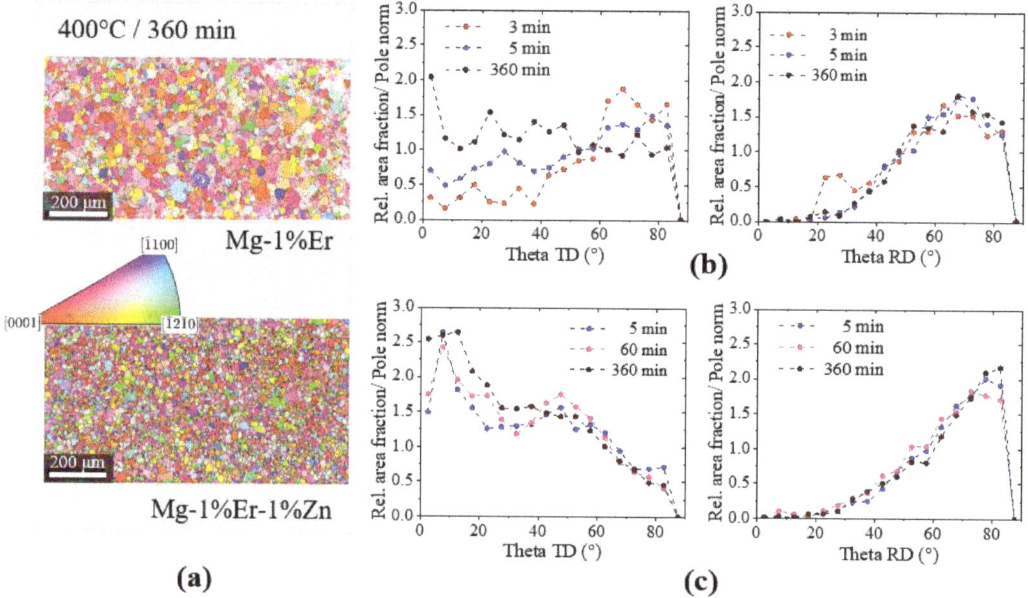

Figure 5. (**a**) EBSD ND-IPF maps of binary (top) and ternary Mg-1%Er-(1%Zn) (bottom) alloys annealed at 400 °C for 360 min; (**b**,**c**) angle distribution profiles between the c-axis and the transverse and rolling directions for different annealing durations.

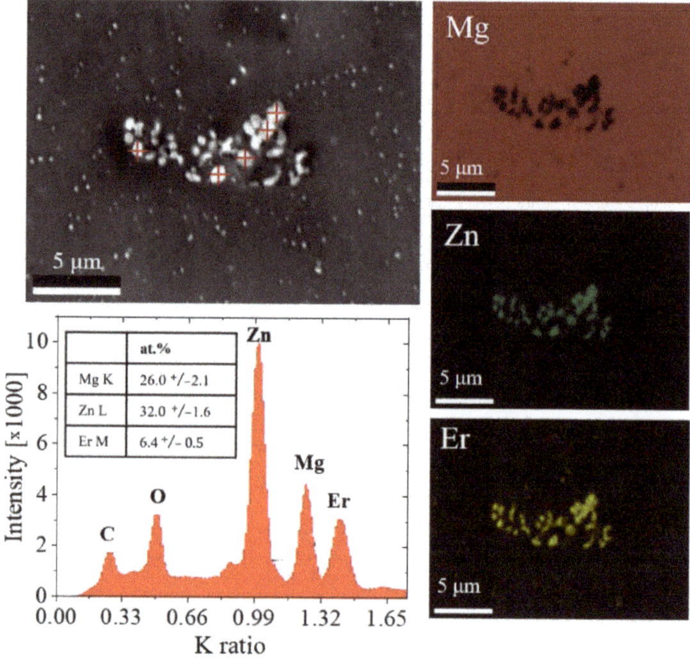

Figure 6. EDS analysis of the Er-containing precipitates observed in the Mg-1%Er-1%Zn alloy.

3.2. Mechanical Properties

Figure 7a shows the true stress-strain curves of the uniaxial tensile tests in RD, 45° and TD for the investigated alloys annealed at 400 °C for 15 and 60 min, respectively. The samples of both alloys exhibited a similar annealing grain size (hence the different times) but a drastically different texture (Figure 7b). As evident from the flow curves, the addition of zinc leads to higher ductility, as well as an increase of the ultimate tensile strength, which are indicative of an enhanced work hardening capability.

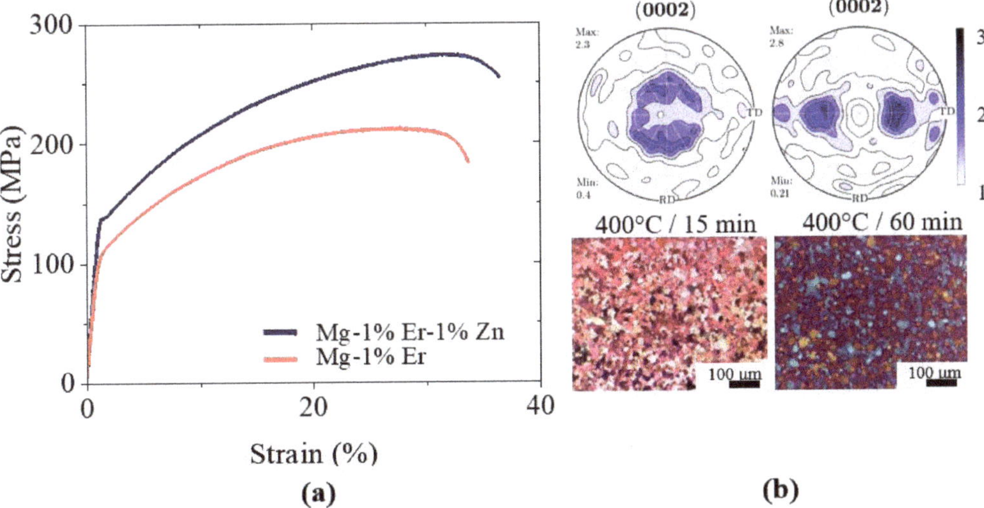

Figure 7. Uniaxial stress–strain response (**a**), optical micrographs and (0002) pole figures of Mg-1%Er annealed for 15 min and Mg-1%Er-1%Zn annealed for 60 min at 400 °C (**b**). The grain size prior to the tensile test was in the range of 14 to 16 µm for both samples.

4. Discussion

The present study revealed distinct differences in the microstructure evolution and mechanical properties between a dilute Mg-1%Er alloy with and without the addition of Zn. During annealing, both alloys developed a significant basal pole spread away from the ND. For the ternary alloy, a titanium rolling texture with off-basal components at ±40° TD developed during recrystallization, and was retained during subsequent grain growth. By contrast, the recrystallization texture of the binary alloy with off-basal peaks at ±20° RD was further modified with longer annealing resulting in a much larger grain size and increased texture scatter. The stable grain size in the Mg-1%Er-1%Zn alloy in contrary to the excessive growth in Mg-1%Er can be explained by the presence of a fine particle dispersion and the resulting drag force (Zener drag). This is known to have a significant impeding effect on the grain boundary motion and can therefore lead to a stable grain size [19,20]. Figure 8 shows a BSE image of the observed precipitates in Mg-1%Er-1%Zn after 60 min of annealing at 400 °C and the corresponding particle size distribution. The precipitate characteristics including particle size, volume fraction, and number density are listed in Table 1.

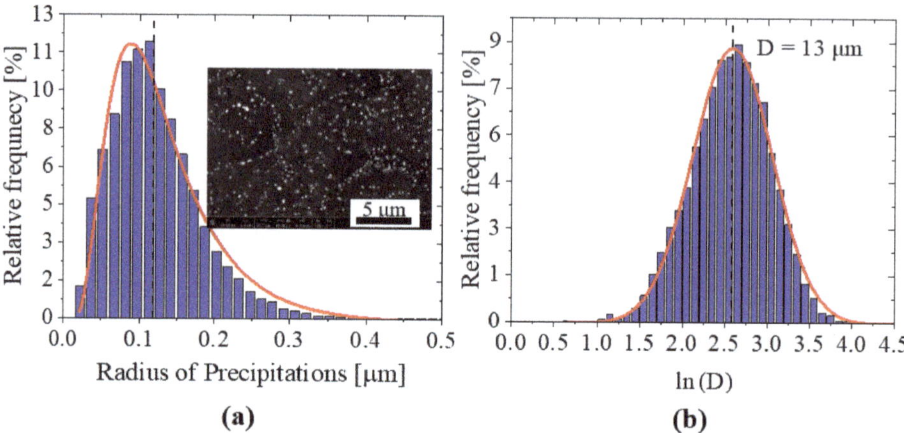

Figure 8. Particle size distribution after 60 min of annealing at 400°C visualized using a histogram with linear *x*-axis (**a**) and the corresponding grain size distribution obtained by optical microscopy (**b**).

The critical grain size according to classical Zener theory is given by Equation (2), assuming that the radius of curvature equals the critical grain diameter [20].

$$D_c = \frac{4r}{3f} \tag{2}$$

The calculated critical grain size of 12 μm, obtained by using the values in Table 2, is similar to the measured grain size (13.28 μm) under the same annealing conditions (cf. Figure 8b). This supports the assumption made earlier that the fine precipitates stabilize the recrystallization microstructure during annealing. Similarly, the absence of precipitates in the binary alloy (Mg-1%Er) could explain the excessive growth rate observed after an initial slow growth stage during the first 15 min of annealing (cf. Figure 3). The drag force during this short stage is believed to be caused by solutes inhibiting the motion of grain boundaries [21–23], which were obviously able to break free after 15 min annealing time. Theoretical studies revealed that accelerated grain growth in a system under the influence of solute pinning will occur when the average grain size becomes close to the critical grain size for which the solute drag force and the capillary driving force are equal. Such state allows the pinned boundaries with a slightly higher curvature than the average to break free from their solute atmosphere, while the other boundaries remain pinned [24]. This seems to be in accordance with the observed growth behavior of the Mg-1%Er alloy and the resulting growth of non-basal texture components (cf. Figure 5b) after a critical grain size has been reached.

Table 2. Characteristics of the Er and Zn-containing precipitates in Mg-1%Er-1%Zn obtained by means of statistical analysis of BSE images

Volume Fraction f:	0.0131
Average Particle Size r:	0.122 μm (from dist. 0.1178)
Number density:	0.2246 1/μm^2

In the ternary alloy, the process of recrystallization was responsible for the RD → TD transition of the basal pole spread seen in Figure 4, which can be contributed to either oriented nucleation or selective nucleus growth. Selective growth of texture components can be caused by anisotropic segregation behavior and/or an equalization of different grain boundary mobilities annihilating a prior growth advantage of certain orientations, e.g., of basal oriented grains as suggested by Barrett et al. [25]. Figure 9 presents EBSD maps (ND-IPF) of the microstructure of the ternary alloy at 5 and 10 min annealing at 350 °C. The orientation of recrystallized grains was evaluated separately from the orientation of the overall microstructure in order to discern the effect of recrystallization on the development of the ±40° TD texture. For this, the common criterion of grain orientation spread (GOS) was used, where recrystallized grains typically depicted GOS < 1°. The overall texture revealed that the RD-tilted components were a feature of the deformation microstructure. They did not exist in the recrystallization texture that was solely composed of the TD-tilted components, readily after 5 min of annealing (Figure 9a). After completion of recrystallization at 10 min of annealing, the RD-tilted components were no longer a feature of the overall texture that was only dictated by the TD- recrystallization texture (cf. Figure 9b).

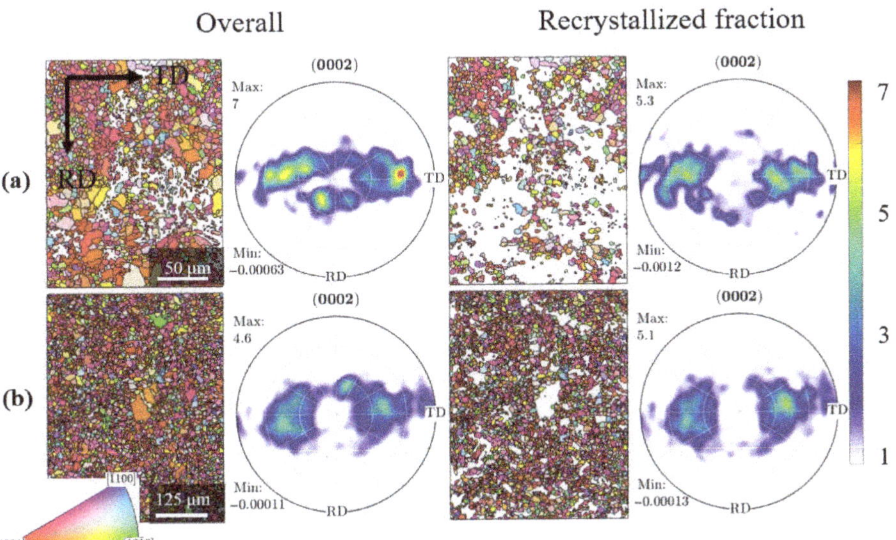

Figure 9. EBSD IPF maps (with respect to ND) and their (0002) pole figures of Mg-1%Er-1%Zn annealed at 350 °C for 5 min (**a**) and 10 min (**b**). The left column features the overall map data, and the right column only the recrystallized fraction with GOS < 1°.

Additional grain size vs. texture analysis of the recrystallized structure in Figure 9b (400 °C/10 min) revealed that the ±40° TD texture in the ternary alloy intensifies during recrystallization growth of the nuclei. A similar effect was not observed in the binary system (Figure 10b), as no significant size dependence of texture during recrystallization was found. While there was a progressive texture change in the binary alloy during annealing, which involved both recrystallization and grain growth, the texture change in the counterpart ternary alloy was much more rapid, and in that regard, limited to recrystallization. As the drag force exerted by particles of a given size and volume fraction is too small to match the capillary driving force of grains significantly smaller than the critical grain size D_c, particle pinning should only play a minor role during recrystallization and early grain growth. Accordingly, selective growth at the recrystallization stage is likely to be governed by solutes. In comparison to the binary alloy, specific boundaries in the ternary alloy seem to break free from their solute atmosphere much sooner enabling texture selection readily during recrystallization. This may be a result of a lower solute content due to the large amount of fine Er/Zn-rich precipitates, whereas in the binary alloy the Er content is fully dissolved in the matrix.

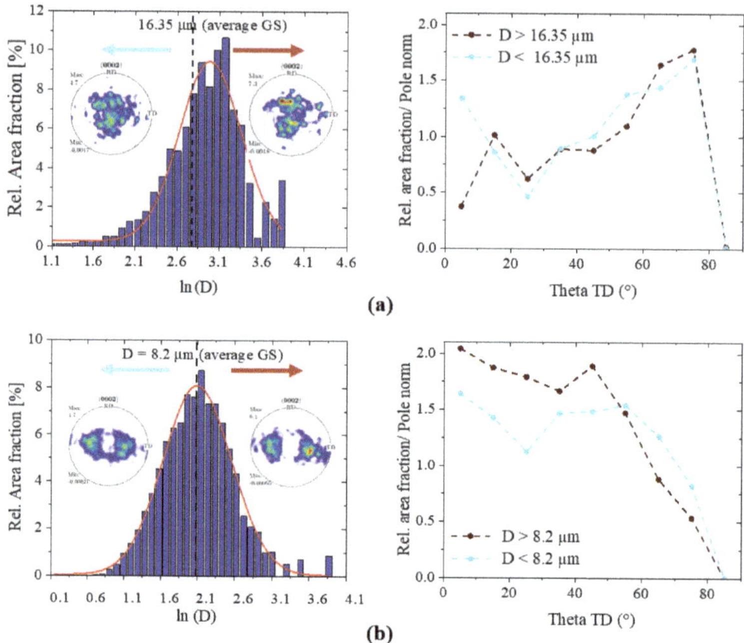

Figure 10. Grain size distributions from EBSD measurements of Mg-1%Er annealed at 400 °C for 4 min (**a**) and Mg-1%Er-1%Zn annealed at 350 °C for 10 min (**b**) along with the orientation distribution of the c-axis with respect to TD for small and large recrystallized grains. The threshold used in each case was the average grain size obtained from EBSD.

The observed recrystallization kinetics of the two alloys further indicates a difference in the drag force during recrystallization. Figure 11 shows the recrystallized volume fraction as a function of annealing time and the corresponding JMAK (Johnson-Mehl-Avrami-Kolmogorow) fit for both Mg-1%Er and Mg-1%Er-1%Zn alloys. A Comparison of the recrystallization curves reveals a quicker recrystallization in the ternary alloy (~210 s) compared to the counterpart binary alloy (~500 s). Additionally, the initial stage (linear portion of the curve) until the onset of stable nuclei growth is extended in Mg-1%Er compared to the other alloy, indicating a longer incubation time in the first. As the incubation time is related to reduced boundary mobility, a stronger solute drag effect in Mg-1%Er may cause a longer incubation period.

Mg-1%Er-1%Zn exhibited an elevated yield strength, strain hardening capability, and failure ductility. The elevated yield stress is partially caused by precipitation hardening, which contributed by approximately 5 MPa to the 20 MPa offset (approximation by Orowan). Furthermore, the addition of zinc might have magnified the impact of solid solution strengthening by means of synergetic solute effects. To investigate the role of texture and microstructure in the activation of deformation modes during tension, EBSD-assisted slip trace analysis of at least 120 grains per strained sample was employed in the current study. A slip trace appears as a straight line as a result of the intersection of an active slip plane with the free surface. Figure 12 shows the frequency of slip traces corresponding to the active slip systems during 5% tension in RD, for the binary and ternary alloys. Additionally, the frequency of twinning was evaluated by a visual investigation of each grain. Basal, pyramidal I and <c + a> pyramidal II, as well as $\{10\bar{1}2\}$ tensile twining were detected with almost equal shares of approximately 22% in the ternary system. In the binary alloy, basal slip prevailed with about 64%, while all other slip systems accounted for less than 20% of the overall observed slip traces. A large contribution of basal slip to deformation could justify the lower yield point in the flow curve of the Mg-1%Er alloy.

This is because basal slip, even in rare-earth containing Mg alloys, requires a much lower critical resolved shear stress (CRSS) than non-basal slip [26]. The observed enhanced slip activity on non-basal slip planes in the strained Mg-1%Er-1%Zn sample can be attributed to both a favorable texture and a change of the CRSS ratio between basal and non-basal slip, evoked by the present Zn and Er solutes. The relatively low contribution of prismatic slip to deformation might be attributed to the investigated low strain of 5% being insufficient to activate cross-slip of basal screw dislocations onto the prismatic plane [27,28].

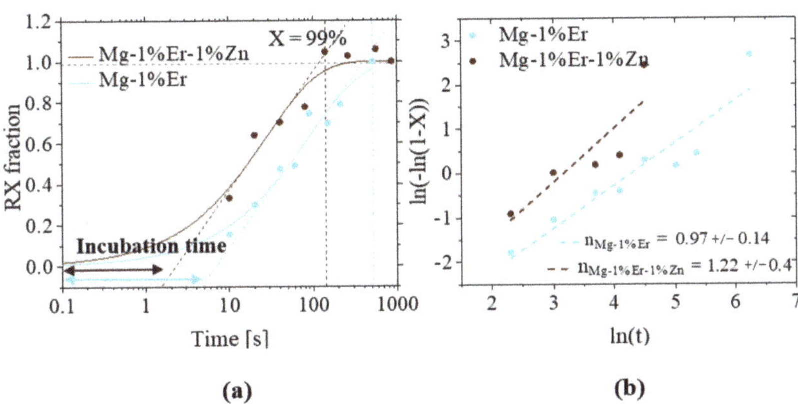

Figure 11. (**a**) Recrystallization kinetics of binary and ternary Mg-1%Er-(1%Zn) annealed at 350 °C and the corresponding Avrami exponents (**b**).

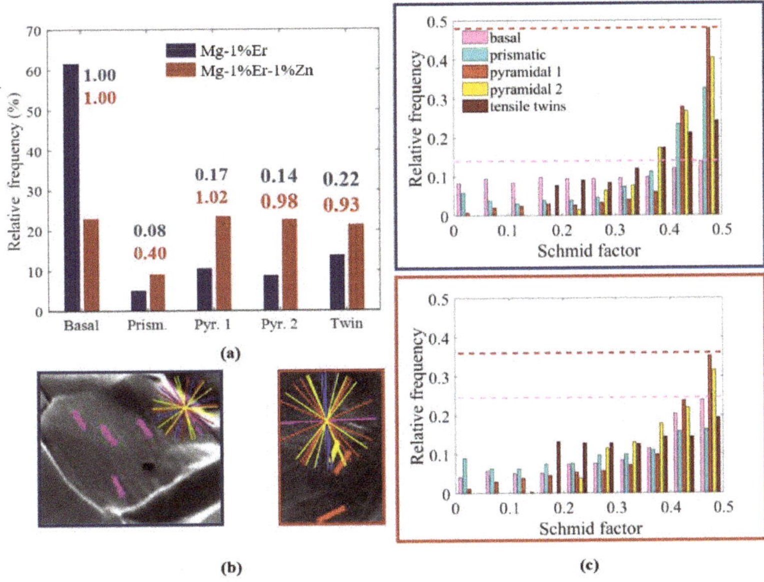

Figure 12. (**a**) Distribution of slip traces corresponding to the active slip systems in Mg-1%Er and Mg-1%Er-1%Zn during tension to 5% strain. The initial condition of the sample corresponded to annealing at 400 °C for 60 min. (**b**) Exemplary SE images of one grain exhibiting the morphology of slip traces with overlaid possible slip systems obtained from EBSD for the determination of the active slip system. (**c**) Schmid factor distribution calculated from the initial texture of the grains, in which the slip traces were detected in both alloys.

Figure 12 also shows the distribution of Schmid factors (SF) of both alloys with respect to the applied external stress corresponding to different slip traces detected. Generally, non-basal slip (particularly pyramidal) seems to be the main plasticity carrier in both alloys. It also seems to comply principally well with the Schmid law, i.e., corresponding slip traces are preferentially detected in grains with the highest SF, and their frequency drops consecutively with decreasing SF. In the binary alloy, it is evident that the activation of basal slip in grains with a very low SF deviates from the Schmid behavior. This is likely due to the naturally low CRSS of basal slip but also to local stress concentrations at grain boundaries that could cause non-Schmid activation of different slip modes. Another important finding from the SF distribution in Figure 12 is the large difference in the contribution of basal vs. non-basal slip to deformation between the two alloys. In the Mg-Er sample at SF = 0.5, non-basal slip (including prismatic and both <a> and <c + a> pyramidal slip) was ~2–3 times more active than basal slip. In contrast, the level of contribution of these slip modes to deformation at SF = 0.5 in the Mg-1%Er-1%Zn sample was more comparable. Both types of pyramidal slip were still preferential with a frequency ratio of ~1.2–1.5 to basal slip. Prismatic slip was no longer favorable compared to basal slip (frequency ratio of ~0.7).

The present findings certainly highlight the effect of solutes on the deformation behavior, particularly the non-basal slip behavior. There are also indications of solute-induced basal slip strengthening in the current alloys in accordance with previous experimental and computational reports [29–31]. The addition of Zn that alters the substitutional solute chemistry of the Mg-1%Er alloy would obviously lead to complex interaction of solute species within the matrix. From energetic perspectives, Zn and Er can cluster in the lattice to relieve the misfit strains arising from the solute size mismatch. They would also cosegregate to defects in the microstructure, and can be hence expected to have a stronger interaction with grain boundaries and dislocations leading to strikingly different slip system activation during deformation and boundary migration characteristics during annealing. This is indeed demonstrated in the current work by the unique texture development upon rolling and upon annealing of the Mg-1%Er-1%Zn alloy.

5. Conclusions

1. The combination of Zn and Er in a dilute substitutional Mg-1%Er-1%Zn alloy (wt %) led to unique sheet texture development upon rolling and subsequent annealing characterized by pronounced basal pole peaks at ±40° TD. This type of sheet texture is only common for hcp metals with c/a < 1.633, such as titanium but was nevertheless observed in the current study.
2. The same texture was not observed in a binary version of the alloy, i.e., Mg-1%Er subjected to similar processing, which suggests that synergetic effects of multiple solute species, in this case Zn/Er, are crucial in terms of providing the ±40°TD orientations for recrystallization nuclei during deformation and their selective growth during subsequent annealing.
3. Mg-1%Er-1%Zn alloy obtained a stable final texture (±40° TD) and grain size (<20 μm) upon completion of static recrystallization. By contrast, the binary Mg-1%Er alloy revealed a continuous modification of its microstructure throughout the annealing process suggesting an important role of grain growth following recrystallization.
4. The microstructural stability of the ternary alloy during longer annealing durations originates from a fine dispersion of dense nanosized particles in the matrix that were found to impede grain growth by Zener drag.
5. Recrystallization nuclei demonstrated selective growth behavior favoring TD-tilted texture components, which was the main mechanism for the ±20° RD → ±40° TD texture transition taking place during recrystallization of the deformed Mg-1%Er-1%Zn alloy. This was likely influenced by solute drag on specific boundaries during early recrystallization.

6. With respect to the mechanical properties in tension, the addition of zinc, the present precipitates, solute strengthening effects and a favorable soft texture led to a remarkable enhancement in the yield strength, strain hardening capability, and failure ductility as compared with the binary Mg-1%Er alloy.

7. EBSD-assisted slip trace analysis at 5% strain unraveled promoted non-basal slip behavior and obvious basal slip strengthening owing to solute/dislocation interaction that requires extended advanced experimental and computational efforts to better understand the interaction of multiple solute species and solute strengthening.

Author Contributions: Conceptualization, F.-Z.M. and T.A.-S.; Methodology, F.-Z.M. and R.P.; Software, R.P.; Validation, F.-Z.M. and F.S.; Formal analysis, F.-Z.M. and F.S.; Investigation, F.S., R.M., and F.-Z.M.; Resources, S.K.-K. and T.A.-S.; Data curation, F.-Z.M. and T.A.-S.; Writing—original draft preparation, F.-Z.M.; Writing—review and editing, T.A.-S.; Visualization, F.-Z.M.; Supervision, S.K.-K. and T.A.-S.; Project administration, T.A.-S. and S.K.-K.; Funding acquisition, T.A.-S. All authors have read and agreed to the published version of the manuscript.

Funding: This research was funded by Deutsche Forschungsgemeinschaft (DFG), grant number grant no. AL 1343/7-1.

Institutional Review Board Statement: Not applicable.

Informed Consent Statement: Not applicable.

Acknowledgments: The financial support of the Deutsche Forschungsgemeinschaft (DFG) under grant no. AL 1343/7-1 is gratefully acknowledged. We also thank Burak Erol for his assistance during the experimental work.

Conflicts of Interest: The authors declare no conflict of interest. The funders had no role in the design of the study; in the collection, analyses, or interpretation of data; in the writing of the manuscript, or in the decision to publish the results.

References

1. Friedrich, H.E.; Mordike, B.L. Technology of Magnesium and Magnesium Alloys. In *Magnesium Technology: Metallurgy, Design Data, Applications*; Springer: Berlin/Heidelberg, Germany, 2006.
2. Gehrmann, R.; Frommert, M.M.; Gottstein, G. Texture effects on plastic deformation of magnesium. *Mater. Sci. Eng. A* **2005**, *395*, 338–349. [CrossRef]
3. Hirsch, J.; Al-Samman, T. Superior light metals by texture engineering: Optimized aluminum and magnesium alloys for automotive applications. *Acta Mater.* **2013**, *61*, 818–843. [CrossRef]
4. Imandoust, A.; Barrett, C.D.; Al-Samman, T.; Inal, K.A.; El Kadiri, H. A review on the effect of rare-earth elements on texture evolution during processing of magnesium alloys. *J. Mater. Sci.* **2016**, *52*, 1–29. [CrossRef]
5. Hadorn, J.P.; Hantzsche, K.; Yi, S.; Bohlen, J.; Letzig, D.; Wollmershauser, J.A.; Agnew, S.R. Role of Solute in the Texture Modification During Hot Deformation of Mg-Rare Earth Alloys. *Metall. Mater. Trans. A* **2011**, *43*, 1347–1362. [CrossRef]
6. Hantzsche, K.; Bohlen, J.; Wendt, J.; Kainer, K.U.; Yi, S.B.; Letzig, D. Effect of rare earth additions on microstructure and texture development of magnesium alloy sheets. *Scr. Mater.* **2010**, *63*, 725–730. [CrossRef]
7. Al-Samman, T.; Li, X. Sheet texture modification in magnesium-based alloys by selective rare earth alloying. *Mater. Sci. Eng. A* **2011**, *528*, 3809–3822. [CrossRef]
8. Robson, J.D. Effect of Rare-Earth Additions on the Texture of Wrought Magnesium Alloys: The Role of Grain Boundary Segregation. *Metall. Mater. Trans. A* **2013**, *45*, 3205–3212. [CrossRef]
9. Barnett, M.R.; Nave, M.D.; Bettles, C.J. Deformation microstructures and textures of some cold rolled Mg alloys. *Mater. Sci. Eng. A* **2004**, *386*, 205–211. [CrossRef]
10. Ball, E.A.; Prangnell, P.B. Tensile-compressive yield asymmetries in high strength wrought magnesium alloys. *Scr. Metall. Mater.* **1994**, 31. [CrossRef]
11. Mackenzie, L.; Pekguleryuz, M. The recrystallization and texture of magnesium–zinc–cerium alloys. *Scr. Mater.* **2008**, *59*, 665–668. [CrossRef]
12. Basu, I.; Al-Samman, T. Triggering rare earth texture modification in magnesium alloys by addition of zinc and zirconium. *Acta Mater.* **2014**, *67*, 116–133. [CrossRef]
13. Jiang, M.G.; Xu, C.; Yan, H.; Lu, S.H.; Nakata, T.; Lao, C.S.; Chen, R.S.; Kamado, S.; Han, E.H. Correlation between dynamic recrystallization and formation of rare earth texture in a Mg-Zn-Gd magnesium alloy during extrusion. *Sci. Rep.* **2018**, *8*, 16800. [CrossRef] [PubMed]

14. Hielscher, R.; Schaeben, H. A novel pole figure inversion method: Specification of theMTEXalgorithm. *J. Appl. Crystallogr.* **2008**, *41*, 1024–1037. [CrossRef]
15. Humphreys, F.J.; Hatherly, M. Recrystallization and Related Annealing Phenomena. In *Recrystallization and Related Annealing Phenomena*; Chapter 9; Elsevier: Amsterdam, The Netherlands, 2004; pp. 285–319. [CrossRef]
16. Lee, H.P.; Esling, C.; Bunge, H.J. Development of the Rolling Texture in Titanium. *Textures Microstruct.* **1988**, *7*, 317–337. [CrossRef]
17. Li, M.R.; Deng, D.W.; Kuo, K.H. Crystal structure of the hexagonal (Zn, Mg)4Ho and (Zn, Mg)4Er. *J. Alloy. Compd.* **2006**, *414*, 66–72. [CrossRef]
18. Fisher, I.R.; Islam, Z.; Panchula, A.F.; Cheon, K.O.; Kramer, M.J.; Canfield, P.C.; Goldman, A.I. Growth of large-grain R-Mg-Zn quasicrystals from the ternary melt (R = Y, Er, Ho, Dy and Tb). *Philos. Mag. B* **2009**, *77*, 1601–1615. [CrossRef]
19. Hadorn, J.P.; Hantzsche, K.; Yi, S.; Bohlen, J.; Letzig, D.; Agnew, S.R. Effects of Solute and Second-Phase Particles on the Texture of Nd-Containing Mg Alloys. *Metall. Mater. Trans. A* **2012**, *43*, 1363–1375. [CrossRef]
20. Humphreys, F.J.; Hatherly, M. Grain Growth Following Recrystallization. In *Recrystallization and Related Annealing Phenomena*; Chapter 11; Elsevier: Amsterdam, The Netherlands, 2004; pp. 333–378. [CrossRef]
21. Lücke, K.; Detert, K. A quantitative theory of grain-boundary motion and recrystallization in metals in the presence of impurities. *Acta Metall.* **1956**, *5*, 628–637. [CrossRef]
22. Hillert, M. Solute drag, solute trapping and diffusional dissipation of gibbs energy. *Acta Mater.* **1999**, *47*, 4481–4505. [CrossRef]
23. Hutchinson, C.R.; Brechet, Y. Solute Drag: A review of the 'Force' and 'Dissipation' approaches to the efffect of solute on grain and interphase boundary motion. In *Termodynamic, Microstructures and Plasticity, Alphonse Finel, Dominique Maziere, Muriel Veron*; Springer: Berlin/Heidelberg, Germany, 2003.
24. Kim, S.G.; Park, Y.B. Grain boundary segregation, solute drag and abnormal grain growth. *Acta Mater.* **2008**, *56*, 3739–3753. [CrossRef]
25. Barrett, C.D.; Imandoust, A.; El Kadiri, H. The effect of rare earth element segregation on grain boundary energy and mobility in magnesium and ensuing texture weakening. *Scr. Mater.* **2018**, *146*, 46–50. [CrossRef]
26. Barnett, M.R. A Taylor Model Based Description of the Proof Stress of Magnesium AZ31 during Hot Working. *Metall. Mater. Trans. A* **2003**, *34*, 1799–1806. [CrossRef]
27. Chino, Y.; Kado, M.; Mabuchi, M. Enhancement of tensile ductility and stretch formability of magnesium by addition of 0.2wt%(0.035at%)Ce. *Mater. Sci. Eng. A* **2008**, *494*, 343–349. [CrossRef]
28. Sandlöbes, S.; Friák, M.; Neugebauer, J.; Raabe, D. Basal and non-basal dislocation slip in Mg–Y. *Mater. Sci. Eng. A* **2013**, *576*, 61–68. [CrossRef]
29. Wang, L.; Huang, Z.; Wang, H.; Maldar, A.; Yi, S.; Park, J.-S.; Kenesei, P.; Lilleodden, E.; Zeng, X. Study of slip activity in a Mg-Y alloy by in situ high energy X-ray diffraction microscopy and elastic viscoplastic self-consistent modeling. *Acta Mater.* **2018**, *155*, 138–152. [CrossRef]
30. Wu, J.; Si, S.; Takagi, K.; Li, T.; Mine, Y.; Takashima, K.; Chiu, Y.L. Study of basal <a> and pyramidal <c + a> slips in Mg-Y alloys using micro-pillar compression. *Philos. Mag.* **2020**, *100*, 1–22. [CrossRef]
31. Tehranchi, A.; Yin, B.; Curtin, W.A. Solute strengthening of basal slip in Mg alloys. *Acta Mater.* **2018**, *151*, 56–66. [CrossRef]

Article

Adjustment of the Mechanical Properties of Mg2Nd and Mg2Yb by Optimizing Their Microstructures

Jonas Schmidt [1,*], Irene J. Beyerlein [2], Marko Knezevic [3] and Walter Reimers [1]

[1] Institute of Material Science and Technology, Technische Universität Berlin, Ernst-Reuter-Platz 1, 10587 Berlin, Germany; walter.reimers@physik.tu-berlin.de
[2] Department of Mechanical Engineering, Materials Department, University of California, Santa Barbara, CA 93106, USA; beyerlein@ucsb.edu
[3] Department of Mechanical Engineering, University of New Hampshire, Durham, NH 03824, USA; marko.knezevic@unh.edu
* Correspondence: jonas.schmidt@tu-berlin.de; Tel.: +49-030-314-26715

Abstract: The deformation behavior of the extruded magnesium alloys Mg2Nd and Mg2Yb was investigated at room temperature. By using in situ energy-dispersive synchrotron X-ray diffraction compression and tensile tests, accompanied by Elasto-Plastic Self-Consistent (EPSC) modeling, the differences in the active deformation systems were analyzed. Both alloying elements change and weaken the extrusion texture and form precipitates during extrusion and subsequent heat treatments relative to common Mg alloys. By varying the extrusion parameters and subsequent heat treatment, the strengths and ductility can be adjusted over a wide range while still maintaining a strength differential effect (SDE) of close to zero. Remarkably, the compressive and tensile yield strengths are similar and there is no mechanical anisotropy when comparing tensile and compressive deformation, which is desirable for industrial applications. Uncommon for Mg alloys, Mg2Nd shows a low tensile twinning activity during compression tests. We show that heat treatments promote the nucleation and growth of precipitates and increase the yield strengths isotopically up to 200 MPa. The anisotropy of the yield strength is reduced to a minimum and elongations to failure of about 0.2 are still achieved. At lower strengths, elongations to failure of up to 0.41 are reached. In the Mg2Yb alloy, adjusting the extrusion parameters enhances the rare-earth texture and reduces the grain size. Excessive deformation twinning is, however, observed, but despite this the SDE is still minimized.

Keywords: Mg-RE alloys; extrusion; mechanical properties; microstructure; in situ diffraction; crystal plasticity; deformation twinning; texture

Citation: Schmidt, J.; Beyerlein, I.J.; Knezevic, M.; Reimers, W. Adjustment of the Mechanical Properties of Mg2Nd and Mg2Yb by Optimizing Their Microstructures. *Metals* **2021**, *11*, 377. https://doi.org/10.3390/met11030377

Academic Editor: Dmytro Orlov

Received: 15 January 2021
Accepted: 19 February 2021
Published: 25 February 2021

Publisher's Note: MDPI stays neutral with regard to jurisdictional claims in published maps and institutional affiliations.

1. Introduction

Magnesium is one of the lowest density ($\rho = 1.7$ g/cm^3) metallic structural materials. Due to its high specific strengths, it is suitable for lightweight construction applications in the electronic industry, the biomedical industry, and the sports equipment sector [1–4]. However, the poor formability at room temperature limits the processing possibilities of magnesium alloys [5,6].

The low ductility and formability is caused by the hexagonal crystal system of magnesium, which only offers primary deformation systems in the basal plane. The <a> basal slip contributes to plastic deformation at low applied stresses due to its small critical resolved shear stresses (CRSS). Together with the <a> prismatic slip system, which requires higher activation energies, deformation can only can be realized in the <a> direction. Deformation in the <c> direction is only possible with <c + a> pyramidal slip systems, which are hard to activate [7–10]. Recently published studies show that <c + a> pyramidal slip is important for high elongations of magnesium alloys [11].

Plastic deformation can also be realized by twinning. In twinning, the crystal structure is reoriented by a defined angle along a plane of symmetry [12,13]. This induces a microscopic change in length and contributes to macroscopic deformation [14,15]. The twin systems are divided into {10.1} tension twinning (TTW-ing) and {10.2} compression twinning (CTW-ing), depending on the stress applied along the c-axis of the unit cell. TTWs are formed when applying either tension along the c-axis or compression perpendicular to it and feature an 86° rotation about a <11.0> axis [12,13,16]. TTWs usually have a low CRSS and are, therefore, more easily activated. In contrast, CTWs with a higher CRSS feature a 56° rotation about a <11.0> axis when compression stress is applied parallel to the c-axis.

Twin formation is a unidirectional deformation mechanism that depends on the orientation of the crystals and the applied stress. In both monocrystalline and polycrystalline magnesium products, the properties depend on the orientation of the crystals and are affected by the production parameters. Typically extruded AZ31-alloys show fiber textures after extrusion [17] in which the basal planes are oriented parallel to the direction of extrusion. When compressive stress is applied along the extrusion direction, the perpendicular aligned c-axes of the grains are subjected to tensile stress, preferably forming tensile twins [18]. Due to the low CRSS of tensile twins, the compressive yield strength (CYS) is much lower than the tensile yield strength (TYS) in the extrusion direction. This asymmetric mechanical behavior is described by the Strength Differential Effect (SDE) [19,20]. For practical applications of structural materials, it is desirable to have a symmetrical material behavior, equivalent to an SDE close to 0.

The texture can be modified by changing the production/extrusion parameters and alloying elements. Studies have shown that rare earth [RE] elements can have a strong influence on the recrystallization texture [17,21,22]. They alter the grain orientations and weaken the texture during hot deformation/extrusion. Depending on the resulting orientations, various deformation mechanisms are active during subsequent cold deformation. The CRSS and the contribution to the total deformation of the individual systems have an influence on the resulting macroscopic mechanical properties.

RE solutes are used in the Mg alloys of the WE-series, containing 7 to 9% RE, mainly yttrium (Y) and neodymium (Nd) [23,24]. Although the alloys are characterized by high strength, high costs make it desirable to reduce the RE content while maintaining the advantageous properties. Neodymium in particular is suitable, as it achieves good properties and strongly influences the texture [25]. Seitz et al. [26] extruded an Mg2wt.%Nd alloy and found a small asymmetry in the tension and compression.

For this study, two magnesium alloys with 2 wt.% rare earth content, Mg2Nd and Mg2Yb, were investigated with regard to their microstructure, deformation behavior, and mechanical properties. By varying the process parameters, the microstructure can be changed so that the strength and ductility can be adjusted over a wide range. In particular, a significant reduction in the SDE can be achieved.

In a first step, two casted billets of each alloy were extruded, varying the cooling conditions. The extruded bars were investigated by electron microscopy (SEM, TEM), laboratory X-ray texture measurements, and mechanical testing (compression, tension) to determine the mechanical properties, the microstructure, and their changes during deformation. To gain further insight into the different deformation behavior, a combination of in situ energy-dispersive X-ray synchrotron diffraction and simulations with the elasto-plastic self-consistent (EPSC) model was used [27]. The particular version of the model used in the present work is from [28]. These methods allow an analysis of the active deformation mechanisms as a function of load. Furthermore, the parameters of a dislocation density-based strain hardening model of the different deformation mechanisms are obtained. By adding Nd, an advantageous texture can be achieved. The texture activates similar deformation systems at the beginning of plastic preforming and, in particular, effectively suppresses the formation of tensile twins under compressive stress. The addition of Yb has a lower impact on the texture and a more pronounced dynamic recrystallization can be observed. This results in low strengths and high SDEs.

In the next step, heat treatments and variations in extrusion parameters were used to modify the microstructure and thus the mechanical properties. The grain size was reduced while maintaining or even improving the advantageous texture. Due to the low solubility of Nd and ytterbium (Yb) in Mg, subsequent heat treatments were used to further increase the strength. In the Mg2Nd alloy, subsequent heat treatments lead to the formation of fine precipitates, generating a significant hardening effect. In the case of the Mg2Yb alloy, heat treatments for precipitation hardening are less effective. However, by adjusting the process parameters the grain size can be reduced and the texture improved as well, so that, despite differences in plastic deformation, the yield strengths in compression are almost equal to the yield strengths in tension. This resulted in high strengths and low SDEs.

2. Materials and Methods

2.1. Material, Microstructure

The extrusion billets had a nominal composition of Mg2wt.%Nd (Mg2Nd) and Mg2wt.%Yb (Mg2Yb) with a diameter of 123 mm and a length of 115 mm. They were casted at the Helmholtz-Zentrum Geesthacht (HZG). The billets were homogenized and solution annealed in the single-phase region of the phase diagram [29]. Depending on the composition and phase diagram, the temperature was 500 °C (6 h) for the Mg2Nd alloy and 450 °C (10 h) for Mg2Yb with an 8 h heating period. After the heat treatment, the billets were quenched in water. Prior to extrusion, the billets were inductively heated rapidly to extrusion temperature and then immediately extruded. The indirect extrusion process was carried out at the Technische Universität Berlin, Extrusion Research and Development Center, varying some of the extrusion parameters including billet temperature (T_B), cooling method, extrusion ratio (R), and product speed (v_P), as shown in Table 1.

Table 1. Extrusion parameters.

Alloys	Series	T_B (°C)	Cooling	R	v_P (m/min)
	A	400	Air	61:1	0.5
Mg2Nd	B	400	Water	61:1	0.5
	C	400	Water	61:1	0.25
	A	400	Air	61:1	1.3
Mg2Yb	B	400	Water	61:1	1.6
	C	300	Water	61:1	1.8

To improve the mechanical properties with regard to technological application, the extruded bars were subsequently heat-treated at different temperatures (150, 204, and 300 °C). To determine the heat treatment duration, hardness tests were performed on samples of different ageing times. Samples with the highest hardness were further selected for mechanical testing with tensile and compression tests.

For optical microscopy and grain size analyses, the as-extruded specimens were prepared by grinding and polishing and subsequent chemical polishing with a solution of 12 mL hydrochloric acid (37%), 8 mL nitric acid (65%), and 100 mL ethanol. Etching with a picric etching solution (4.2 g picric acid, 10 mL acetic acid, 10 mL H_2O, and 70 mL ethanol) revealed the grain structure. To examine precipitates in the SEM, polished samples in the as-extruded and heat-treated states were used.

Thin TEM foils were cut, mechanically polished, and electrolytically thinned by a twin-jet TenuPol 3 with a solution of 5.3 g lithiumchloride, 11.16 g magnesiumperchlorate, 500 mL methanol, and 100 mL 2-butoxy-ethanol.

HR-TEM images were recorded on a FEI Tecnai G2 20 S-TWIN TEM and a FEI Titan 80–300 Berlin Holography Special TEM (Technische Universität Berlin, Zentraleinrichtung Elektronenmikroskopie).

For energy-dispersive synchrotron in situ tests, compression specimens with a diameter of 7 mm and a length of 15 mm and tensile specimens with a diameter of 6 mm and a length of 36 mm were used.

The texture of the polished specimens in the as-extruded conditions and deformed state were measured with the laboratory X-ray diffraction method using CoKα radiation and a 3 mm collimator. The experimental pole figures for the (10.0), (0002), (10.1), (10.2) and (11.0)-reflections were measured and the inverse pole figures were calculated from the experimental data using the MTEX software package [30].

Compression samples (D = 10 mm, l_0 = 20 mm) and tensile samples (D = 6 mm, l_0 = 36 mm) were machined from the extruded bar. Compression samples were compressed up to engineering strains of −0.01, −0.04, −0.08, −0.15, and to fracture, while tensile samples were deformed to fracture. The quasi-static tension and compression tests were carried out with a universal testing machine (MTS810) with strain rates of 2.5×10^{-4} s^{-1}.

For extruded magnesium alloys, the yield strength in tension (TYS) has been reported to be higher than the yield strength in compression (CYS). This difference in strength is called the strength differential effect (SDE) and can be expressed as:

$$\text{SDE} = \frac{|\text{CYS}| - |\text{TYS}|}{|\text{CYS}| + |\text{TYS}|}. \tag{1}$$

2.2. In Situ Energy Dispersive Synchrotron X-Ray Diffraction

In situ energy-dispersive synchrotron X-ray diffraction experiments were carried out at the 7T-MPW-EDDI-beamline at the BESSY-II synchrotron [31]. The beamline is equipped with a superconducting 7T multipole wiggler, which provides a white beam with a usable range of about 8 to 150 keV. For the experiments, an energy range of 20 to 60 keV and the Bragg angle 2θ of 10.34° were used. The beam was limited by slit systems to 1×2 mm^2 on the incoming and 0.1×7 mm^2 on the detector side. The in situ compression and tensile tests were carried out with a tensile-compressive loading device designed by Fa. Walter+Bai AG, which is mounted on a 4-axes positioner to allow x-y-z translation and Ψ rotation around the axis of the beam.

In the unloaded sample state, the lattice spacings d_0^{hkil} were measured in 11 Ψ angles between 0 and 89.9° with an exposure time of 60 s. These data were used for the determination of d_0^{hkil}. During mechanical testing, the d^{hkil} were measured at several uniaxial tensile and compression steps at 0° and 89.9° to determine the elastic lattice strains ε^{hkil} in the axial and transversal direction by Equation (2) [32]:

$$\varepsilon^{hkil} = \frac{d^{hkil}}{d_0^{hkil}} - 1 = \frac{E_0^{hkil}}{E^{hkil}} - 1. \tag{2}$$

Here, d^{hkil}, d_0^{hkil}, and E^{hkil}, E_0^{hkil} are the lattice spacings and their corresponding diffraction lines at a load step and prior loading. The in situ measurements were stress controlled in the linear elastic region. After reaching the yield strength, displacement control was used.

2.3. EPSC Modeling

The elasto-plastic self-consistent (EPSC) model [27] is a polycrystal plasticity model which simulates the constitutive response of a material based on single crystal data, a hardening model, given texture, and initial grain shape and size. A recent description of the formulation used here can be found in [28,33,34].

To describe hardening, separate models are employed for slip and twinning. For slip, a dislocation density-based hardening model for the CRSS values in different crystallographic slip systems is employed [35,36]. For twinning, a model for the domain reorientation and twin expansion as well as the twin boundary effect on slip is used, as described by the composite grain model [37]. To activate twinning, we employ the finite initial fraction (FIF) assumption [7], which states that twin nucleation is accompanied by twin growth to a finite initial fraction. For the simulations here, an FIF of 0.01 is used.

The material parameters used in the model pertain to elasticity, the initial stress to activate slip, the rates of dislocation storage, and the critical stress for twin growth. These are listed in Table 3. For the first of these, we used single-crystal elastic constants for Mg, which are C_{11} = 59.5 GPa, C_{12} = 26.1 GPa, C_{13} = 21.8 GPa, C_{33} = 65.6 GPa, and C_{44} = 16.3 GPa for both alloys. The remaining parameters were identified iteratively and specifically for each alloy until the simulation results matched all the experimental results. The simulation results enable da direct comparison with the synchrotron X-ray diffraction measurements from the compression and tensile tests. The average stress–strain response, the texture evolution, and the elastic lattice strains allow to check the EPSC predictions by the experimental diffraction data. The EPSC simulation further gives information on the active deformation modes and their CRSS as a function of the strain. For the final simulation, a grain set of 10,000 grains, which represent the as-extruded texture, was used.

3. Results

3.1. Microstructure

The microstructure of the as-extruded material is presented in Figure 1 via optical micrographs to show the grain structure and inverse pole figures to show the texture. The microstructures of all three series are dominated by globular recrystallized grains. The average grain size of the Mg2Nd alloy decreases from 13 ± 1 μm for the air-cooled A-series to 7 ± 1 μm for the water-quenched B-series. This is due to hindered grain growth after the dynamic recrystallization caused by the rapid drop in temperature. An additional decrease to an average grain size of 3 ± 1 μm for the C-series was achieved by reducing the extrusion speed. The Mg2Yb alloy shows an average grain size of 49 ± 4 μm for the air-cooled A-series and 26 ± 2 μm for the water-quenched B-series. Extrusion at a reduced temperature decreases the grain size of the C-series to 8 ± 2 μm. In the slower extruded C-series of the Mg2Nd alloy and both water-quenched series of the Mg2Yb alloy, elongated grains that are not fully recrystallized are occasionally found in the longitudinal sections (Figure 1h,n,q).

The X-ray diffraction texture measurements reveal pronounced RE-fiber-textures with overall low maximum intensities (Figure 1). The texture of the A-series of the Mg2Nd alloys has a maximum at the <11.2>pole and the B-series shows additional intensity between the <10.2> and <10.1>pole. The C-series has its maximum at the <10.0> pole with additional intensities at the <10.1> and<11.2> pole. The A- and C-series of the Mg2Yb alloys show a very weak extrusion texture with of maximum of just 1.3 multiples of random distribution (mrd) and 1.5 mrd, respectively, and intensities at the <10.2> and <11.2> pole. The B-series, in contrast, shows the highest observed intensity of 2.3 mrd at the <10.0> pole. The texture of the C-series is similar to the A-series, with an additional intensity at the <10.0> pole.

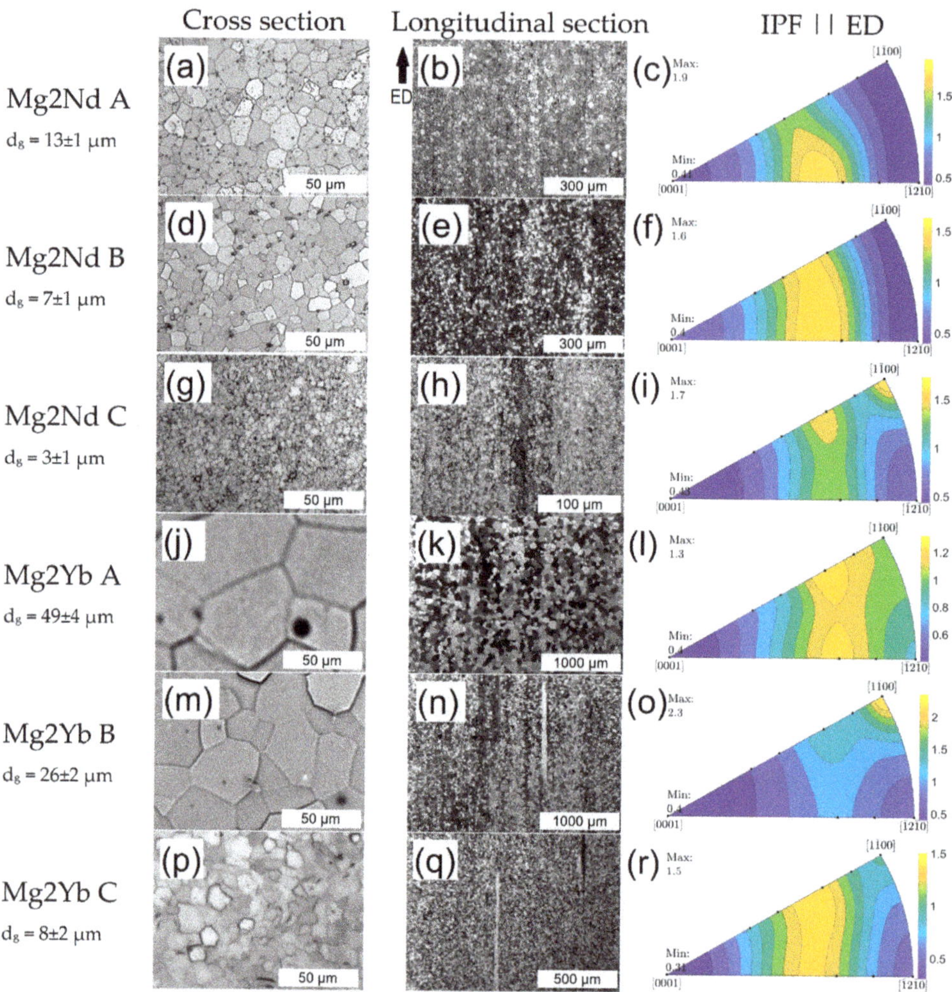

Figure 1. Cross section, longitudinal section, and inverse polfigures in the extrusion direction (ED) of the Mg2Nd alloy (**a–c**) A-series, (**d–f**) B-series, (**g–i**) C-series, and of the Mg2Yb alloy (**j–l**) A-series, (**m–o**) B-series, (**p–r**) C-series.

3.1.1. Mg2Nd

All series of the Mg2Nd alloy show coarse precipitates in different shapes and diameters up to 4 μm. These precipitates are randomly distributed in the grains. In the slower-cooled A-series, additional precipitates can be seen on the grain boundaries (Figure 2a). The area fraction of the precipitates was determined to approx. 3.2% for the A- and B-series and 3.5% for the C-series. Due to the high number of precipitates, analyses of the solution content of Nd in the Mg-matrix were carried out using a high-resolution electron probe micro-analyzer (EPMA). These showed that a Nd concentration of 1.6 wt.% was present in the matrix. Further measurements show that these are stable Nd-rich β_e (Mg41Nd5) precipitates [38]. Heat treatments at 150 °C and 204 °C promote the formation of coherent Guinier–Preston zones and β'' precipitates with diameters of 10 to 20 nm (Figure 2b) [29,38,39]. Heat treatments at higher temperatures produce stable β_e-precipitates that either do not lead to precipitation hardening or, even worse, reduce the strength. These samples were, therefore, not investigated further.

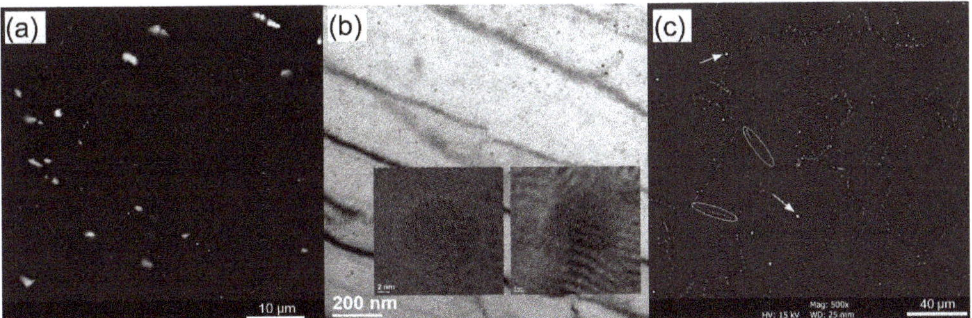

Figure 2. (**a**) SEM Mg2Nd A-series, (**b**) TEM images of precipitates in the heat-treated (150 °C, 7 h) Mg2Nd B-series, <0002> zone axis, (**c**) SEM Mg2Yb A-series.

3.1.2. Mg2Yb

In the water-quenched B and C series, only the larger precipitates with diameters of 1 to 2 µm are found in the as-extruded state. They are located within the grains and are homogeneously distributed. In samples of the A series, the same precipitates are found within the grains. In addition, predominantly smaller precipitates (Ø~0.5 to 1 µm) are observed at the grain boundaries (Figure 2c). These precipitates are the β'-phase (Mg$_2$Yb) precipitates [40], and formed after extrusion during slow cooling. Heat treatments at 150 °C have little effect on the microstructure and mechanical properties of the Mg2Yb alloy. In contrast, an increase in the phase fraction of the β' precipitates can be observed after a heat treatment for 15 min at 300 °C. No changes in grain size and texture were observed during all heat treatments of the Mg2Nd and Mg2Yb alloys.

3.2. Mechanical Properties

The yield strengths (CYS, TYS) and elongations to failure (ε_{fc}, ε_{ft}), as determined from the stress–strain curves, are summarized in Table 2. For the Mg2Nd alloy, the flow curves of the A-series look similar in compression (Figure 3a) and tension (Figure 3b), and achieve elongations to failure above 0.32. We find that the B-series shows similar behavior (Supplementary Materials Figure S3). In contrast, the shape of the flow curves of the C-series differ between the compression and tensile tests (Figure S3). The air-cooled A-series yields at −105 and 107 MPa in the uniaxial compression and tensile tests, respectively, resulting in nearly zero SDE (Table 2). Reducing the grain size to 7 µm by adjusting the cooling method (B-series) after extrusion increased the CYS and TYS to −125 and 122 MPa, respectively. In the case of the C-series alloy, the associated additional grain size reduction to 3 µm, by the slower extrusion rate, increased the CYS and TYS further to −176 and 175 MPa, resulting in an SDE of 0. However, in this case the flow curve of the compression test exhibits a sigmoidal shape (Figure S3), which is a well-known signature of tension twinning [18]. Still, the increased activity of TTW-ing significantly reduces the elongation to failure (ductility) in the compression test. The tensile test had similarly good values, just as with the A- and B-series.

Table 2. Mechanical properties of the investigated alloys (the error of compressive yield strength (CYS) and tensile yield strength (TYS) is below 5 MPa, the error of ε_{fc} and ε_{ft} is below 0.02).

Alloy	Series	Heat Treatment	CYS (MPa)	ε_{fc} (-)	TYS (MPa)	ε_{ft} (-)	SDE
	A	-	−105	−0.32	107	0.36	−0.02
	A + HT	150 °C, 6 h	−123	−0.24	129	0.20	−0.05
	B	-	−125	−0.29	122	0.41	0.02
Mg2Nd	B + HT	150 °C, 7 h	−160	−0.24	133	0.17	0.18
	C	-	−176	−0.22	175	0.34	0.01
	C + HT	150 °C, 7 h	−197	−0.18	203	0.34	−0.03
	C + HT	204 °C, 3 h	−191	−0.2	197	0.33	−0.03
	A	-	−58	−0.3	89	0.3	−0.42
	A + HT	300 °C, 15 min	−65	−0.29	104	0.21	−0.47
Mg2Yb	B	-	−77	−0.22	100	0.32	−0.26
	B + HT	300 °C, 15 min	−83	−0.23	110	0.25	−0.28
	C	-	−111	−0.22	115	0.32	−0.04

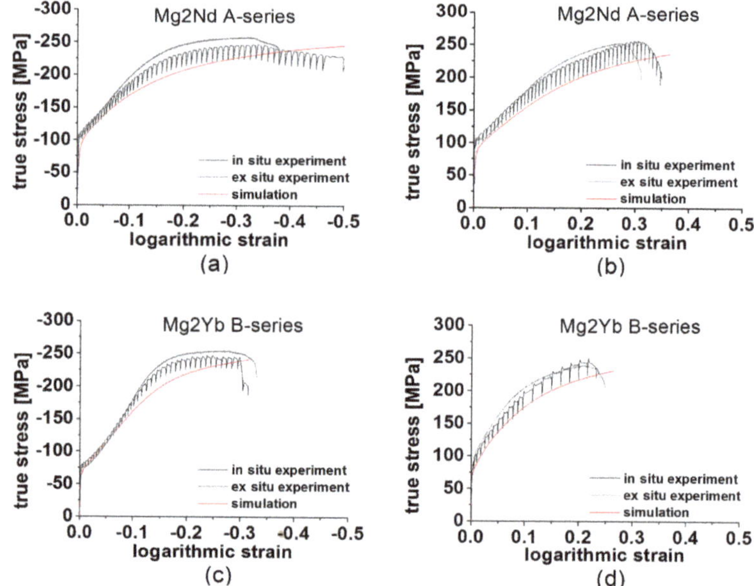

Figure 3. Comparison of the flow curves from the in situ experiment, ex situ experiment and the simulation of the Mg2Nd A-series (**a**) compression and (**b**) tensile tests and the Mg2Yb B-series (**c**) compression and (**d**) tensile tests.

The Mg2Yb alloy is significantly more anisotropic in its plastic deformation and mechanical properties than the Mg2Nd alloy, especially in the B-series. The flow curves of the compression tests (Figure 3c and Figure S3) show a strong sigmoidal shape and the CYS are significantly lower than the TYS. This leads to SDEs of high magnitude (Table 2). Nevertheless, a strong influence of the grain size can be observed in both the CYS and TYS. As grain size decreases, the yield strengths increase and this effect is more pronounced in compression than in tension. Therefore, the C-series ultimately achieves a low SDE of −0.04.

Next, we heat-treated the series of the Mg2Nd alloys at 150 °C. We find that heat treating increases the TYS of the A-series slightly more than the CYS, leading to a larger SDE than without heat treatment. The yield strengths become similar to those of the B-series in the as-extruded state. In the case of the B-series, heat treating causes the CYS and TYS to increase asymmetrically, resulting in a strongly increased SDE. The strength of the

C-series can be increased in a similar way as that seen in the A-series. A 3 h heat treatment at 204 °C leads to a CYS and a TYS of -191 and 195 MPa (SDE = −0.03) and a 7 h heat treatment at 150 °C to −197 and 203 MPa (SDE = −0.03), respectively. While the ductility of the A series decreases significantly, the ductility of the heat-treated C-samples remains unchanged in the tensile tests. However, it decreases slightly in the compression tests.

Asymmetric precipitation hardening is even more pronounced in the Mg2Yb alloy. For both the A- and B-series, the TYS increases more than the CYS. Therefore, an increase in the SDE can be observed in both cases. While the ductility during compression deformation remains nearly the same after heat treatment, it decreases significantly during tensile deformation.

3.3. Deformed Microstructure

In order to investigate the development of the microstructure during deformation, compression tests were stopped at certain deformation steps. Figure 4 shows the grain structure and texture for the Mg2Nd A-series and Mg2Yb B-series for some of these intermediate strain levels (the microstructure and textures for the other series and at other strain levels can be found in Figures S4 and S5).

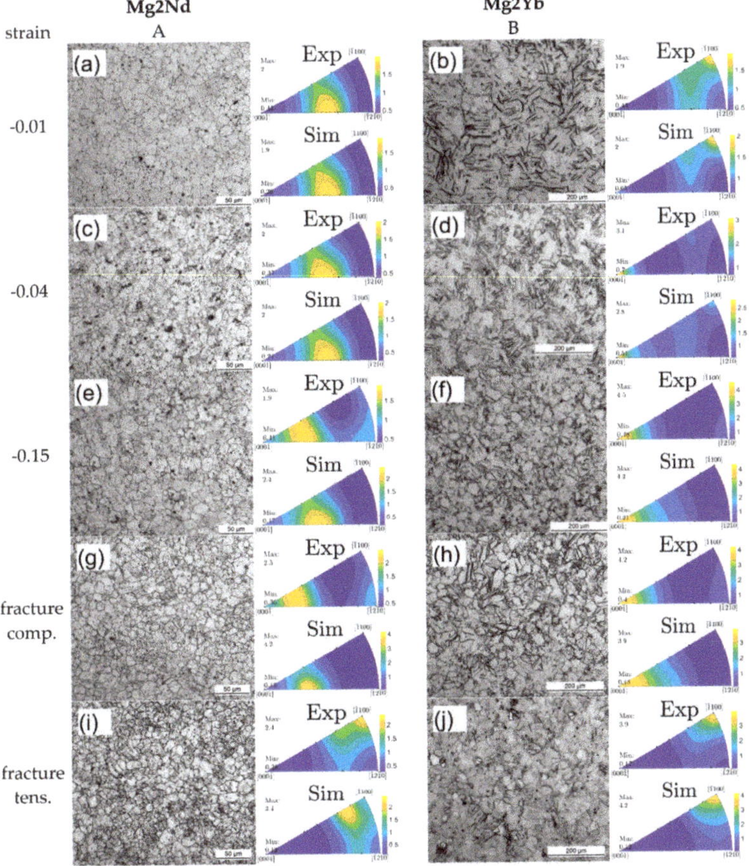

Figure 4. Cross sections and comparison of the measured and simulated texture at (**a**,**b**)−0.01, (**c**,**d**)−0.04, (**e**,**f**)−0.15 strain and elongation to failure in (**g**,**h**) compression and (**i**,**j**) tension for the Mg2Nd A-series and the Mg2Yb B-series, respectively.

3.3.1. Mg2Nd

No twins are observed in the optical micrographs of the A- and B-series of the Mg2Nd alloy. While the flow curves of the Mg2Nd alloy show a similar shape, there is a clear difference in the texture evolution between the compression and tensile tests (Figure 4g,i). The inverse pole figures show a continuous shift in texture during deformation. During compression, the maximum intensity is close to the <11.2>-pole in as-extruded state, gradually shifting towards the <0002>-basal pole with increasing deformation (Figure 4a,c,e,g). During tensile testing, however, the initial <11.2>-texture with low maximum intensities changes to a sharp <10.0>-texture (Figure 4i) with increased maxima of 3.9 and 4.4 mrd for the A- and B-series, respectively (B-series shown in Figure S4).

3.3.2. Mg2Yb

On the contrary, the excessive twinning of the Mg2Yb alloy is observed in the compressive deformation of the A- and B-series. For as little as −0.01 strain, thin twin lamellae can be observed in the optical micrographs (Figure 4b). However, these fine twins have no effect on the texture. With increasing the strain to −0.04, the area fraction of the twinned grains increases and the existing twin lamellae widen. Together the resulting rise in twin activity leads to a noticeable change in texture towards a <0002>-basal texture (Figure 4d). Specifically, due the formation and growth of twins, the maximum intensity at the <0002>-pole increases, while the initial texture slowly depletes. The intensity at the <10.0>-pole decreases both in the A- and B-series and disappears at −0.04 strain. While the <10.1> and <11.2>-texture components from the A-series can still be recognized, they too disappear at −0.08 and a strong <0002>-basal texture remains (Figure S5 and Figure 4f). With increasing deformation, the maximum intensity at the <0002>-pole further increases, an increase that is more pronounced in the B-series than in the A-series.

3.4. EPSC Simulation

To understand the deformation behavior of the investigated alloy, in situ experiments were performed. Table 3 shows the parameters used for the hardening models in the EPSC simulation. With these parameters, a good agreement was obtained between the simulated and the experimental data in terms of flow curves (Figure 3 and Figure S3) and texture evolution (Figure 4 and Figure S2). As validation, we find that the model is capable of calculating the elastic lattice strains of different crystallographic planes. The calculated elastic lattice strains are compared with those determined experimentally in the in situ tests in Figure 5, Figure S4, and Figure S5 and also show good agreement. As a result of the simulation, further insights into the deformation behavior and the activities of the deformation systems (Figure 6, Figure S6, and Figure S8), as well as their CRSS (Figure 7, Figure S7, and Figure S9), during the deformation process are gained.

Table 3. Hardening parameters for slip and twin systems.

Parameter	Mg2Nd (c/a = 1.623)				Mg2Yb (c/a = 1.626)			
	Prismatic	Basal	Pyramidal	TTW	Prismatic	Basal	Pyramidal	TTW
$\tau_{0,f}^{s}/\tau_{0}^{t}$ (Mpa)	39	11	60	18	21	1	67	2
k_{1}^{s} (m^{-1})	2.2×10^{8}	6.0×10^{7}	4.5×10^{9}	-	4.0×10^{8}	3.0×10^{7}	5.5×10^{9}	-
D^{s} (MPa)	4.5×10^{3}	3.0×10^{3}	3.5×10^{3}	-	4.0×10^{3}	4.0×10^{3}	4.0×10^{3}	-
g^{s}	7.2×10^{-3}	4.3×10^{-3}	7.3×10^{-3}	-	6.5×10^{-3}	4.8×10^{-3}	7.2×10^{-3}	-
H^{s}, H^{t}	70	90	30	110	190	150	40	110
C^{st}	50	50	400	-	50	50	400	-

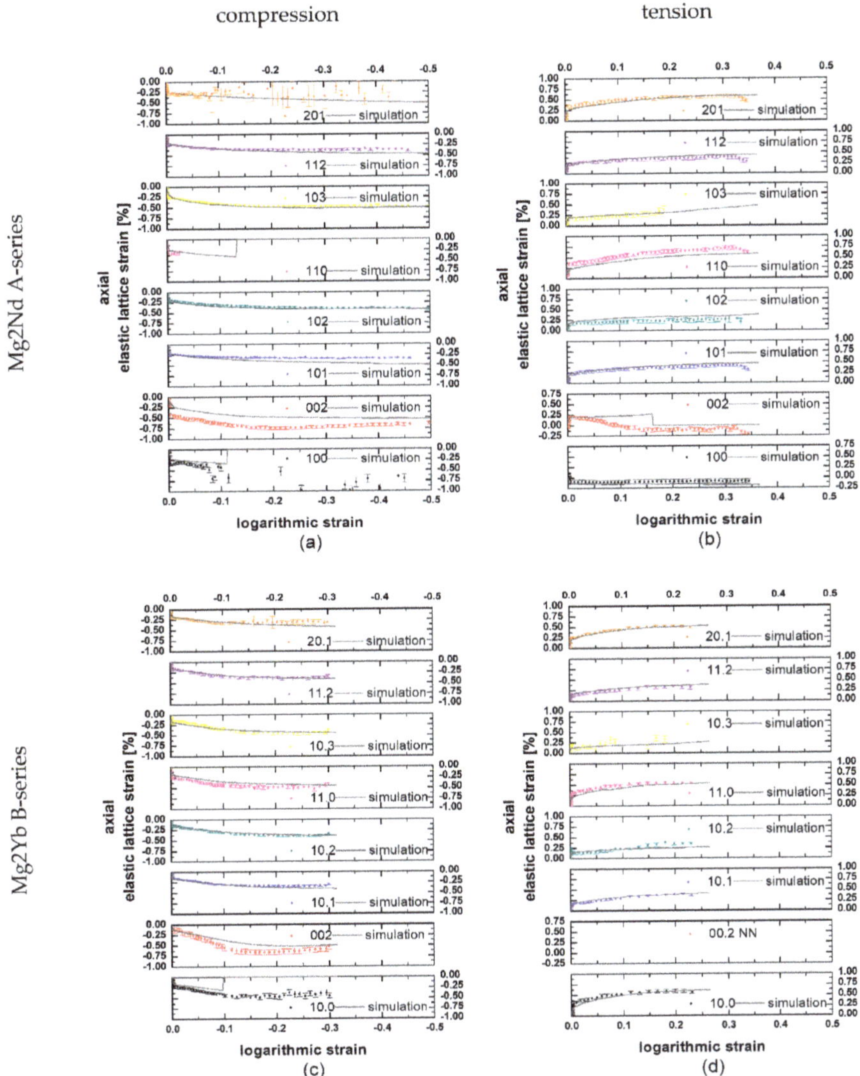

Figure 5. Comparison of the experimentally measured and simulated axial elastic lattice strains of the Mg2Nd A-series during (**a**) compression and (**b**) tensile tests and the Mg2Yb B-series during (**c**) compression and (**d**) tensile tests.

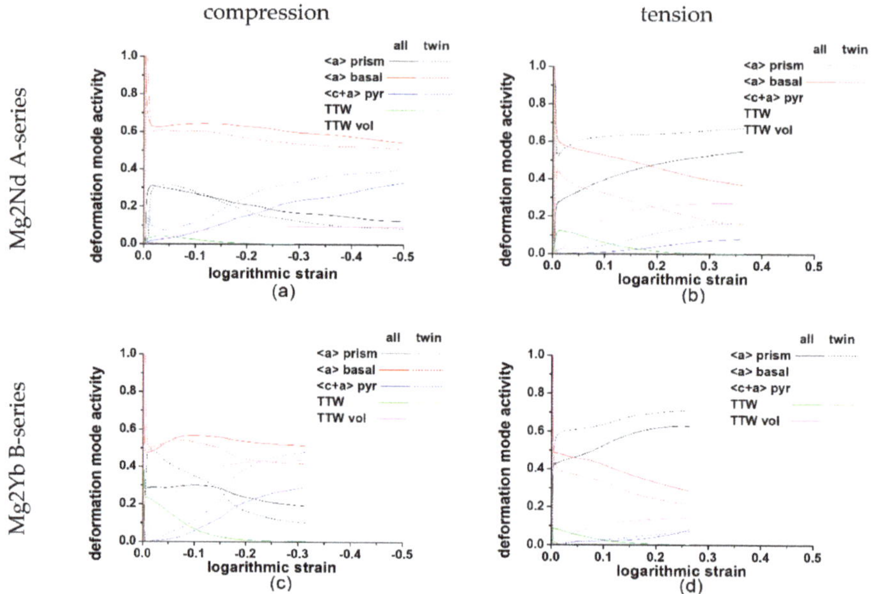

Figure 6. Deformation mode activity as a function of strain of the Mg2Nd A-series during (**a**) compression and (**b**) tensile tests and the Mg2Yb B-series during (**c**) compression and (**d**) tensile tests.

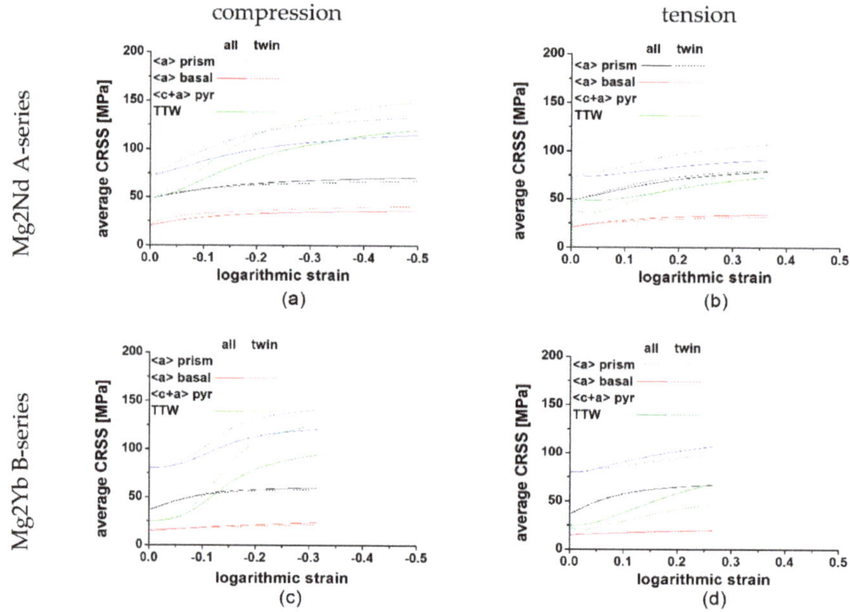

Figure 7. Critical resolved shear stresses (CRSS) of the different deformation systems as a function of strain of the Mg2Nd A-series during (**a**) compression and (**b**) tensile tests and the Mg2Yb B-series during (**c**) compression and (**d**) tensile tests.

3.4.1. Flow Curves

In the mechanical testing, the strains were applied step by step, and at each step EDS-XRD measurements were carried out. During the data acquisition steps, the material rapidly relaxed, leading to a reduced stress. The stress measurement is recorded in the relaxed state for most of the time. The simulated flow curves were matched to the experimental flow curves from the in situ tests at these reduced stresses. It should be noted that an elongated yield point (Lüders strain) was observed in the in situ tensile tests of the Mg2Nd alloy (Figure 3). However, this region was not used in the simulation model, since it does not take into account localized plastic deformation.

3.4.2. Elastic Lattice Strains

The elastic strains were calculated from the data of the in situ compression and tensile tests. These provide further information on the deformation behavior. Figure 6; Figure 7 show the evolution of the elastic lattice strains for the A- and B-series of the Mg2Nd and Mg2Yb alloys in axial direction of the (10.0), (0002), (10.1), (10.2), (11.0), (10.3), (11.2), and (20.1) reflections during the compression and tensile tests.

During deformation, grains with (10.1), (10.2), (10.3), and (11.2) orientations possess high Schmid Factors (SF) for <a> basal slip, making it the most likely active slip system in these grains. Grains with (10.0), (11.0), and (20.1) orientations have a high SF for <a> prismatic slip. Grains oriented in (0002) require <c> axis deformation to accommodate the applied load. They cannot deform by either <a> basal or <a> prismatic slip since the slip direction is perpendicular to the loading direction, and this results in an SF of zero for these <a> slip modes. Consequently, <c + a> pyramidal slip is the only possible slip system. Another mechanism accommodating <c>-axis deformation is mechanical TTW-ing. Since TTW-ing is a unidirectional deformation system accommodating only <c>-axis extension and not compression, the behavior of the alloy in tension and compression tests differs. During compression (10.0) and (11.0) -oriented grains possess a high SF for TTW-ing, whereas in tensile tests (0002) grains have a high SF for TTW-ing.

Mg2Nd

For both compression and tensile tests, the grains associated with <a> basal slip ((10.1), (10.2), (10.3), and (11.2)) show the lowest initial elastic lattice strains in the axial direction. With increasing deformation, only a small increase in elastic lattice strains can be observed. Grains associated with <a> prismatic slip ((10.0), (11.0), and (20.1)) exhibit higher initial elastic lattice strains, which increase with further deformation. During compression, the intensity of these reflections decreases and gradually disappears, leading to greater inaccuracies of the calculated elastic lattice strains. Eventually the (10.0), (11.0), and (20.1) reflections disappear, so the elastic lattice strains can no longer be determined. Equivalent observations were made for the (0002) grains during tensile deformation. During compression, the grains oriented in (0002) initially show the highest elastic lattice strains, which increase the most during plastic deformation. They reach their maximum at a strain of −0.15 and remain almost constant thereafter.

Mg2Yb

For the A- and B-series of the Mg2Yb alloy, a similar evolution of the elastic lattice strains is observed. However, during compression grains associated with <a> prismatic slip ((10.0), (11.0), and (20.1)) can be analyzed at all times. Yet, similar to the Mg2Nd alloy, increasing inaccuracies can be observed with increasing deformation. Small initial elastic lattice strains in these orientations associated with <a> basal slip ((10.1), (10.2), (10.3), and (11.2)) indicate a low CRSS to activate plastic deformation, or, vice versa, higher elastic lattice strains indicate a higher CRSS. In addition, the increase in the elastic lattice strains can be associated with the work hardening of the active slip system. The EPSC model can quantify these qualitative predictions and calculate the evolution of the CRSS of the different deformation systems and their contribution to the total deformation.

3.4.3. Activities and CRSS of Deformation Systems
Mg2Nd

Compression

The deformation during uniaxial compression is mainly carried by <a> basal slip with a constant contribution of approximately 0.6 (Figure 6a) and little work hardening. This can be seen in the low increase in the CRSS (Figure 7a). The maximum SF of <a> basal slip is not affected by the continuous texture change and remains at a high value of 0.49. Initially, the <a> prismatic slip system is the second most active, but its activity decreases continuously. It shows more hardening (increase in the CRSS) than the <a> basal slip system. TTW-ing only has a minor role at the beginning. As previously seen in Figure 4, there are only few twinned grains at a strain of −0.08. After a total strain of −0.1 to −0.15, the activity of TTW-ing is negligible. As the activity of <a> prismatic slip and TTW-ing decreases, the <c + a> pyramidal slip system becomes more important at higher strains. The elastic lattice strain of the (0002) plane increases continuously up to a deformation of approximately −0.15 true strain. Afterwards, they remain at a constant level. This means that up to this strain, the deformation of the grains in (0002) or c-direction is mainly carried out elastically. Subsequently, an increase in the deformation activity of the <c+a> pyramidal slip can be observed, accompanied by a significant increase in the CRSS.

Tension

As with compression deformation, the <a> basal slip system is the most important deformation system in the early stages of straining. However, the contribution decreases continuously from 0.55 to 0.4, while the share of the <a> prismatic slip system increases and exceeds the share of the <a> basal slip system at a total strain of about 0.15 (Figure 6b). The CRSS of the <a> prismatic slip system behaves in the same way as the <c + a> pyramidal slip system under compressive stress. The increased activity of the <a> prismatic slip system is accompanied by a much stronger increase in CRSS (Figure 7b). Compared to the compression test too, an increased activity of TTW-ing is observed in the tensile test, thus leading to a higher final total volume of twinned grains.

Mg2Yb

Compression

As suggested in the texture evolution, TTW-ing contributes to deformation at small strains. The B-series alloys achieve a higher volume fraction of TTW than the A-series alloys. The B-series also has a higher proportion of TTW volume fraction than the A-series. This can be attributed to the more pronounced <10.0> texture component. For the slip systems, the <a> basal slip is the most active. Its volume fraction increases up to 0.6 with decreasing TTW-ing (Figure 6c). The portion of <a> prismatic slip decreases slightly with increasing deformation. In the range of ×0.1 to −0.15 deformation an increase in <c + a> pyramidal slip can be observed. The twinned grains, which are oriented in (0002), can now only deform further by <c + a> pyramidal slip. The CRSS of the <c + a> pyramidal slip increases highly for the twinned grains (Figure 7c). In contrast, the CRSS of the <a> basal and <a> prismatic slip shows a weaker increase. The CRSS of the TTW-ing increases more than the CRSS of the slip systems.

Tension

In the tensile tests, <a> prismatic slip is the dominant deformation system (Figure 6d). The share of the total deformation increases with increasing strain, whereas the contribution of the <a> basal slip continuously decreases. Since there are only a few grains oriented in (0002) in the original texture, a low activity of TTW-ing can only be observed at the beginning. After completion, there are no more grains with this orientation. Therefore, the proportion of <c + a> pyramidal slip is also low. Just as in compression tests, the CRSS of TTW-ing increases strongly as soon as all favorably oriented grains are twinned (Figure 7d).

4. Discussion

4.1. Strength

Plastic deformation can be carried out by different deformation systems such as crystallographic slip and mechanical twinning. A deformation system is activated if the shear stress τ exceeds the CRSS of the respective deformation system. The magnitude of the shear stress depends on the orientation of the planes with respect to the direction of loading. Thus, the texture has a significant influence on which deformation systems can be activated to which extent. The activity of the respective deformation systems and their CRSS, in turn, have a significant influence on the mechanical properties. Additionally, the CRSS of polycrystalline materials depends on the grain size.

From the yield strengths displayed in Table 2, the following dependencies (Table 4), regarding grain size can be calculated using the Hall–Petch relationship:

$$YS = \sigma_0 \frac{K}{\sqrt{d}}. \tag{3}$$

Table 4. Calculated Hall–Petch parameters.

Alloy	Mg2Nd		Mg2Yb	
HP Parameter	**CYS**	**TYS**	**CYS**	**TYS**
K (MPa $\mu m^{0.5}$)	239 ± 14	232 ± 29	243 ± 28	116 ± 22
σ_0 (MPa)	-37 ± 6	39 ± 12	-25 ± 7	74 ± 6

As seen in Table 4, the Hall–Petch parameters for the Mg2Nd alloy show similar values for compression and tension, with both having large values for the K-factor. This indicates the direct dependence on the grain size. The initial texture of the A- and B-series activates similar slip systems at the beginning of the compression and tensile deformation. As a result, the CYS and TYS are also similar, resulting in a low SDE. In contrast, the texture of the C-series is altered and grains elongated in ED are observed (Figure 1h). This favors TTW-ing, as seen in the flow curves (Figure S3) [16,41]. However, this does not lead to a reduced CYS (-176 MPa) compared to the TYS (175 MPa). This is due to the strong dependence of the $CRSS_{ttw}$ on grain size (Table 5), maintaining an SDE of 0.

Table 5. Initial critical resolved shear stresses (CRSS) for the different slip and twinning modes.

Alloy	Series	GS (μm)	$CRSS_{ba}$	$CRSS_{pr}$	$CRSS_{py}$	$CRSS_{ttw}$	CRSS Ratio ba:pr:py:ttw
Mg2Nd	A	13	20.7	49.2	73.4	49.4	1: 2.4: 3.6: 2.4
	B	7	23.2	51.1	74.7	61.6	1: 2.2: 3.2: 2.7
Mg2Yb	A	49	12.7	34.4	79.5	19.7	1: 2.7: 6.3: 1.6
	B	26	15.0	37.3	80.4	25.6	1: 2.5: 5.4: 1.7

On the other hand, the Hall–Petch parameters of the Mg2Yb alloy for CYS and TYS are different. For the CYS, a stronger dependence on grain size can be observed than for the TYS. This can be explained by the fact that during compression tests of the Mg2Yb alloy, a high activity of TTW-ing and <a> basal slip carry the plastic deformation initially (Figure 7b). The CRSS values for TTW-ing and <a> basal slip have a strong dependence on grain size (Table 5). In contrast, the <a> prismatic slip system is more active during tensile deformation. With a decreasing grain size from 49 μm to 26 μm for the A- and B-series, the CRSS of TTW-ing and <a> basal slip increases by 30% and 18%, respectively. On the other hand, the CRSS of <a> prismatic and <c + a> pyramidal slip only increases by 8% and 1%, respectively. The sigmoidal shape of the flow curve in the compression test (Figure S3) indicates that TTW-ing is still an important deformation system in the C-series.

Due to further reductions in grain size, however, the $CRSS_{ttw}$ has most likely continued to increase. As a result, the CYS increases relatively more than the TYS, which was also observed in the mechanical data (Table 2). According to this, the SDE was reduced from −0.42 for the A-series to −0.04 for the C-series. Subsequent heat treatments at 300 °C further increased the yield strengths of the A- and B-series. Since the CYS increases less than the TYS, it can be assumed that the precipitates have a stronger hardening effect on the initially dominant <a> basal and <a> prismatic slip systems in tensile deformation than on the TTW-ing, which occurs during compression.

4.2. Plastic Deformation

Under compressive load, the slip plane normal of an active slip system rotates in the direction of the load. During tensile deformation, the crystal also rotates continuously in such a way that the slip direction approaches the loading direction.

4.2.1. Mg2Nd

Since the texture shifts continuously toward the <0002> basal pole, it can be associated with <a> basal slip. Figure 8 shows that the initial texture (Figure 1) also exhibits high Schmid Factors for the <a> basal slip system, making it even more favorable for activation. The slow rotation of the grains toward the <0002> orientation results in a smaller SF for <a> prismatic slip and a larger SF for <c + a> pyramidal slip. In simulation, a continuous decrease in the activity of the <a> prismatic slip is observed with increasing deformation and texture shift, while the activity of the <c + a> pyramidal slip increases (Figure 6a). During tensile deformation, the texture shifts toward the <10.0> pole; therefore, <a> slip systems must be responsible for this as well (Figure 4i). Initially, the <a> basal slip system has a higher SF than the <a> prismatic slip system. Yet, at higher strains, this changes and <a> prismatic slip becomes the favored deformation system, resulting in an increase in the deformation mode activity. The increased activity of <c + a> pyramidal slip in compression tests and <a> prismatic slip in tensile test leads to higher dislocation densities and work hardening. This can be observed in the different increases in their CRSS (Figure 7b). The CRSS of the <c + a> pyramidal slip system shows the highest increase during compression, while <a> prismatic slip hardens the most under tension.

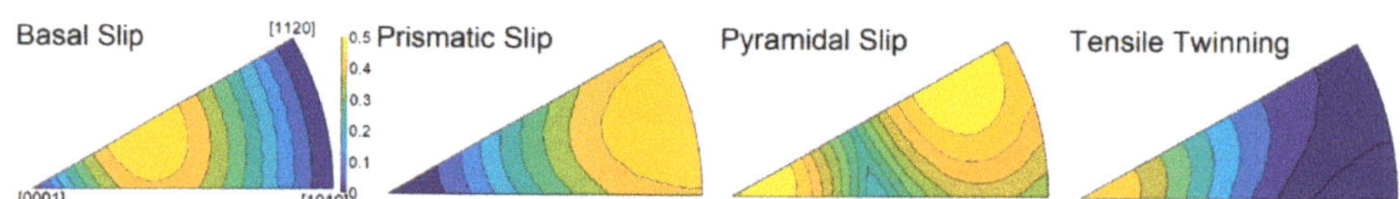

Figure 8. Theoretical calculated Schmid Factors as a function of grain orientation for uniaxial tension, displayed as inverse polfigures.

4.2.2. Mg2Yb

Compared to the continuous texture shift in the compression and tensile tests of the Mg2Nd series, the compression tests of the Mg2Yb series show an abrupt texture change (Figure 4d). This is related to the formation of TTWs. After approx. −0.1 deformation, the TTW-ing is mostly completed. The twinned grains initially also deform with <a> slip systems, but as the deformation increases <c + a> pyramidal slip becomes the dominant deformation system. This leads to grain rotation in the direction of the <11.3> slip direction, so that the <0002> texture becomes less pronounced and the intensity is spread further around the basal pole. In tensile tests, a similar behavior to the series of the Mg2Nd alloy is observed. Yet, since the initial texture is different and <a> prismatic slip is favored, a higher activity can be observed at the beginning of the deformation. As a result, it becomes the dominant deformation system even at lower strains compared to the Mg2Nd alloy series.

4.3. Ductility

Mg and most Mg alloys have a low ductility at room temperature. This is mainly due to the hexagonal crystal structure, which bears only a small number of easily activated slip systems. At room temperature, the <a> basal slip system usually has significantly lower CRSS values than the non-basal slip systems, which is why deformation at low temperatures is mainly realized by <a> basal slip. The <a> basal slip system, however, has only two independent slip systems and thus does not meet the von Mises criterion of deformation compatibility, which states that at least five independent slip systems are required for uniform plastic deformation.

By modifying the texture, the deformation behavior and thus the mechanical properties can be improved. Additions of rare earth elements showed texture changes that may have had a positive effect on the deformation behavior and especially on the ductility at low temperatures. Investigations by Stanford et al. [21] showed that the extrusion product of a binary Mg-Gd alloy (1.5wt.%Gd) with modified Mg extrusion texture achieves twice the elongation to failure of an extrusion product of pure magnesium with a comparable grain size and typical Mg extrusion texture (<10.0>/<11.0> double fiber texture). Due to the comparable grain size, the increase in ductility could be directly attributed to the texture modification.

These findings and the presented microstructure and texture suggest good ductility and good mechanical properties of the extruded products. In these modified textures, a significantly higher number of grains shows a favorable orientation for the activation of the <a> basal slip system. At the same time, fewer grains are favorably oriented for the activation of the TTW system, which further can reduce the asymmetry of the tensile and compressive yield strength (SDE). The higher activity of the basal slip system results in a higher ductility at room temperature.

The A- and B-series of the Mg2Nd alloy show <11.2> textures (Figure 1). This means that the c-axis of the majority of grains is tilted by 45–50° to the extrusion direction. The grains are in a favorable orientation for the activation of the <a> basal slip system. The higher activity of the <a> basal slip system results in a higher ductility at room temperature. Meanwhile the C-series has an additional <10.0> texture component and therefore increased TTW activity in compression tests, as seen in the flow curves (Figure S3). This leads to reduced elongation to failure in compression tests, compared to the A- and B-series, while the elongation to failure in tensile tests reaches similar values (Table 2). In all cases, heat treatments for precipitation hardening result in reduced elongation at failure for the Mg2Nd alloy. Considering this, the grain size reduction by adapting the cooling conditions, as seen in the B-series, is preferable to a subsequent heat treatment of the A-series. Similar yield strengths can be achieved by both methods. However, the ductility of the B-series is better than that of the A-series, suggesting that the additional process step of heat treatment is not necessary.

In contrast, the A- and B-series of Mg2Yb alloy have different textures. The A-series has an extraordinarily weak texture with a maximum intensity of just 1.3 mrd, nearly completely random, with intensities at <11.2> and <10.1> poles, while the B-series has a distinct <10.0> texture. This results in a difference in the ductility. Despite the unfavorable texture, the A-series deformed at lot due to TTW-ing in the compression test due to the large grain size (as explained in 4.1). The TTW are reoriented by 89°, so that they are not exactly oriented in <0002> but slightly tilted, depending on the initial orientation. This results in a wider spread compared to the B-series, which has a strong <0002> texture with a higher intensity (Figure S5). The TTW in the A-series can deform by <a> basal and <a> prismatic slip due to the tilt. While the majority of the twinned grains in the B-series are unfavorably oriented for the <a> basal slip system, the activity of the <a> basal slip system is, therefore, correspondingly low. As a result, the twinned grains deform primarily due to <c+a> pyramidal slip (Figure 6c). This leads to limited ductility and crack initiation due to stress concentrations.

5. Conclusions

Three approaches were used to modify the microstructure to adjust the mechanical properties and reduce the asymmetric yield behavior and anisotropy of the investigated magnesium alloys: texture weakening, grain size reduction, and precipitation strengthening.

As a result, the Mg2Nd series, which had high yield strengths and was symmetrical in compression and tension, resulting in an SDE close to 0, could be extruded. Thus, products with a wide variety of yield strengths (CYS −105 to −197 MPa, TYS 107 to 203 MPa) and good ductility, with elongation to failures of −0.18 to −0.32 in compression and 0.17 to 0.41 in tension tests, were obtained. The YSs of the Mg2Yb alloy are below those of the Mg2Nd alloy and range from −58 to −111 MPa for the CYS and 89 to 115 MPa for the TYS, with similarly high elongations to failure between −0.22 and −0.3 in compression and 0.21 and 0.32 in tensile tests.

By adjusting the extrusion parameters, the grain size of the Mg2Nd alloy can be gradually reduced, thereby significantly increasing the strength. The texture of the A- and B-series effectively suppresses TTW-ing. At the beginning of compressive and tensile deformation, similar deformation mechanisms are active. This leads to a symmetrical yielding behavior and SDEs close to zero. A slower extrusion speed decreases the grain size and weakens the texture. The appearance of a <10.0> texture component in the C-series promotes TTW-ing during compression. However, because of the small grain size, CYS and TYS are similar and the SDE remains at 0. The precipitation state can be changed by subsequent heat treatment, which, in turn, further increases the yield strengths while maintaining good ductility.

Depending on the extrusion parameters, the Mg2Yb series tend to have larger grain sizes and less pronounced rare earth textures. This leads to intensive TTW-ing under compressive stress, resulting in high SDEs. Subsequent heat treatments can increase the YS. Since the effect on the slip systems is stronger than on the formation of TTWs, this leads to further increased SDEs. However, by reducing the extrusion temperature the grain size of the Mg2Yb alloy is also reduced and an advantageous texture can be achieved. This is accompanied by a significant increase in YS. Since $CRSS_{ttw}$ in particular is very sensitive to grain size, the CYS increases more than the TYS. Therefore, it is also possible to obtain an SDE close to 0 for the Mg2Yb alloy.

Supplementary Materials: The following are available online at https://www.mdpi.com/2075-4701/11/3/377/s1: Figure S1: Cross sections of the A- and B-series of the Mg2Nd and Mg2Yb alloy at different strains and elongation to failure. Figure S2: Comparison of the measured and simulated texture development during deformation of the A- and B-series of the Mg2Nd and Mg2Yb alloy at different strains. Figure S3: Comparison of the flow curves from compression and tensile tests of the A-, B- and C-series of the Mg2Nd and Mg2Yb alloy. Figure S4: Comparison of the experimentally measured and simulated axial elastic lattice strains during (a, b) compression and (c, d) tensile tests of the A- and B-series of the Mg2Nd alloy. Figure S5: Comparison of the experimentally measured and simulated axial elastic lattice strains during (a, b) compression and (c, d) tensile tests of the A- and B-series of the Mg2Yb alloy. Figure S6: Deformation mode activity as a function of strain of the A- and B-series of the Mg2Nd alloy. Figure S7: CRSS of the different deformation systems as a function of strain of the A- and B-series of the Mg2Nd alloy. Figure S8: Deformation mode activity as a function of strain of the A- and B-series of the Mg2Yb alloy. Figure S9: CRSS of the different deformation systems as a function of strain of the A- and B-series of the Mg2Yb alloy.

Author Contributions: Conceptualization, J.S., I.J.B., W.R.; methodology, J.S., I.J.B.; software, J.S., I.J.B., M.K.; validation, J.S., I.J.B., W.R.; formal analysis, J.S., I.J.B.; investigation, J.S.; resources, I.J.B., W.R.; data curation, J.S., I.J.B.; writing—original draft preparation, J.S.; writing—review and editing, I.J.B., M.K., W.R.; visualization, J.S, I.J.B.; supervision, W.R.; project administration, J.S., I.J.B., W.R.; funding acquisition, W.R. All authors have read and agreed to the published version of the manuscript.

Funding: This research was funded by Deutsche Forschungsgemeinschaft (DFG), grant number AOBJ595087.

Acknowledgments: The authors would like to thank Katrin Böttcher, Mateus Dobecki, Felix Hohlstein, Alexander Poeche, Tim Plöger, Jakob Schröder (TU Berlin, Metallische Werkstoffe), and Manuela Klaus (HZB) for the support of the in situ experiments; Sören Selve (TU Berlin, Zelmi) for conducting the HR-TEM analysis; Jörg Nissen (TU Berlin, Zelmi) for conducting the EPMA analysis; and Christoph Fahrenson (TU Berlin, Zelmi) for conducting the EBSD analysis. I.J.B. acknowledges financial support from the National Science Foundation Designing Materials to Revolutionize and Engineer our Future (DMREF) program (NSF CMMI-1729887). M. K. gratefully acknowledges support from the U.S. National Science Foundation (NSF) under grant no. CMMI-1727495.

Conflicts of Interest: The authors declare no conflict of interest. The funders had no role in the design of the study; in the collection, analysis, or interpretation of data; in the writing of the manuscript; or in the decision to publish the results.

References

1. Zeng, Z.; Stanford, N.; Davies, C.H.J.; Nie, J.-F.; Birbilis, N. Magnesium extrusion alloys: A review of developments and prospects. *Int. Mater. Rev.* **2019**, *64*, 27–62. [CrossRef]
2. Mordike, B.; Ebert, T. Magnesium. *Mater. Sci. Eng. A* **2001**, *302*, 37–45. [CrossRef]
3. Gupta, M.; Sharon, N.M.L. *Magnesium, Magnesium Alloys, and Magnesium Composites*; Wiley: Hoboken, NJ, USA, 2011; ISBN 9780470494172.
4. Friedrich, H.E.; Mordike, B.L. *Magnesium Technology*; Springer: Berlin/Heidelberg, Germany, 2006; ISBN 3-540-20599-3.
5. Arul Kumar, M.; Beyerlein, I.J.; Tomé, C.N. A measure of plastic anisotropy for hexagonal close packed metals: Application to alloying effects on the formability of Mg. *J. Alloy. Compd.* **2017**, *695*, 1488–1497. [CrossRef]
6. Chelladurai, I.; Miles, M.P.; Fullwood, D.T.; Carsley, J.E.; Mishra, R.K.; Beyerlein, I.J.; Knezevic, M. Microstructure Correlation with Formability for Biaxial Stretching of Magnesium Alloy AZ31B at Mildly Elevated Temperatures. *JOM* **2017**, *69*, 907–914. [CrossRef]
7. Clausen, B.; Tomé, C.N.; Brown, D.W.; Agnew, S.R. Reorientation and stress relaxation due to twinning: Modeling and experimental characterization for Mg. *Acta Mater.* **2008**, *56*, 2456–2468. [CrossRef]
8. Hutchinson, W.B.; Barnett, M.R. Effective values of critical resolved shear stress for slip in polycrystalline magnesium and other hcp metals. *Scr. Mater.* **2010**, *63*, 737–740. [CrossRef]
9. Ion, S.E.; Humphreys, F.J.; White, S.H. Dynamic recrystallisation and the development of microstructure during the high temperature deformation of magnesium. *Acta Metall.* **1982**, *30*, 1909–1919. [CrossRef]
10. Yoo, M.H. Slip, twinning, and fracture in hexagonal close-packed metals. *MTA* **1981**, *12*, 409–418. [CrossRef]
11. Liu, B.-Y.; Liu, F.; Yang, N.; Zhai, X.-B.; Zhang, L.; Yang, Y.; Li, B.; Li, J.; Ma, E.; Nie, J.-F.; et al. Large plasticity in magnesium mediated by pyramidal dislocations. *Science* **2019**, *365*, 73–75. [CrossRef]
12. Yoo, M.H.; Lee, J.K. Deformation twinning in h.c.p. metals and alloys. *Philos. Mag. A* **1991**, *63*, 987–1000. [CrossRef]
13. Beyerlein, I.J.; Zhang, X.; Misra, A. Growth Twins and Deformation Twins in Metals. *Annu. Rev. Mater. Res.* **2014**, *44*, 329–363. [CrossRef]
14. Kumar, M.A.; Beyerlein, I.J. Local microstructure and micromechanical stress evolution during deformation twinning in hexagonal polycrystals. *J. Mater. Res.* **2020**, *35*, 217–241. [CrossRef]
15. Knezevic, M.; Beyerlein, I.J. Multiscale Modeling of Microstructure-Property Relationships of Polycrystalline Metals during Thermo-Mechanical Deformation. *Adv. Eng. Mater.* **2018**, *20*, 1700956. [CrossRef]
16. Christian, J.W.; Mahajan, S. Deformation twinning. *Prog. Mater. Sci.* **1995**, *39*, 1–157. [CrossRef]
17. Stanford, N.; Atwell, D.; Beer, A.; Davies, C.; Barnett, M.R. Effect of microalloying with rare-earth elements on the texture of extruded magnesium-based alloys. *Scr. Mater.* **2008**, *59*, 772–775. [CrossRef]
18. Barnett, M.R. Twinning and the ductility of magnesium alloys. *Mater. Sci. Eng. A* **2007**, *464*, 1–7. [CrossRef]
19. Hirth, J.P.; Cohen, M. On the strength-differential phenomenon in hardened steel. *Metall. Trans.* **1970**, *1*, 3–8. [CrossRef]
20. Spitzig, W.; Sober, R.; Richmond, O. Pressure dependence of yielding and associated volume expansion in tempered martensite. *Acta Metall.* **1975**, *23*, 885–893. [CrossRef]
21. Stanford, N.; Barnett, M.R. The origin of "rare earth" texture development in extruded Mg-based alloys and its effect on tensile ductility. *Mater. Sci. Eng. A* **2008**, *496*, 399–408. [CrossRef]
22. Stanford, N.; Barnett, M. Effect of composition on the texture and deformation behaviour of wrought Mg alloys. *Scr. Mater.* **2008**, *58*, 179–182. [CrossRef]
23. Lentz, M.; Klaus, M.; Coelho, R.S.; Schaefer, N.; Schmack, F.; Reimers, W.; Clausen, B. Analysis of the Deformation Behavior of Magnesium-Rare Earth Alloys Mg-2 pct Mn-1 pct Rare Earth and Mg-5 pct Y-4 pct Rare Earth by In Situ Energy-Dispersive X-ray Synchrotron Diffraction and Elasto-Plastic Self-Consistent Modeling. *Metall. Mater. Trans. A* **2014**, *45*, 5721–5735. [CrossRef]
24. Lentz, M.; Klaus, M.; Wagner, M.; Fahrenson, C.; Beyerlein, I.J.; Zecevic, M.; Reimers, W.; Knezevic, M. Effect of age hardening on the deformation behavior of an Mg–Y–Nd alloy: In-situ X-ray diffraction and crystal plasticity modeling. *Mater. Sci. Eng. A* **2015**, *628*, 396–409. [CrossRef]

25. Bohlen, J.; Yi, S.; Letzig, D.; Kainer, K.U. Effect of rare earth elements on the microstructure and texture development in magnesium–manganese alloys during extrusion. *Mater. Sci. Eng. A* **2010**, *527*, 7092–7098. [CrossRef]
26. Seitz, J.-M.; Eifler, R.; Stahl, J.; Kietzmann, M.; Bach, F.-W. Characterization of MgNd2 alloy for potential applications in bioresorbable implantable devices. *Acta Biomater.* **2012**, *8*, 3852–3864. [CrossRef] [PubMed]
27. Turner, P.A.; Tomé, C.N. A study of residual stresses in Zircaloy-2 with rod texture. *Acta Metall. Et Mater.* **1994**, *42*, 4143–4153. [CrossRef]
28. Zecevic, M.; Knezevic, M.; Beyerlein, I.J.; Tomé, C.N. An elasto-plastic self-consistent model with hardening based on dislocation density, twinning and de-twinning: Application to strain path changes in HCP metals. *Mater. Sci. Eng. A* **2015**, *638*, 262–274. [CrossRef]
29. Rokhlin, L.L. *Magnesium Alloys Containing Rare Earth Metals*; CRC Press: London, UK, 2003; ISBN 9780429179228.
30. Bachmann, F.; Hielscher, R.; Schaeben, H. Texture Analysis with MTEX—Free and Open Source Software Toolbox. *Solid State Phenom.* **2010**, *160*, 63–68. [CrossRef]
31. Klaus, M.; Garcia-Moreno, F. The 7T-MPW-EDDI beamline at BESSY II. *JLSRF* **2016**, *2*. [CrossRef]
32. Genzel, C.; Denks, I.A.; Coelho, R.; Thomas, D.; Mainz, R.; Apel, D.; Klaus, M. Exploiting the features of energy-dispersive synchrotron diffraction for advanced residual stress and texture analysis. *J. Strain Anal. Eng. Des.* **2011**, *46*, 615–625. [CrossRef]
33. Beyerlein, I.J.; Knezevic, M. Mesoscale, Microstructure-Sensitive Modeling for Interface-Dominated, Nanostructured Materials. In *Handbook of Materials Modeling*; Andreoni, W., Yip, S., Eds.; Springer International Publishing: Cham, Switzerland, 2020; pp. 1111–1152. ISBN 978-3-319-44676-9.
34. Wang, J.; Zecevic, M.; Knezevic, M.; Beyerlein, I.J. Polycrystal plasticity modeling for load reversals in commercially pure titanium. *Int. J. Plast.* **2020**, *125*, 294–313. [CrossRef]
35. Beyerlein, I.J.; Tomé, C.N. A dislocation-based constitutive law for pure Zr including temperature effects. *Int. J. Plast.* **2008**, *24*, 867–895. [CrossRef]
36. Knezevic, M.; Beyerlein, I.J.; Brown, D.W.; Sisneros, T.A.; Tomé, C.N. A polycrystal plasticity model for predicting mechanical response and texture evolution during strain-path changes: Application to beryllium. *Int. J. Plast.* **2013**, *49*, 185–198. [CrossRef]
37. Proust, G.; Tomé, C.N.; Kaschner, G.C. Modeling texture, twinning and hardening evolution during deformation of hexagonal materials. *Acta Mater.* **2007**, *55*, 2137–2148. [CrossRef]
38. Nie, J.-F. Precipitation and Hardening in Magnesium Alloys. *MTA* **2012**, *43*, 3891–3939. [CrossRef]
39. Lefebvre, W.; Kopp, V.; Pareige, C. Nano-precipitates made of atomic pillars revealed by single atom detection in a Mg-Nd alloy. *Appl. Phys. Lett.* **2012**, *100*, 141906. [CrossRef]
40. Dobromyslov, A.V.; Kaigorodova, L.I.; Sukhanov, v.d.; Dobatkina, T.V. Decomposition of a supersaturated solid solution in the Mg-3.3 wt% Yb alloy. *Phys. Met. Met.* **2007**, *103*, 64–71. [CrossRef]
41. Dobroň, P.; Chmelík, F.; Yi, S.; Parfenenko, K.; Letzig, D.; Bohlen, J. Grain size effects on deformation twinning in an extruded magnesium alloy tested in compression. *Scr. Mater.* **2011**, *65*, 424–427. [CrossRef]

Article

Hot Rolling of Magnesium Single Crystals

José Antonio Estrada-Martínez [1], David Hernández-Silva [1,*] and Talal Al-Samman [2]

[1] Department of Metallurgical Engineering, Instituto Politécnico Nacional-ESIQIE, 07738 Mexico City, Mexico; jestradam1202@alumno.ipn.mx
[2] Institute for Physical Metallurgy and Materials Physics, RWTH Aachen, 52056 Aachen, Germany; alsamman@imm.rwth-aachen.de
* Correspondence: dhernandez@ipn.mx

Abstract: To analyze the effect of the initial orientation in the activity of twinning and texture development, magnesium single crystals were rolled at 400 °C (nominal furnace temperature) in two specific orientations. In both orientations, the rolling direction of the sheet (RD) was parallel to the *c*-axis. For orientation 1, the $[11\bar{2}0]$ direction was parallel to the normal direction (ND), and for orientation 2, it was parallel to the $[10\bar{1}0]$ direction. The samples were rolled at 30%, 50% and 80% of thickness reduction. After rolling, all the samples were quenched in water to retain the microstructure. The microstructure and texture evolution were characterized by X-ray diffraction and Electron Backscatter Diffraction (EBSD). The initial single crystals were turned into polycrystals, where most grains had their *c*-axis almost parallel to the ND, and this reorientation was explained by extension twinning. The active twin variants in orientation 1 aligned the basal plane ~30° from the sheet plane and caused a weaker basal texture compared to orientation 2, where the twin variants aligned the basal plane almost parallel to the sheet plane. Strain localization inside contraction twins was observed, and consequently, non-basal grains nucleated inside these twins and weakened the final basal texture only in orientation 1.

Keywords: magnesium single crystal; texture; twinning

Citation: Estrada-Martínez, J.A.; Hernández-Silva, D.; Al-Samman, T. Hot Rolling of Magnesium Single Crystals. *Metals* **2021**, *11*, 443. https://doi.org/10.3390/met11030443

Academic Editor: Andrey Belyakov

Received: 30 January 2021
Accepted: 3 March 2021
Published: 8 March 2021

Publisher's Note: MDPI stays neutral with regard to jurisdictional claims in published maps and institutional affiliations.

1. Introduction

Magnesium has a hexagonal closely packed structure, which causes high mechanical anisotropy and low formability at room temperature, owing to the limited number of easily activate slip systems. In addition, it is well known that magnesium develops a basal texture after deformation [1,2]. The above has given rise to many investigations seeking to improve the mechanical properties and formability of this material, which requires a vast understanding of the deformation mechanisms present in the forming processes.

Studies on deformation of single crystals have several advantages compared to deformation of polycrystals. The use of specifically oriented single crystals with respect to the loading axes allow to isolate and identify the deformation mechanisms depending on the initial orientation of the crystal, since the operating deformation mechanisms, at least at early stages of deformation, can be determined using the Schmid's Law [3–6].

First studies on magnesium single crystals were performed by Wonziewicz et al. [5] and Kelly et al. [6]. In these studies, the crystals were subjected to plane strain compression (PSC) tests in different orientations, which showed profuse deformation by twinning and recrystallization. However, these investigations were limited to small strains and focused on the effect of temperature on the activity of the deformation mechanisms.

More recently, Chapuis et al. [7] evaluated the temperature dependency of critical resolved shear stresses (CRSS) of the slip and twinning systems of Mg single crystals by PSC tests and concluded that basal slip and $\{10\bar{1}2\}$ twinning are not temperature-dependent mechanisms, prismatic and pyramidal II slip were identified only at about 300 °C and contraction twinning systems $\{10\bar{1}1\}$ and $\{10\bar{1}3\}$ were identified as temperature-dependent systems because their CRSS decreased with the temperature increase.

Molodov et al. [8–14] characterized the deformation behavior and microstructure evolution of Mg single crystals in plane strain compression, using Electron Backscatter Diffraction (EBSD) and X-ray pole figure measurements. They focused on the activation of twinning and dynamic recrystallization mechanisms and its influence on the texture evolution.

Currently, no works on rolling of magnesium single crystals have been reported, and considering the contributions of the previously described works, single crystals could help us to understand the deformation behavior of magnesium during rolling deformation, which allows a certain broadening of the sample, contrary to PSC, where broadening is usually suppressed by the walls of the channel die. Therefore, the aim of this work is to provide a basic understanding of the deformation mechanisms involved in hot rolling of magnesium single crystals at 400 °C and its contribution to the final texture.

2. Materials and Methods

The single crystals of commercially pure magnesium (min. 99.95%) were fabricated by crystal growth using the Bridgman method in a vertical configuration. To manufacture monocrystalline rolling samples with their geometric axes coinciding with specific crystallographic directions, each grown single crystal was oriented by means of the Laue X-ray back diffraction method [15], then, slabs with dimensions of 40 mm × 20 mm × 6 mm were cut by electrical discharge machining (EDM). Figure 1 shows the schemes of the rolling samples and their orientation relationship with the crystal directions. The *c*-axis of the crystals were aligned parallel to the rolling direction in both orientations. In the sample labeled "orientation 1" (O-1), the normal direction (ND) was aligned parallel to $[11\bar{2}0]$ crystal direction and the transversal direction (TD) corresponds to $[10\bar{1}0]$. In the sample of "orientation 2" (O-2), the normal direction (ND) was aligned parallel to $[10\bar{1}0]$ crystal direction and the transversal direction (TD) corresponds to $[11\bar{2}0]$. The misalignment between the crystallographic directions and the specimen axes was less than 0.5°.

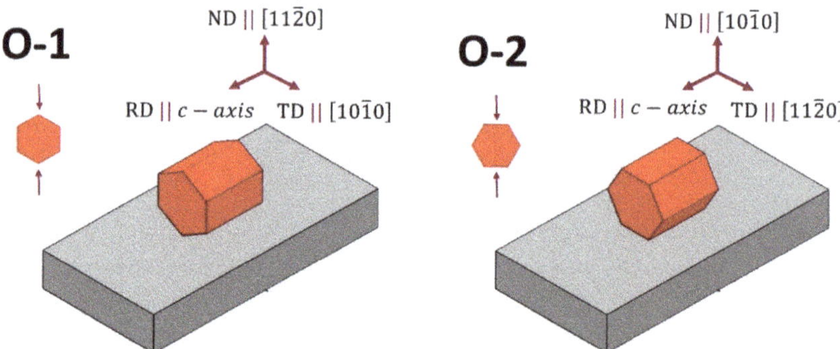

Figure 1. Single crystal orientations (O) for the rolling experiments.

It is well known that above about 250 °C, additional slip systems become operative in the deformation of pure magnesium and fulfill the Von Mises criterion, which states that in a polycrystalline material, five independent sliding systems must be activated to perform any possible deformation without fail. That is why the typical temperature range for magnesium rolling is between 300 and 500 °C [16–19]. In the current work, considering that the specimens were single crystals, and it might be a little harder to deform them successfully compared to polycrystalline material, the slabs of both orientations were heated at 400 °C for 30 min to ensure the deformation ability of the material and avoid cracking during deformation. Subsequently, each slab was rolled at 200 mm/s to ~30%, ~50% and ~80% of thickness reductions, using 7, 11 and 16 rolling passes, respectively. The complete rolling schedule is presented in Table 1. Between each rolling pass, the samples were returned to the furnace for reheating at 400 °C for 10 min. After the final pass, the sheets were quenched in water to freeze the microstructure and prevent post-mortem static recrystallization. For the microstructural and texture analysis, several samples were cut from the sheets mid-plane using EDM.

Table 1. Rolling schedule.

Rolling Pass	Deformation (φ)	Re-Heating Time	Thickness Reduction
1	0.05	10	4.88
2	0.05	10	9.52
3	0.05	10	13.93
4	0.05	10	18.13
5	0.05	10	22.12
6	0.06	10	26.66
7	0.06	10	30.93
8	0.10	10	34.95
9	0.10	10	41.14
10	0.12	10	47.80
11	0.12	10	53.70
12	0.15	10	60.15
13	0.15	10	65.15
14	0.17	10	71.06
15	0.17	10	75.59
16	0.20	–	80.01

The sample preparation for X-ray and EBSD measurements consisted of soft grinding with 2000 and 4000 grinding paper of silicon carbide during 10 min for each particle size, followed by 3 and 1 μm diamond suspension polishing for 15 and 20 min, respectively. After that, electropolishing in a solution of 125 mL ethanol and 75 mL H_3PO_4 was performed at 2.0 V, submerging the specimens in the electrolyte for 40 min. In the case of microstructural characterization by optical microscopy, a chemical color-etching with a freshly prepared solution of 10 mL H_2O, 10 mL CH_3-COOH and 70 mL picral 4% was applied.

To characterize the crystallographic texture, X-ray pole figure measurements were conducted in a Bruker D8 Advance diffractometer (Bruker, Billerica, MA, USA), equipped with a HI-STAR multi-wire area detector (Bruker, Billerica, MA, USA) with a circular beryllium window, operating at 30 kV and 25 mA. A set of six incomplete pole figures of the $\{10\bar{1}0\}$, (0002), $\{10\bar{1}1\}$, $\{10\bar{1}2\}$, $\{11\bar{2}0\}$ and $\{10\bar{1}3\}$ families were measured and used to determine the orientation density function (ODF) using the MTEX 5.1.1 toolbox [20].

Electron Backscatter Diffraction (EBSD) measurements were conducted on a JEOL JSM-6701F (JEOL Ltd., Tokyo, Japan) scanning electron microscope equipped with a field emission gun ($LaB6$ filament) and an HKL-Nordlys II EBSD detector (Oxford Instruments PLC, Abingdon, UK). An acceleration voltage of 30 kV was used for all measurements. The HKL Channel5 5.11 (Oxford Instruments PLC, Abingdon, UK) software was used for the post-processing of the raw data.

3. Results

The sheets of pure Mg single crystals rolled at 400 °C are shown in Figure 2, which reveal a smooth surface without visible cracks for both orientations. There were no significant changes in the final dimensions (as a broadening) between the sheets of two orientations, which could indicate small changes in the deformation behavior caused by the difference of the initial crystallographic orientation.

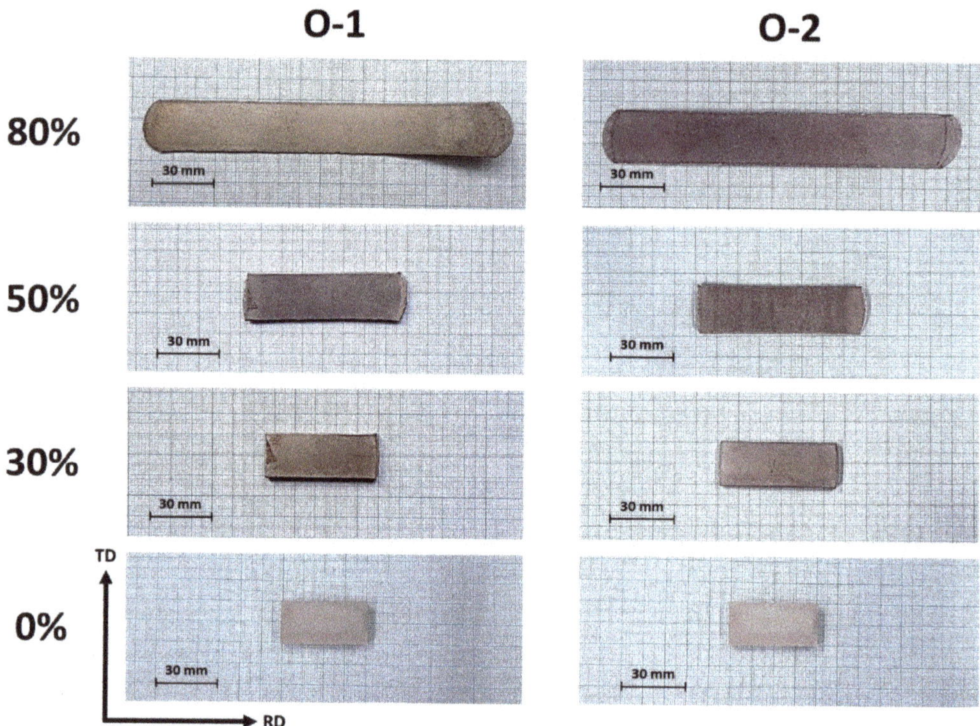

Figure 2. Pictures of magnesium hot-rolled sheets of orientation 1 and orientation 2 at 0% (raw material) 30%, 50% and 80% of thickness reduction.

3.1. Microstructure Evolution during Rolling Process

The microstructure of the rolled samples was examined by optical microscopy and the results are presented in Figure 3. The single crystal has been totally turned into a polycrystal of relatively large grains with a particular characteristic, featured by their straight grain boundaries. Thus, they can be described as polygonal grains. In both orientations, many twins and recrystallized twin bands, consisting of fine dynamic recrystallized grains along the twin bands, were observed. The network of twins distributed homogeneously along the microstructure formed symmetric sets, and an inclination angle of ~60° can be observed between them.

At 30% of reduction, many lenticular twins were observed, in O-1 as well as in O-2. In the case of O-2, some recrystallized grains inside the twins can be observed, suggesting the occurrence of dynamic recrystallization (DRX) process. In the O-2, coarser grains compared to O-1 can be observed, but this feature is not so evident in subsequent reductions.

With further deformation (at 50%), the microstructure of O-1 was characterized by the presence of more and finer twins compared to 30%. In O-2, fewer twins were observed, but in this case, they were coarser than those observed in this orientation at 30% and tended to form clusters. In both orientations, fine grains were observed inside the twins and some twin-free grains were identified.

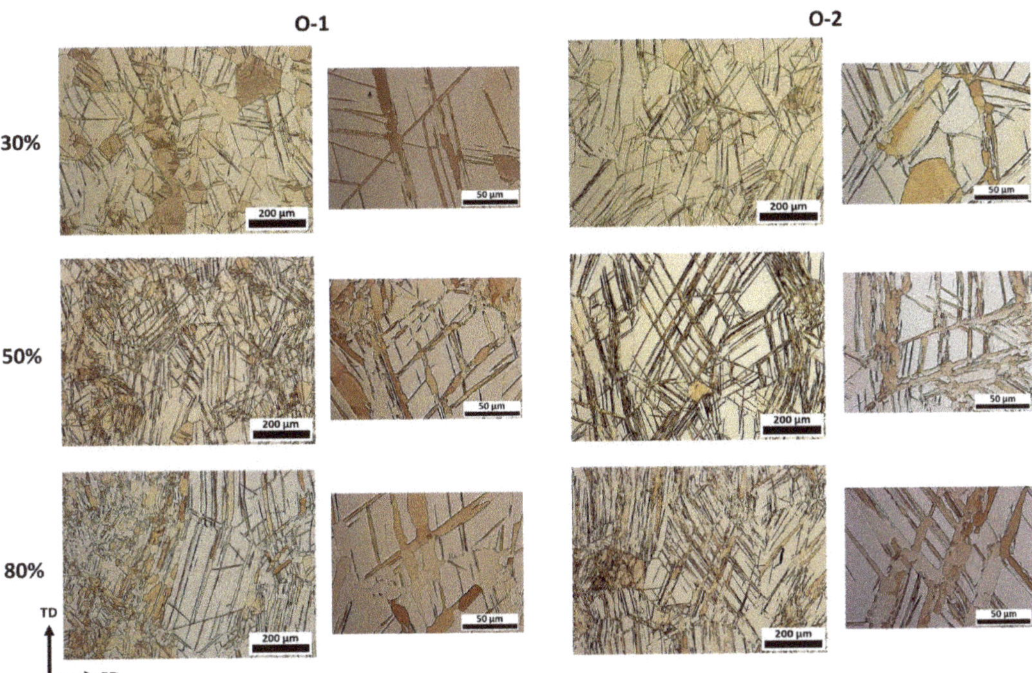

Figure 3. Optical micrographs at 200× and 500× magnifications of the rolling direction (RD)-transversal direction (TD) plane after magnesium single crystal rolling at different strains of both studied orientations.

Finally, at $\varepsilon = 80\%$, the O-1 exhibited a partially recrystallized and heterogeneous microstructure described by two zones. In the first, to the left of the micrograph, recrystallized grains almost consumed all the matrix (there was a grain growth considering the size of recrystallized grains observed in previous strains). In the second, at the center, some recrystallized and coarse twin bands as well as recrystallization-free twin bands were found, suggesting that the grains initially recrystallize within the twins, grow, and consume the surrounding matrix.

It can be observed that the fraction of recrystallized grains increases as the rolling progresses, especially inside the twins, and consumes them (O-1 at 80%). These observations indicate that the DRX process plays a dominant role in the microstructure evolution during rolling, giving rise to new grains and, in consequence, new orientations that could modify the texture, and it is interesting to analyze them.

3.2. Texture Evolution during Hot Rolling

Texture measurements were carried out along the sheets and the average texture was reported by means of pole figures (PF) in Figure 4. For O-1 at the early stage of 30%, the (0002) pole figure exhibits a strong basal texture with an intensity value of 17.8. On the other hand, the O-2 presents a stronger basal texture with a maximum of 23. The corresponding $\{10\bar{1}0\}$ and $\{2\bar{1}\bar{1}0\}$ pole figures of both orientations reveal that the prismatic plane distributes homogeneously along the transversal and rolling direction of the sheet. Other components can be clearly detected at the center of these; in the $\{10\bar{1}0\}$, a weak pick indicates that some prismatic planes are parallel to the ND of the sheet. A texture fiber component can be observed at ~30° from the ND in this PF, which is the same component shown by the weak peak at ND of the $\{2\bar{1}\bar{1}0\}$ pole figure.

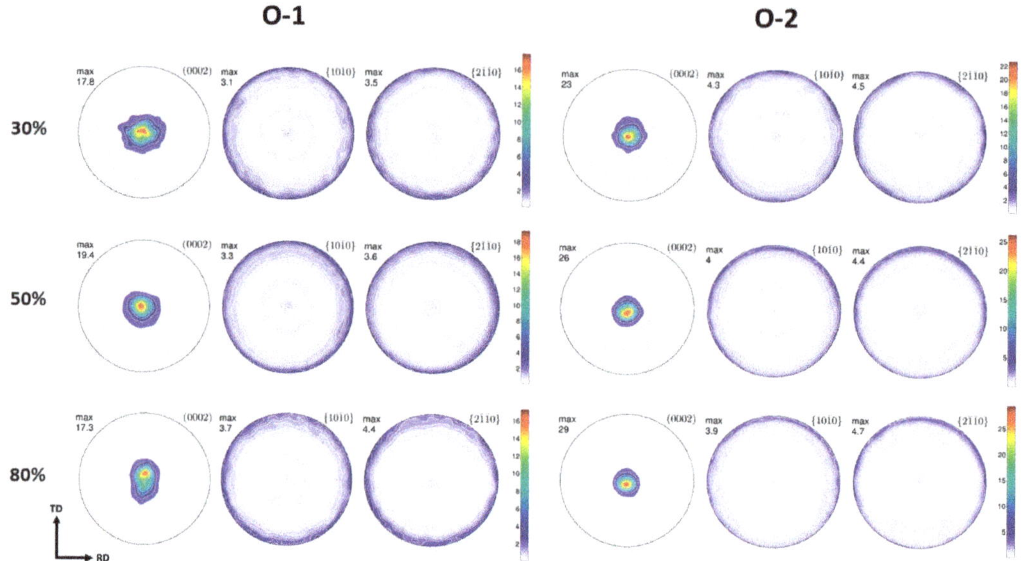

Figure 4. Pole Figures of texture measurements of magnesium sheets at different thickness reductions for both orientations.

With increasing strain at 50% of reduction thickness, in the O-1 texture, the basal texture intensity became strengthened and reached a value of 19.4. Meanwhile, (0002) PF of O-2 shows an intensity of 26. The increment in the intensity is slightly higher for O-2 than for O-1 with respect to the PF at 30%. Regarding the $\{10\bar{1}0\}$ and $\{2\bar{1}\bar{1}0\}$ pole figures, the same components described at 30% can also be observed at 50%.

At the final reduction thickness of 80%, the (0002) PF of O-1 reveals the emergence of a double-pick component slightly deviated from ND with a maximum of 17.3, weakening the basal texture compared with previous stages. In the case of O-2, the maximum of the basal PF increased until 29.0 and no changes in the components that have already been described in previous deformations were found.

According to the microstructure evolution, it is certain that the texture development is strongly associated with the changes on the microstructure during rolling. In general, a strong basal texture with a non-preferential orientation of the a-axis was observed in both orientations since 30% until 80% of thickness reduction, but this basal texture is weaker in O-1 than in O-2. Also, at 80% in O-1, there is a decrease in the intensity of basal texture and the presence of a new double-pick component, the opposite to O-2, where there is always a strengthening of the texture and the components stay the same.

3.3. EBSD Measurements

The microstructure evolution was characterized by EBSD to reveal the twin types during magnesium single crystal rolling. The orientation maps are presented in terms of inverse pole figure coloring with respect to the ND. A step size of 0.7 μm was used for all measurements. High-angle grain boundaries (HAGBs, >15°) were highlighted in black and low-angle grain boundaries (LAGBs, 2–15°) in gray. Black areas show remaining zero solutions (non-indexed EBSD patterns) correlating with high local strain areas. Noise reduction was carefully applied to improve the original indexing using a minimum of five neighbors. In the twin boundary misorientation map, the types of twins were determined based on their particular misorientation angles and rotation axes [21]. The deviation of angle and axis to identify the twin boundaries was within 5° of the ideal values. It is important to mention that the Schmid Factor (SF) maps and misorientation profiles will be analyzed in Section 4.

Figure 5 shows the EBSD results of the interrupted rolling at 30% of reduction of O-1. The original indexing for the mapping was 87% and was improved to 97% by noise reduction, and the map size consists of a grid of 818 × 1063 points. The area fraction of extension twins was 0.019 and 0.048 for contraction twins. In the orientation map (Figure 5a), yellow squares point to high-angle grain boundaries (HAGBs), and these belong to grains that recrystallize at the triple junctions (TJ). Blue boxes enclose low-angle grain boundaries that bulged at the triple junctions (LAGBs-TJ). The nucleation of new grains at the triple junction could be related to the discontinuous dynamic recrystallization (DDRX) mechanism occurrence [22], but it is not the purpose of this paper to address this analysis. Meanwhile, green boxes point out zones where it is possible to observe the transition from a LAGB to HAGB (gray-grain boundary to black-grain boundary), but not at the triple junctions. Cyan rectangles enclose LAGBs that completely cross an extension twin, indicating that the dislocations can slip into the parent grain, completely cross the twin, and move onto the parent grain again.

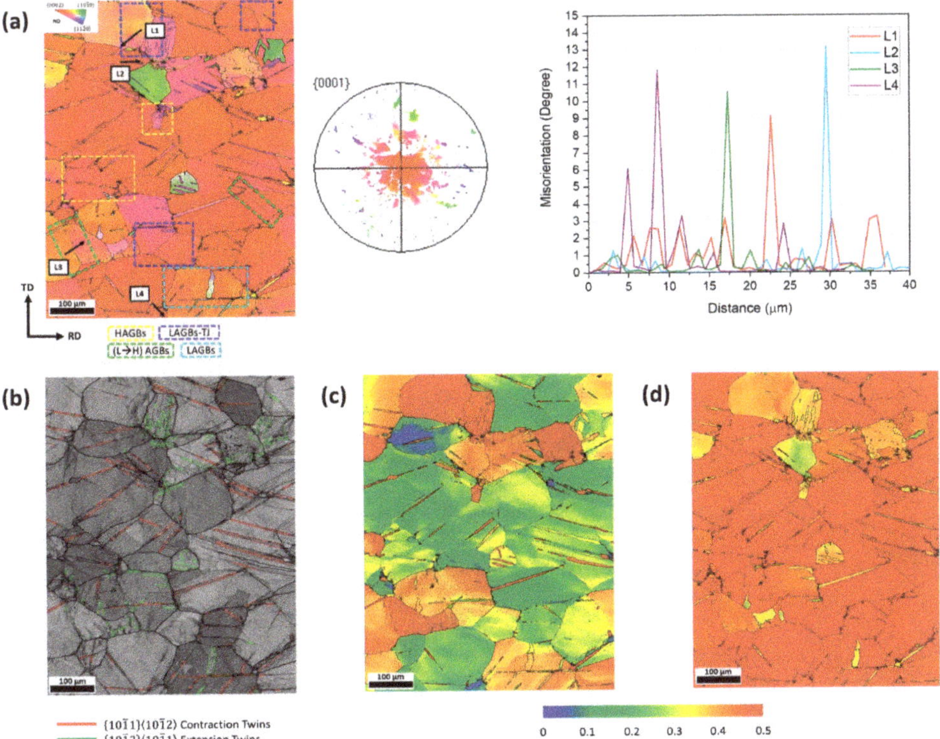

Figure 5. Electron Backscatter Diffraction (EBSD) results obtained from an O-1 specimen at 30% of thickness reduction: (**a**) EBSD map in inverse pole figure (IPF) coloring relative to normal direction (ND), corresponding scattered data (0002) PF and misorientation profiles, (**b**) twin boundary misorientation map, (**c**) Schmid Factor map for $(0002)\langle11\bar{2}0\rangle$ slip system, (**d**) Schmid Factor map for $\{10\bar{1}1\}\langle11\bar{2}3\rangle$ slip system.

The corresponding orientation map of O-2 at 30% is presented in Figure 6: the original indexing was 88% and was improved to 97% by noise reduction, and the map size consists of a grid of 545 × 926 points. The area fraction of extension twins was 0.009 and for contraction twins it was 0.076. In the orientation map (Figure 6a), yellow squares point to high-angle grain boundaries (HAGBs), and these belong to grains that recrystallize at the triple junctions (TJ). Blue boxes enclose low-angle grain boundaries that bulged at the triple junctions (LAGBs-TJ). Cyan rectangles enclose nets of LAGBs that do not crowd at triple junctions. Green frames show zones where the transition from a LAGB to HAGB can be observed. As in O-1, the new matrix has basal-oriented grains, and consequently, compression twins were formed inside. Compared to O-1, more contraction twins (CTs) were observed (Figure 6b), and this can be explained if we analyze the EBSD map where grains of more solid red coloration were observed, i.e., the basal plane in this orientation was aligned almost parallel to the sheet, rendering the *c*-axis almost parallel to the compression load induced by rolling. This can also be observed in the PFs of Figure 4, where the intensity of the basal component of O-2 is higher and sharper than O-1.

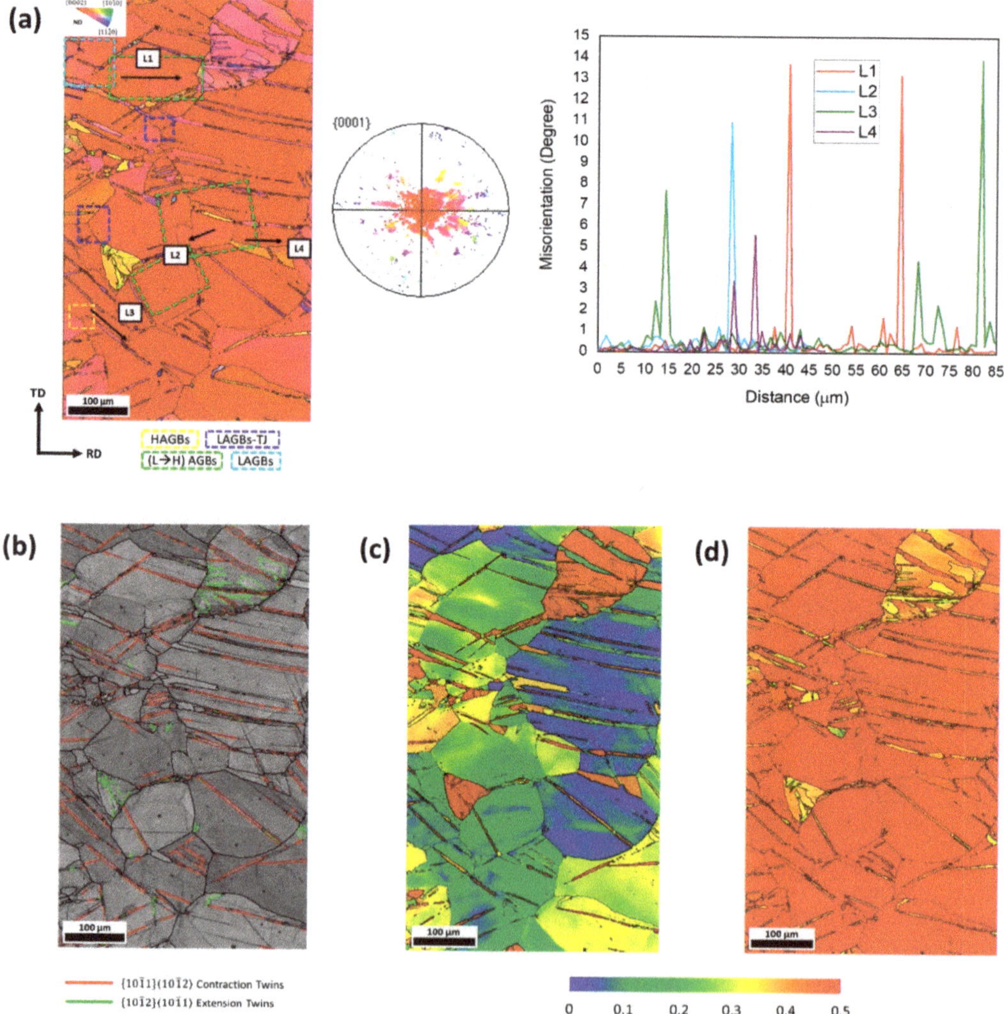

Figure 6. Electron Backscatter Diffraction (EBSD) results obtained from an O-2 specimen at 30% of thickness reduction: (**a**) EBSD map in inverse pole figure (IPF) coloring relative to normal direction (ND), corresponding scattered data (0002) PF and misorientation profiles, (**b**) twin boundary misorientation map, (**c**) Schmid Factor map for $(0002)\langle 11\bar{2}0\rangle$ slip system, (**d**) Schmid Factor map for $\{10\bar{1}1\}\langle 11\bar{2}3\rangle$ slip system.

Figure 7 shows the EBSD result of O-1 at 50%, where the original indexing of the map was 78% and was improved to 84% by noise reduction, and the map size consists of a grid of 505 × 788 points. The area fraction of extension twins of this map was 0.008 and 0.07 for contraction twins. Twin bands embedded in a 'hard' basal orientation matrix (*c*-axis nearly parallel to ND) were observed. The orientation map of Figure 7a shows thicker twin bands (some of them were recrystallized) and more low-angle grain boundaries (LAGBs) than those observed at 30%. Yellow boxes point out grains that recrystallize inside the contraction twin. There are many areas with zero solution along the bands of contraction twins, which indicate a high deformation inside them. Despite the low index ratio in these zones, fine grains can be observed inside the twin-like morphologies derived from the high concentration of plastic deformation within the twins and its subsequent partial recrystallization. Cyan rectangles enclose nets of LAGBs that do not crowd at triple junctions.

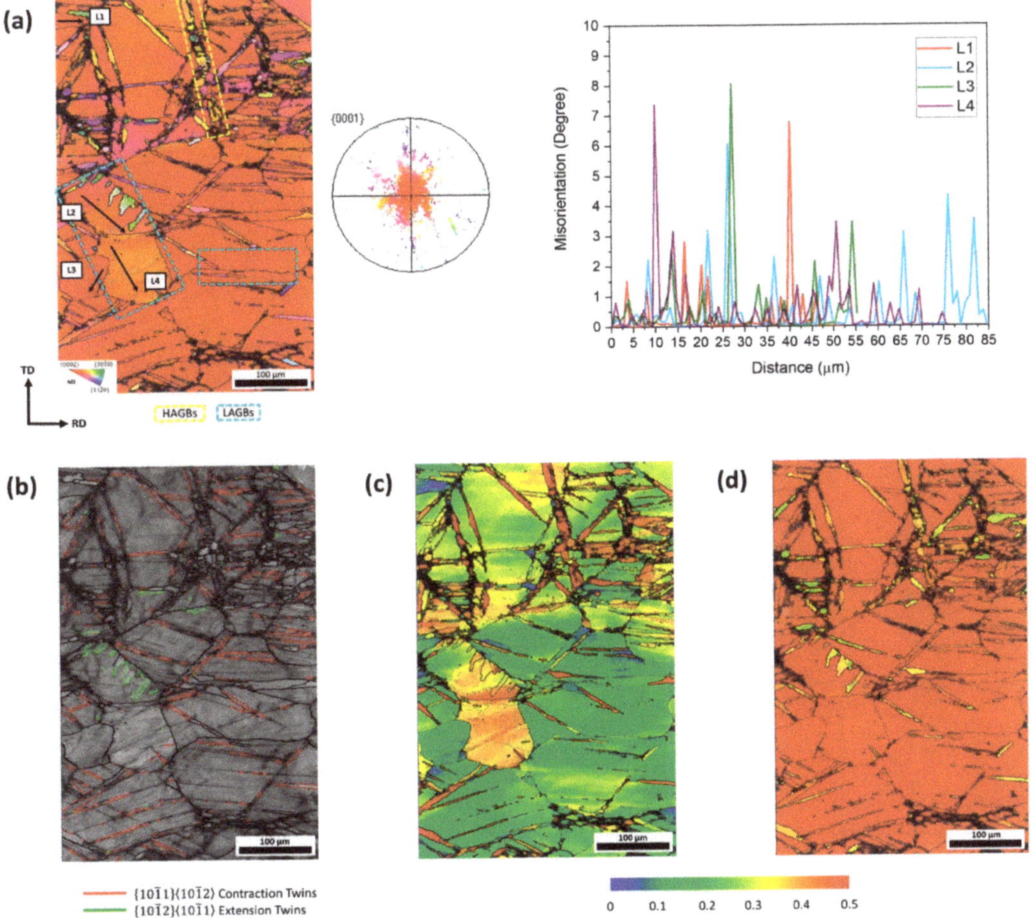

Figure 7. Electron Backscatter Diffraction (EBSD) results obtained from an O-1 specimen at 50% of thickness reduction: (**a**) EBSD map in inverse pole figure (IPF) coloring relative to normal direction (ND), corresponding scattered data (0002) PF and misorientation profiles, (**b**) twin boundary misorientation map, (**c**) Schmid Factor map for $(0002)\langle11\bar{2}0\rangle$ slip system, (**d**) Schmid Factor map for $\{10\bar{1}1\}\langle11\bar{2}3\rangle$ slip system.

In the orientation map of O-2 at 50%, shown in Figure 8, the original indexing for the mapping was 81% and was improved to 88% by noise reduction, and the map size consists of a grid of 511 × 743 points. The area fraction of extension twins of this map was 0.12 and 0.018 for contraction twins. In the orientation map (Figure 8a), yellow squares point out high-angle grain boundaries (HAGBs) that belong to recrystallized grains subdividing the pre-existing grains. Cyan rectangles enclose nets of LAGBs that do not crowd at triple junctions. Green frames show zones where the transition from a LAGB to HAGB can be observed. Also, in this case, most of the grains have the basal plane almost parallel to the sheet plane and therefore many lenticular $\{10\bar{1}1\}\langle10\bar{1}2\rangle$ twins were located inside, but were thinner than those observed in O-1, and recrystallized grains are not clearly observed inside the twinned area.

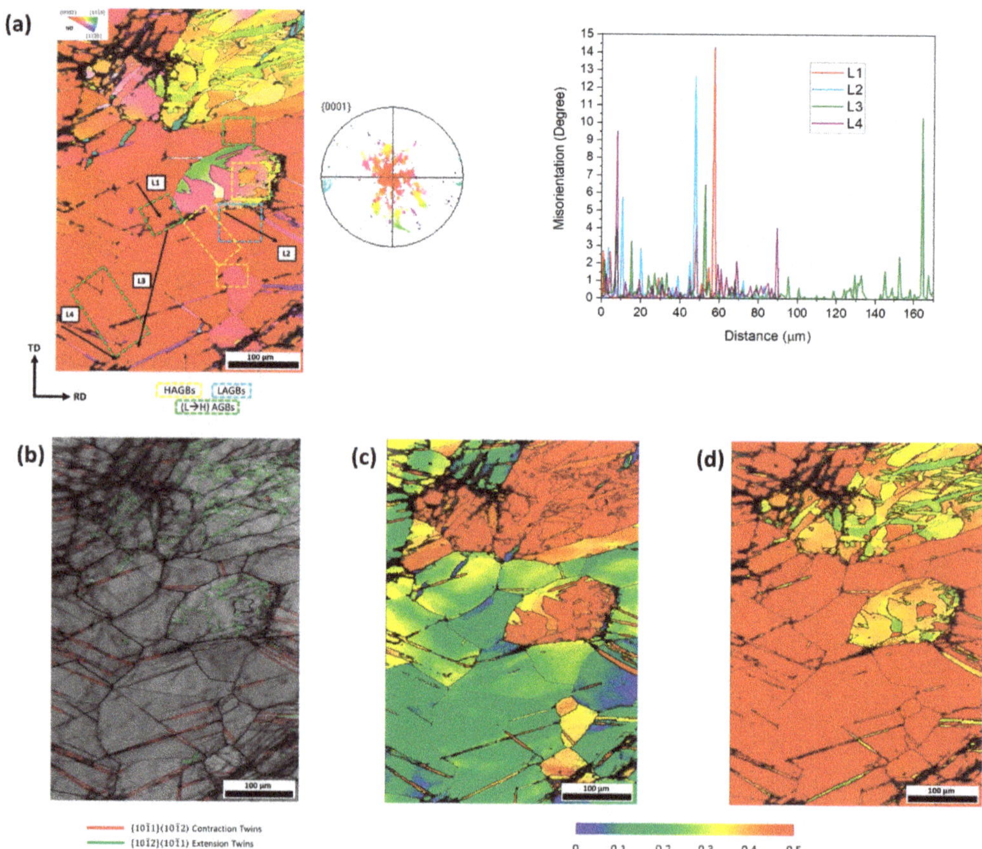

Figure 8. Electron Backscatter Diffraction (EBSD) results obtained from an O-2 specimen at 50% of thickness reduction: (**a**) EBSD map in inverse pole figure (IPF) coloring relative to normal direction (ND), corresponding scattered data (0002) PF and misorientation profiles, (**b**) twin boundary misorientation map, (**c**) Schmid Factor map for $(0002)\langle11\bar{2}0\rangle$ slip system, (**d**) Schmid Factor map for $\{10\bar{1}1\}\langle11\bar{2}3\rangle$ slip system.

Figure 9 shows an EBSD map for O-1 at 80% of thickness reduction. The original indexing for the mapping was 89% and was improved to 98% by noise reduction, and the map size consists of a grid of 850 × 1145 points. The microstructure is not homogenous, and a typical bimodal microstructure consisting of fine and coarse grains was observed. The coarse grains belong to the basal matrix observed in the previous strains. On the other hand, most of the fine grains were originated by the DRX process inside the twins. Practically, all grains have a basal orientation or very close to it. However, some grains near to $\langle 10\bar{1}1 \rangle$, $\langle 10\bar{1}3 \rangle$, $\langle 2\bar{1}\bar{1}1 \rangle$, $\langle 2\bar{1}\bar{1}2 \rangle$ and $\langle 2\bar{1}\bar{1}4 \rangle$ components can also be observed inside the twin-like morphologies, although in a much smaller proportion. The map does not show twinning activity despite the same temperature of deformation and similar texture compared to 30% and 50%, but an evident smaller grain size, indicating the susceptibility of the twinning to the grain size compared to other parameters.

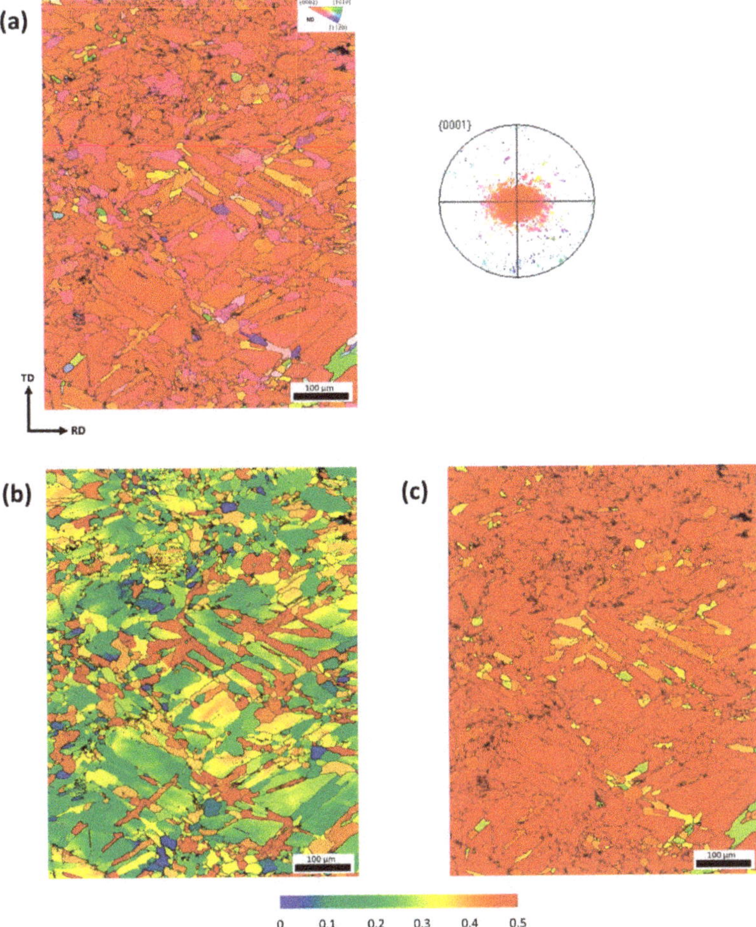

Figure 9. Electron Backscatter Diffraction (EBSD) results obtained from an O-1 specimen at 80% of thickness reduction: (**a**) EBSD map in inverse pole figure (IPF) coloring relative to normal direction (ND) and corresponding scattered data (0002) PF, (**b**) Schmid Factor map for $(0002)\langle 11\bar{2}0 \rangle$ slip system, (**c**) Schmid Factor map for $\{10\bar{1}1\}\langle 11\bar{2}3 \rangle$ slip system.

Finally, Figure 10 shows the EBSD map of O-2 at reduction of 80%, where the original indexing for the mapping was 71% and was improved to 78% by noise reduction, and the map size consists of a grid of 375 × 622 points. This presents partially recrystallized twin bands, leading to grains with orientations slightly deviated from $\langle 10\bar{1}0 \rangle$, $\langle 10\bar{1}1 \rangle$, $\langle 10\bar{1}2 \rangle$, $\langle 2\bar{1}\bar{1}0 \rangle$ and $\langle 2\bar{1}\bar{1}4 \rangle$ components. These bands could also be grains elongated by deformation and then recrystallized, since some of them were observed parallel to the RD in particular cases, however due to their morphology (some bands are almost perpendicular to the RD), they were considered as recrystallized twins. There is still a prominent matrix with a basal orientation, which has many LAGBs, but any twin could be found inside.

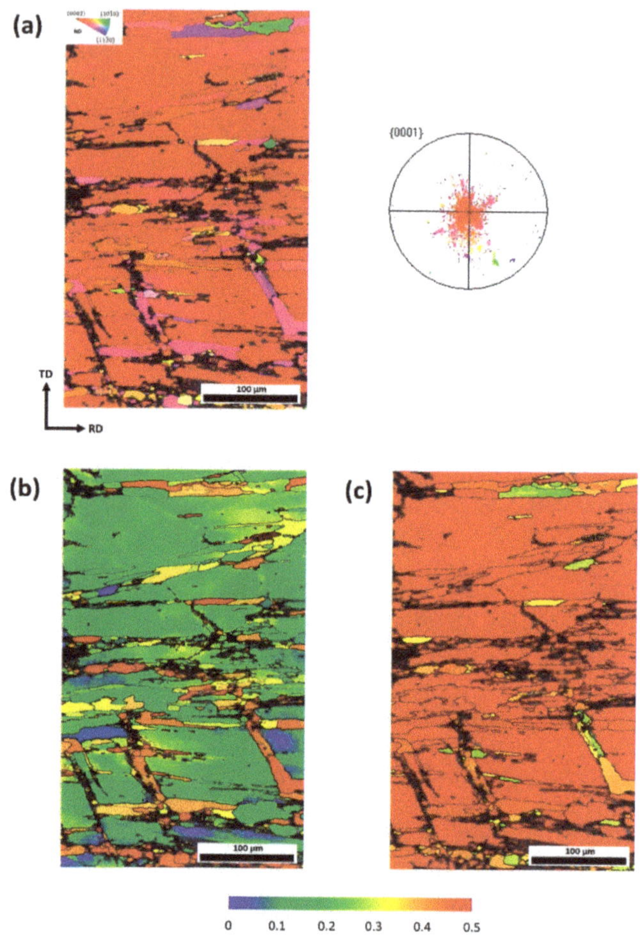

Figure 10. Electron Backscatter Diffraction (EBSD) results obtained from an O-2 specimen at 80% of thickness reduction: (**a**) EBSD map in inverse pole figure (IPF) coloring relative to normal direction (ND) and corresponding scattered data (0002) PF, (**b**) Schmid Factor map for $(0002)\langle11\bar{2}0\rangle$ slip system, (**c**) Schmid Factor map for $\{10\bar{1}1\}\langle11\bar{2}3\rangle$ slip system.

4. Discussion

4.1. Extension Twinning and Its Influence on the Basal Texture

There are different reports in the literature regarding the twinning activity during deformation of magnesium. Some works report that this activity is negligible during deformation at high temperatures [23–25], while others report that twinning is active at high deformation rates regardless of deformation temperature [5,26]. There are also other works that relate twinning activity at high temperatures to large grain sizes [27,28]. In the current work, twinning was an important deformation mechanism, where both high deformation temperature and very large grain size (single crystal) are investigated.

In O-1, the starting orientation was aligned for $\{10\bar{1}2\}$ tensile twinning (Schmid factor of 0.37, compression perpendicular to *c*-axis). Blue areas from the original orientation

($\{11\bar{2}0\}$) can still be observed but in a very marginal proportion, and it is important to note that most of these blue areas correspond to $\{10\bar{1}2\}$ extension twins (ETs) according to the twin boundary misorientation map (Figure 5b), and according to the (002) PF of scattered data, these areas appear close to the RD, as in the case of initial orientation, which could indicate that these zones are leftovers of initial single crystal. In this way, we can explain how the single crystal was almost completely rotated from $\langle 11\bar{2}0 \rangle // ND$ (blue grains) to $\langle 0001 \rangle // ND$ (red grains) by means of nucleation and growth of extension twinning.

The twinned red grains now constitute a new matrix wherein contraction twinning was favorable. This is owing to the c-axis being at 30° from ND (Schmid factor of 0.49, compression load almost parallel to c-axis). At this strain, according to the orientation map, most of the grains have basal planes deviated ~30° from the sheet plane (red-orange grains), so they were subject to compression on the c-axis during rolling; consequently, they show compression twins inside.

As the O-1, the O-2 was aligned for extension twinning (SF = 0.49, compression perpendicular to c-axis), but the active variants according to Schmid's Law were slightly different in each case. The initial orientation 2 must be colored in green according to the standard IPF, but no grains with this characteristic were detected.

It is possible that $\{10\bar{1}2\}$ twinning explains this lattice rotation from the c-axis parallel to RD to the c-axis almost parallel to ND, and the consequent basal texture in both orientations. The activation of $\{10\bar{1}2\}$ twinning depends on the strain path, compression perpendicular to the c-axis or tension parallel to the c-axis, and their activity is governed by the Schmid law. Usually, the twin variant with the higher Schmid Factor (SF) takes place favorably [29–33], but there are, of course, cases where twinning does not obey a Schmid law [11,34,35]. Therefore, in the following, the $\{10\bar{1}2\}$ twinning behavior was analyzed with respect to the loading and crystal orientation. Its corresponding effect on the evolution of grain orientation was also discussed.

Under rolling, the initial single crystals with the c-axis parallel to the RD were favorably aligned for $\{10\bar{1}2\}$ twinning (compression perpendicular to the c-axis). Theoretically, $\{10\bar{1}2\}$ twinning is possible on six twin variants, $(10\bar{1}2)\,[\bar{1}011]$, $(01\bar{1}2)\,[0\bar{1}11]$, $(\bar{1}102)\,[1\bar{1}01]$, $(\bar{1}012)\,[10\bar{1}1]$, $(0\bar{1}12)\,[01\bar{1}1]$ and $(1\bar{1}02)\,[\bar{1}101]$ [29–31]. Figure 8a,b shows basal pole figures depicting the calculated $\{10\bar{1}2\}$ extension twin orientations for the initial specimen orientations. To explain the lattice rotation during single crystals' rolling, the following can be assumed: $\{10\bar{1}2\}$ twins gradually consumed its parent grain (single crystal) and reoriented it by ~86.3°, with basal planes almost parallel to the RD since the early stage of deformation, contributing to the formation of the basal texture but in two very different ways:

For O-1, $(01\bar{1}2)\,[0\bar{1}11]$, $(\bar{1}102)\,[\bar{1}101]$, $(\bar{1}102)\,[1\bar{1}01]$ and $(0\bar{1}12)\,[01\bar{1}1]$ variants have a SF value of 0.37, as shown in Figure 11a, making them equally active considering the Schmid law. These twin variants transformed the matrix into twin orientations with the c-axis aligned at an angle of ~30° from ND, as shown in Figure 11c. These orientations have a non-zero Schmid factor for basal, prismatic and pyramidal slip, but $\{10\bar{1}1\}$ twinning continues to have higher SF. According to Chapuis et al. [7], the CRSS for contraction twins is estimated between 30 and 150 MPa, and it is therefore higher compared to CRSS of basal (1.5–3.6 MPa), prismatic (6–15 MPa) and pyramidal slip (20–25 MPa), and the variations depend on the temperature. Despite the high CRSS (but also a high SF), contraction twins are visible in the twin boundary misorientation map (Figure 5b).

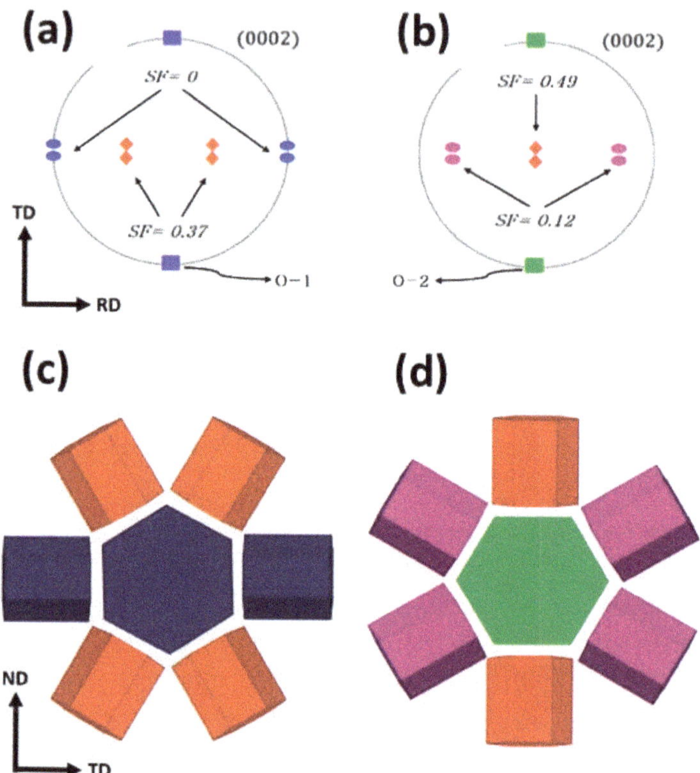

Figure 11. (0002) calculated PF showing the initial orientations (**a**) O-1 and (**b**) O-2, as well as their corresponding extension twin variants. Schematic representation of twin variants for (**c**) O-1 and (**d**) O-2.

On the other hand, regarding O-2, only $(10\bar{1}2)[\bar{1}011]$ and $(\bar{1}012)[10\bar{1}1]$ variants have a SF = 0.49 (Figure 11b) and lay the basal plane almost parallel to the sheet plane, which strengthens the basal texture (Figure 11d), converting the initial matrix into a 'hard' orientation, in which slip was suppressed. According to Molodov [36], the specimens under plane strain compression (PSC) with the starting orientation D (O-2) were again almost single-crystalline, with only low-angle grain boundaries after the conversion of the initial matrix by twinning, so, the crystal rotation from $\langle 11\bar{2}0 \rangle // ND$ to $\langle 0001 \rangle // ND$ by extension twinning is not baseless.

As a result of this reorientation, the intensity of the basal component in the O-2 was higher than O-1. The broadening along the RD in the texture component of (0002) PF for O-1 (Figure 4) points out that these variants: $(01\bar{1}2)[0\bar{1}11]$, $(1\bar{1}02)[\bar{1}101]$, $(\bar{1}102)[1\bar{1}01]$ and $(0\bar{1}12)[01\bar{1}1]$, accommodated the basal plane ~30° rotated from the sheet plane according to the predicted poles of high SF twin variants (Figure 11a). Unlike orientation 2, where the basal component of (0002) PF is sharper (Figure 4) and again agrees with the predicted pole figures (Figure 11b).

4.2. Shear Localization on $\{10\bar{1}1\}$ Twins and Their Influence on the DRX

At 30% of thickness reduction, the misorientation profiles along the black arrows in the EBSD maps (Figures 5 and 6) reveal a continuous change of orientation inside the grains. At some points, this misorientation reaches 13°, revealing that dislocations were also active in addition to the twinning. In O-1 (Figure 5a), the line profiles L1 and

L3 are inside grains with high SF for basal slip according to Figure 5c, meanwhile L2 and L4 cross areas with low SF for basal slip; however, these show a high misorientation, which indicates that not only the basal slip was active. The maps of SF for other slip systems of <a> type dislocations are not shown because the SF was close to zero in these maps. The slip system that shows a high SF inside the basal orientation grains was $\{10\bar{1}1\}\langle11\bar{2}3\rangle$ (Figure 5d) and this is totally in accordance with the study of Kelvin et al. [37], where the first-order (pyramidal-I) slip $\{10\bar{1}1\}\langle11\bar{2}3\rangle$ was presumably the more dominant slip system in *c*-axis compression at room temperature, with a CRSS of 54 MPa. In O-2 (Figure 6a), a similar behavior was observed, but in this case, all the line profiles (L1–L4) are inside grains with low SF for basal slip (Figure 6c), despite this, they reach misorientations of ~14°.

As mentioned above, the orientation maps of O-1 and O-2 at 50% (Figures 7a and 8a, respectively) show areas with high concentration of deformation and consequently, they were not indexed. From the quality of the EBSD patterns of the matrix, we can deduce that these have not been deformed at the same magnitude as twin bands, implying that deformation is heavily concentrated in the softer twins. If we analyze the SF maps for both orientations (Figure 7c,d and Figure 8c,d), we can observe that basal slip has a high SF inside the contraction twins opposite to first-order pyramidal-I, which has a high SF in the basal-oriented grains. Considering the reported values of CRSS of slip basal ~2 MPa [7] and first-order pyramidal-I slip ~54 MPa [37], it is logical to expect that the basal slip will be the dominant mechanism due to its low CRSS, causing the high concentration of deformation within the twin bands. The line profiles L2 and L4 in Figure 7a are inside grains with high SF for basal slip according to Figure 7c, meanwhile L2 and L4 in Figure 7a and L1–L4 in Figure 8a cross areas with low SF for basal slip, however, these reach a high misorientation of 8° and 14° respectively, indicating the activation of the $\{10\bar{1}1\}\langle11\bar{2}3\rangle$ slip system.

In O-1 at 80% of thickness reduction (Figure 9a), the morphological features of recrystallized regions revealed band arrangements embedded within un-recrystallized zones. These kind of arrangements seem to be recrystallized twin bands (DRXed-TBs), and it is evident that the grains inside them have a non-basal orientation, unveiling the effect of contraction twins (CTs) to nucleate them. The rest of the microstructure (out of the bands) shows that most of the recrystallized grains retain a basal orientation. In large grains that presumably belong to the initial basal orientation matrix, LAGBs can be observed, and it suggests that these kind of grains were divided into recrystallized small grains that retain their basal orientation. The many LAGB and SF maps (Figure 9b,c) suggest that basal slip is active inside the grains with non-basal orientation (inside twin bands) and first-order pyramidal-I slip inside the basal-oriented grains of the matrix.

In the case of O-2 at 80% of thickness reduction (Figure 10), there was not a grain refinement, the recrystallized twin bands (DRXed TBs) were coarser and show grains with a large spread orientation (away from basal orientation), allowing the basal slip. As in O-1, there are many LAGBs and sub-grains inside the matrix and recrystallized grains. SF maps (Figure 10b,c) suggest that basal slip is active inside the grains with non-basal orientation, i.e., recrystallized grains inside twin bands and the first-order pyramidal-I slip system inside the basal matrix.

It has been widely reported that $\{10\bar{1}1\}$ twins act as nucleation sites for DRX due to localized strain accumulation within the twin owing to the slip activity inside twin bands and difficult mobility of the twin boundaries, which is referred to as twin-induced dynamic recrystallization. This accumulation of energy produced dynamic recovery in these zones until the misorientation between the low-angle grain boundaries, by more than 15°, and became high-angle grain boundaries and acquired a definitive non-basal orientation [8,25–28,38–40]. Nevertheless, the grains that nucleate inside $\{10\bar{1}1\}$ twins generally cannot grow outside the twin boundaries, so the efficacy of the twin-induced dynamic recrystallization mechanism on the final texture weakening is limited [26,27].

In this investigation, the contraction twin bands in the O-1 were thicker than O-2 at all thickness reductions. At 80%, the orientation map of O-1 (Figure 9a) clearly shows a net of twin-like band morphologies, as described above. If we compare this feature with the one observed in the orientation map of O-2 (Figure 10a), we can easily notice that there is a greater amount of twin-like band morphologies in O-1 than in O-2, which suggests a profuse compression twinning in O-1 during the last rolling step. The recrystallized grains that nucleate inside the contraction twin band without a non-basal orientation seem to have grown slightly out of the twin boundary in O-1, contrary to O-2, where no profuse twin-like morphologies and recrystallized grains inside them can be observed. This explains very well the weakening in the basal texture of O-1 at 80% of thickness reduction, where the recrystallized grains with non-basal orientation contributed to weakening the basal texture from 19.4 to 17.3 multiples of a random distribution (m.r.d,) only in O-1, and gave rise to a new double-pick component in the PF (Figure 4).

5. Conclusions

In the present study, the deformation mechanisms of commercially pure magnesium single crystal during rolling at 400 °C of 30%, 50% and 80% thickness reduction were observed. The microstructural and textural development was investigated by means of EBSD. The following conclusions were drawn:

(1) The original single crystal of both orientations was almost completely rotated into a new matrix of basal grains with the c-axis almost parallel to ND of the sheet. Observations of extension twinning and Schmid Factor analysis indicate that $\{10\bar{1}2\}$ twinning was responsible for this rotation.

(2) The active extension twin variants determined the intensity of the basal texture: orientation 1 presented the weakest basal texture at all strains, its extension twin variants aligned the c-axis ~30° deviated from ND, compared to orientation 2, for which its extension twins aligned the basal plane almost parallel to the sheet plane.

(3) In the new basal matrix, contraction twins were observed at ε of 30% and 50% in both orientations. These contraction twins suffered shear localization attributed to easier conditions for dislocation basal slip motion inside them compared to the 'hard' basal matrix, leading to twin-induced CDRX, and formed recrystallized twin bands, giving rise to new recrystallized grains with non-basal orientation. These recrystallized grains contributed to the weakening of the basal texture only in orientation 1.

Author Contributions: Conceptualization, J.A.E.-M. and T.A.-S.; methodology, J.A.E.-M.; software, J.A.E.-M.; validation, D.H.-S. and T.A.-S.; formal analysis, J.A.E.-M. and D.H.-S.; investigation, J.A.E.-M.; resources, T.A.-S. and D.H.-S.; data curation, J.A.E.-M.; writing—original draft preparation, J.A.E.-M.; writing—review and editing, D.H.-S. and T.A.-S.; visualization, J.A.E.-M.; supervision, D.H.-S. and T.A.-S.; project administration, D.H-S. and T.A.-S.; funding acquisition, D.H.-S. and T.A.-S. All authors have read and agreed to the published version of the manuscript.

Funding: This research received no external funding.

Data Availability Statement: The data presented in this study are available on request from the corresponding author. The data is not publicly available due to ongoing research based on it.

Conflicts of Interest: The authors declare no conflict of interest.

References

1. Xin, Y.; Wang, M.; Zeng, Z.; Huang, G.; Liu, Q. Tailoring the texture of magnesium alloy by twinning deformation to improve the rolling capability. *Scr. Mater.* **2011**, *64*, 986–989. [CrossRef]
2. Agnew, S.R.; Duygulu, Ö. Plastic anisotropy and the role of non-basal slip in magnesium alloy AZ31B. *Int. J. Plast.* **2005**, *21*, 1161–1193. [CrossRef]
3. Kelley, E.W.; Hosford, W.F. The Deformation Characteristics of Textured Magnesium. *Trans. Metall. Soc. AIME* **1968**, *242*, 654–661.
4. Graff, S.; Brocks, W.; Steglich, D. Yielding of magnesium: From single crystal to polycrystalline aggregates. *Int. J. Plast.* **2007**, *23*, 1957–1978. [CrossRef]
5. Wonsiewicz, B.C. *Plasticity of Magnesium Crystals*; Massachusetts Institute of Technology: Cambridge, MA, USA, 1966.
6. Kelley, E.W.; Hosford, W. Plane-strain compression of magnesium and magnesium alloy crystals. *Trans. Metall. Soc. AIME* **1968**, *242*, 5–13.

7. Chapuis, A.; Driver, J.H. Temperature dependency of slip and twinning in plane strain compressed magnesium single crystals. *Acta Mater.* **2011**, *59*, 1986–1994. [CrossRef]
8. Molodov, K.D.; Al-Samman, T.; Molodov, D.A.; Gottstein, G. On the Ductility of Magnesium Single Crystals at Ambient Temperature. *Metall. Mater. Trans. A* **2014**, *45*, 3275–3281. [CrossRef]
9. Molodov, K.D.; Al-Samman, T.; Molodov, D.A.; Gottstein, G. Mechanisms of exceptional ductility of magnesium single crystal during deformation at room temperature: Multiple twinning and dynamic recrystallization. *Acta Mater.* **2014**, *76*, 314–330. [CrossRef]
10. Molodov, K.D.; Al-Samman, T.; Molodov, D.A. Deformation-Induced Recrystallization of Magnesium Single Crystals at Ambient Temperature. *IOP Conf. Ser. Mater. Sci. Eng.* **2015**, *82*, 12014. [CrossRef]
11. Molodov, K.; Al-Samman, T.; Molodov, D.; Gottstein, G. On the role of anomalous twinning in the plasticity of magnesium. *Acta Mater.* **2016**, *103*, 711–723. [CrossRef]
12. Molodov, K.D.; Al-Samman, T.; Molodov, D.A. On the diversity of the plastic response of magnesium in plane strain compression. *Mater. Sci. Eng. A* **2016**, *651*, 63–68. [CrossRef]
13. Molodov, K.D.; Al-Samman, T.; Molodov, D.A. Profuse slip transmission across twin boundaries in magnesium. *Acta Mater.* **2017**, *124*, 397–409. [CrossRef]
14. Molodov, K.D.; Al-Samman, T.; Molodov, D.A.; Korte-Kerzel, S. On the twinning shear of $\{10\bar{1}2\}$ twins in magnesium—Experimental determination and formal description. *Acta Mater.* **2017**, *134*, 267–273. [CrossRef]
15. Molodov, D.A.; Ivanov, V.A.; Gottstein, G. Low angle tilt boundary migration coupled to shear deformation. *Acta Mater.* **2007**, *55*, 1843–1848. [CrossRef]
16. Bettles, C.; Barnett, M. *Advances in Wrought Magnesium Alloys: Fundamentals of Processing, Properties and Applications*; Elsevier: Amsterdam, The Netherlands, 2012.
17. Kaiser, F.; Kainer, K.U. *Magnesium Alloys and Technology*; John Wiley&Sons: Hoboken, NJ, USA, 2003.
18. Polmear, I.; StJohn, D.; Nie, J.-F.; Qian, M. *Light Alloys: Metallurgy of The Light Metals*; Butterworth-Heinemann: Oxford, UK, 2017.
19. Friedrich, H.E.; Mordike, B.L. *Magnesium Technology*; Springer: New York, NY, USA, 2006; Volume 212.
20. Hielscher, R.; Schaeben, H. A novel pole figure inversion method: Specification of the MTEX algorithm. *J. Appl. Crystallogr.* **2008**, *41*, 1024–1037. [CrossRef]
21. Nave, M.D.; Barnett, M.R. Microstructures and textures of pure magnesium deformed in plane-strain compression. *Scr. Mater.* **2004**, *51*, 881–885. [CrossRef]
22. Jiang, M.G.; Xu, C.; Yan, H.; Fan, G.H.; Nakata, T.; Lao, C.S.; Chen, R.S.; Kamado, S.; Han, E.H.; Lu, B.H. Unveiling the formation of basal texture variations based on twinning and dynamic recrystallization in AZ31 magnesium alloy during extrusion. *Acta Mater.* **2018**, *157*, 53–71. [CrossRef]
23. Barnett, M.R.; Stanford, N.; Cizek, P.; Beer, A.; Xuebin, Z.; Keshavarz, Z. Deformation mechanisms in Mg alloys and the challenge of extending room-temperature plasticity. *J. Miner.* **2009**, *61*, 19–24. [CrossRef]
24. Bettles, C.; Gibson, M. Current wrought magnesium alloys: Strengths and weaknesses. *J. Miner.* **2005**, *57*, 46–49. [CrossRef]
25. Xu, S.W.; Matsumoto, N.; Kamado, S.; Homma, T.; Kojima, Y. Dynamic microstructural changes in Mg–9Al–1Zn alloy during hot compression. *Scr. Mater.* **2009**, *61*, 249–252. [CrossRef]
26. Li, X.; Yang, P.; Wang, L.-N.; Meng, L.; Cui, F. Orientational analysis of static recrystallization at compression twins in a magnesium alloy AZ31. *Mater. Sci. Eng. A* **2009**, *517*, 160–169. [CrossRef]
27. Levinson, A.; Mishra, R.K.; Doherty, R.D.; Kalidindi, S.R. Influence of deformation twinning on static annealing of AZ31 Mg alloy. *Acta Mater.* **2013**, *61*, 5966–5978. [CrossRef]
28. Guan, D.; Rainforth, W.M.; Ma, L.; Wynne, B.; Gao, J. Twin recrystallization mechanisms and exceptional contribution to texture evolution during annealing in a magnesium alloy. *Acta Mater.* **2017**, *126*, 132–144. [CrossRef]
29. Čapek, J.; Máthis, K.; Clausen, B.; Barnett, M. Dependence of twinned volume fraction on loading mode and Schmid factor in randomly textured magnesium. *Acta Mater.* **2017**, *130*, 319–328. [CrossRef]
30. Hong, S.-G.; Park, S.H.; Lee, C.S. Role of {10–12} twinning characteristics in the deformation behavior of a polycrystalline magnesium alloy. *Acta Mater.* **2010**, *58*, 5873–5885. [CrossRef]
31. Park, S.H.; Lee, J.H.; Moon, B.G.; You, B.S. Tension–compression yield asymmetry in as-cast magnesium alloy. *J. Alloys Compd.* **2014**, *617*, 277–280. [CrossRef]
32. Godet, S.; Jiang, L.; Luo, A.A.; Jonas, J.J. Use of Schmid factors to select extension twin variants in extruded magnesium alloy tubes. *Scr. Mater.* **2006**, *55*, 1055–1058. [CrossRef]
33. Jiang, J.; Godfrey, A.; Liu, W.; Liu, Q. Identification and analysis of twinning variants during compression of a Mg–Al–Zn alloy. *Scr. Mater.* **2008**, *58*, 122–125. [CrossRef]
34. Pei, Y.; Godfrey, A.; Jiang, J.; Zhang, Y.B.; Liu, W.; Liu, Q. Extension twin variant selection during uniaxial compression of a magnesium alloy. *Mater. Sci. Eng. A* **2012**, *550*, 138–145. [CrossRef]
35. Wang, F.; Sandlöbes, S.; Diehl, M.; Sharma, L.; Roters, F.; Raabe, D. In situ observation of collective grain-scale mechanics in Mg and Mg–rare earth alloys. *Acta Mater.* **2014**, *80*, 77–93. [CrossRef]
36. Molodov, D.A. *Investigation of Deformation Mechanisms in Magnesium Crystals*; RWTH Aachen: Aachen, Germany, 2017.
37. Xie, K.Y.; Alam, Z.; Caffee, A.; Hemker, K.J. Deformation Behavior of Mg Single Crystals Compressed Along *c*-axis. In *Magnesium Technology 2016*; Springer: New York, NY, USA, 2016; pp. 209–211.

38. Al-Samman, T.; Molodov, K.D.; Molodov, D.A.; Gottstein, G.; Suwas, S. Softening and dynamic recrystallization in magnesium single crystals during *c*-axis compression. *Acta Mater.* **2012**, *60*, 537–545. [CrossRef]
39. Guan, D.; Rainforth, W.M.; Gao, J.; Sharp, J.; Wynne, B.; Ma, L. Individual effect of recrystallisation nucleation sites on texture weakening in a magnesium alloy: Part 1 double twins. *Acta Mater.* **2017**, *135*, 14–24. [CrossRef]
40. Sabat, R.K.; Sahoo, S.K.; Panda, D.; Mohanty, U.K.; Suwas, S. Orientation dependent recrystallization mechanism during static annealing of pure magnesium. *Mater. Charact.* **2017**, *132*, 388–396. [CrossRef]

Article

Effect of Al Content on Texture Evolution and Recrystallization Behavior of Non-Flammable Magnesium Sheet Alloys

Sumi Jo *, Dietmar Letzig and Sangbong Yi *

Magnesium Innovation Centre, Institute of Material and Process Design, Helmholtz-Zentrum Geesthacht, Max-Planck-Str. 1, D-21502 Geesthacht, Germany; dietmar.letzig@hzg.de
* Correspondence: su.jo@hzg.de (S.J.); sangbong.yi@hzg.de (S.Y.); Tel.: +49-4152-87-2002 (S.J.); +49-4152-87-1911 (S.Y.)

Abstract: The effect of Al content on the texture evolution and recrystallization behavior of the non-flammable Mg sheet alloys containing Ca and Y was investigated in this study. With a decrease in the Al content from 3 wt.% to 1 wt.%, the amounts of the other alloying elements dissolved in the matrix, especially Ca, are increased. The increase of the alloying elements in a solid solution brought out the retarded recrystallization and weakened texture with the basal poles tilted toward the sheet transverse direction. Extension twinning activity increased when Al content with decreasing, resulting in the texture broadening towards the sheet transverse direction in the as-rolled sheets. The textures of the AZXW1000 and AZXW2000 sheets weaken uniformly in all sample directions during annealing, while the AZXW3000 sheet shows less weakening of the rolling direction split component. The texture weakening of the alloys with lower Al contents is attributed to the retarded recrystallization caused by the larger amount of the dissolved Ca solutes. Based on the non-basal texture and relatively stable grain structure, the Mg alloy sheet containing a relatively small amount of Al is advantageous to improve the formability.

Keywords: magnesium; sheets; recrystallization; texture; formability; non-flammability

Citation: Jo, S.; Letzig, D.; Yi, S. Effect of Al Content on Texture Evolution and Recrystallization Behavior of Non-Flammable Magnesium Sheet Alloys. *Metals* **2021**, *11*, 468. https://doi.org/10.3390/met11030468

Academic Editor: Håkan Hallberg

Received: 26 February 2021
Accepted: 9 March 2021
Published: 12 March 2021

Publisher's Note: MDPI stays neutral with regard to jurisdictional claims in published maps and institutional affiliations.

1. Introduction

As the lightest structural metallic material, Mg alloys are attractive for application in the transportation sectors to increase energy efficiency. However, a wide application of Mg alloys has been hindered by some drawbacks, such as poor room temperature formability, poor corrosion resistance, and high flammability. While the early studies regarding the formability improvement were conducted using the commercial AZ31 alloy sheet [1–5], it has been well acknowledged that the alloying of rare earth elements (RE), Ca or Sr [6–9] significantly improve the room temperature formability to be comparable to that of 6000 series aluminum alloy sheets [10]. The addition of Ca into Mg alloys has been known to improve high temperature mechanical properties and ignition-proof behavior. Moreover, the simultaneous addition of Ca with Y results in an excellent ignition-proof behavior and improved corrosion resistance [11–13].

A large amount of alloying elements cause difficulties in controlling the microstructure during the sheet processes, including continuous casting. Strong microstructure inhomogeneity caused by high alloying amount, e.g., centerline segregation formed in a strip produced via continuous casting and high quantity of secondary phases due to the high affinity Al with other alloying elements, like RE and Ca, deteriorates the mechanical and corrosion properties. Furthermore, the formation of secondary phases consumes the alloying elements dissolved in the matrix, such that the role of such elements in the texture weakening is diminished by high Al content [10,14–17]. In view of the effects of Al on microstructure and texture evolution, recent studies of highly formable Mg sheet alloys put focus on the development of low-Al containing alloys. It was reported that the Mg alloy sheets containing Al close to 1 wt.%, Mg-1.2Al-0.5Ca-0.4Mn-1.6Zn [18], Mg-1.3Al-0.5Ca-0.7Mn-0.8Zn [19], and Mg-1Zn-1Al-0.5Ca-0.4Mn-0.2Ca [20], exhibit a weak texture, high formability, and improved mechanical properties. In this study, the texture evolution and recrystallization behavior of the Mg sheet alloys having excellent non-flammability containing Al, Ca, and Y, which are essential elements to achieve the non-flammability, as well as texture weakening, were investigated by varying the Al content from 1 wt.% to 3 wt.%. The focus of the present study is laid to clarify the relationship between the Al content and microstructural evolution of the non-flammable alloys in consideration of the amount of the dissolved alloying elements in the matrix.

2. Materials and Methods

The ingots of AZXW1000, 2000, and 3000 alloys, with the nominal compositions of Mg-xAl-1Zn-0.5Ca-0.5Y-0.1Mn, x = 1, 2, 3 (in wt.%), were prepared by gravity casting. The molten metal with the target composition was cast in a steel crucible under a protective atmosphere of Ar and SF_6 mixture. The furnace temperature was set to 780 °C until the melts were poured into steel crucibles preheated at 700 °C. After pouring the melt, the crucible was quenched in water. It is to mention that the protection gas was used to ensure a high safety during the casting experiments, while the examined alloys exhibited an excellent ignition-proof behavior, e.g., no ignition during the pouring of the molten metal. The billets with a thickness of 10 mm were machined from the cast materials, and then hot rolled at 450 °C to a final thickness of 1.2 mm. The step reduction degree was increasing from 10 to 30% with the rolling passes, and the intermediate annealing of the rolled sheet was conducted at 450 °C for 10 min prior to each rolling step. The recrystallization annealing of the rolled sheet was conducted at 400 °C for 10 min. The textures of the as-rolled and annealed sheets were measured using a Panalytical X-ray diffractometer (Malvern Panalytical, Malvern, UK) and Cu K_α radiation. The (0002) pole figures were recalculated using a MATLAB-based toolbox MTEX [21] from six measured pole figures. Stretch formability (Index Erichsen, I.E) of the annealed sheets was tested at room temperature. The tests were carried out with a blank holding force of 10 kN, a punch diameter of 20 mm, and a deformation rate (punch displacement) of 0.5 mm/min. Microstructural analyses were performed using a scanning electron microscope (SEM, Vega 3 SB, TESCAN, Brno, Czech Republic), equipped with an energy-dispersive X-ray spectroscope (EDS, eumeX, Heidenrod, Germany). Electron backscatter diffraction (EBSD) analysis was performed by using an FEG-SEM (Ultra 55, Zeiss, Jena, Germany) equipped with an EBSD detector (Hikari camera (EDAX, Mahwah, NJ, USA)) with TSL OIM software (version 7, EDAX, Mahwah, NJ, USA). Specimens for electron microscopy were prepared by electro-polishing in a Struers AC2 solution at −20 °C and 30 V. The thermodynamic calculation for the corresponding alloy compositions was carried out using Pandat software (version 2017, Compu Therm LLC, Wisconsin, USA) [22] to calculate the amount of the dissolved Ca solutes in the matrix.

3. Results and Discussion

The recalculated (0002) pole figures of the as-rolled and annealed sheets are presented in Figure 1. The maximum pole densities of the (0002) pole figures of the as-rolled AZXW1000, 2000, and 3000 sheets are 5.5, 4.9, and 6.2, respectively. The as-rolled AZXW1000 and AZXW2000 sheets show the basal poles tilted towards the rolling direction (RD) and, simultaneously, the basal pole spread in the transverse direction (TD), whereas the texture of the as-rolled AZXW3000 sheet is presented by the basal pole spread from the normal direction (ND) towards the RD. The annealed AZXW1000 and AZXW2000 sheets show the weak texture with the basal pole split toward the TD contrary to the texture of the annealed AZXW3000 sheet. These textures obtained in the AZXW1000 and AZXW2000 sheets normally develop in the highly formable Mg alloy sheets containing RE, Ca and Zn [14,16,23–28]. The appearances of the Erichsen samples of the examined sheets are shown in Figure 2. The annealed AZXW2000 sheet (IE = 6.2) indeed has higher formability than the AZXW3000 sheet (IE = 4.8), while both sheets have higher IE values than the commercial AZ31 sheet (IE = 2~3). The improved formability can be understood as a result of the texture weakening and the accompanying development of the non-basal type texture.

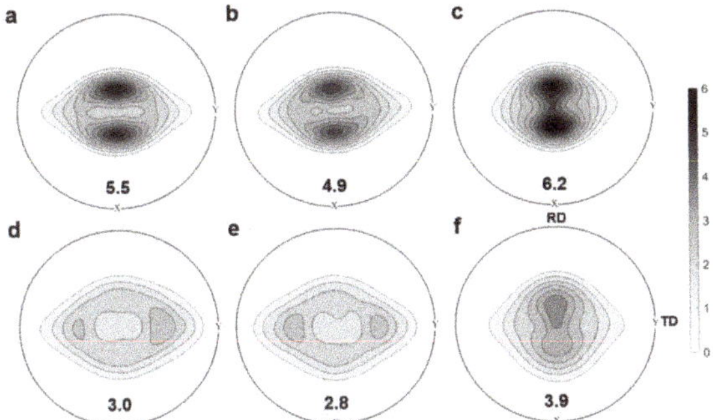

Figure 1. Recalculated (0002) pole figures of the (**a–c**) as-rolled and (**d–f**) annealed AZXW1000, 2000, and 3000 alloy sheets, respectively.

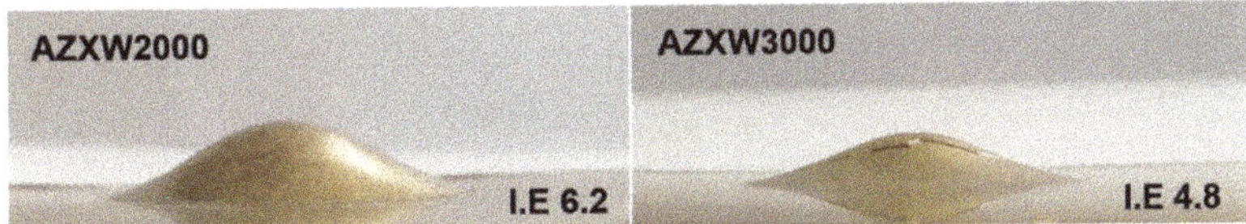

Figure 2. Appearances of the Erichsen samples of the AZXW2000 and AZXW3000 sheets, respectively.

The optical micrographs, which are taken at the longitudinal section, of the as-rolled and annealed sheets are shown in Figure 3a–f, respectively. The microstructure of the as-rolled sheets commonly consists of the deformed grains, deformation twins and secondary phases. The fully recrystallized microstructures were obtained in all examined alloys after annealing treatment at 400 °C for 10 min. The average grain sizes of the recrystallized AZXW1000, 2000, and 3000 sheets are comparable, 5.2, 6.9, and 6.1 μm, respectively.

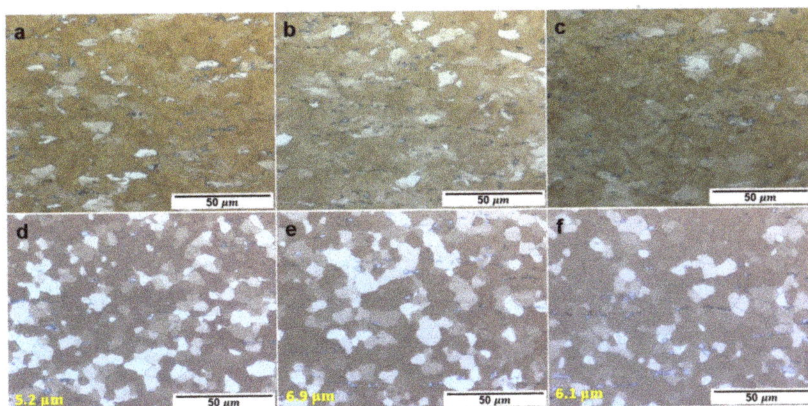

Figure 3. Optical micrographs of the (**a–c**) as-rolled and (**d–f**) annealed AZXW1000, 2000 and 3000 sheets, respectively.

Figure 4 shows the variation of the phase fractions as a function of Al content, calculated by using the Pandat software. The total fraction of the secondary phases of the alloys containing 1, 2, and 3 wt.% of Al increases with the Al content, 1.85, 2.42, and 3.97 wt.%, respectively. The amounts of Al_4MgY and Al_8Mn_5 phases increase when Al content is higher than 1 and 2 wt.%. The formation of Mg_2Ca and Al_2Y phases is found only in the alloy containing 1 wt.% Al, while Al_2Ca phase is formed in the alloys with Al amount higher than 2 wt.%. Considering the calculated fractions of the equilibrium phases, it is plausible that the fine particles with the average size of 0.2 μm observed in the as-rolled and annealed AZXW1000 and AZXW2000 sheets, Figure 3a,b,d,e, are inferred to be Mg_2Ca and Al_2Y phases.

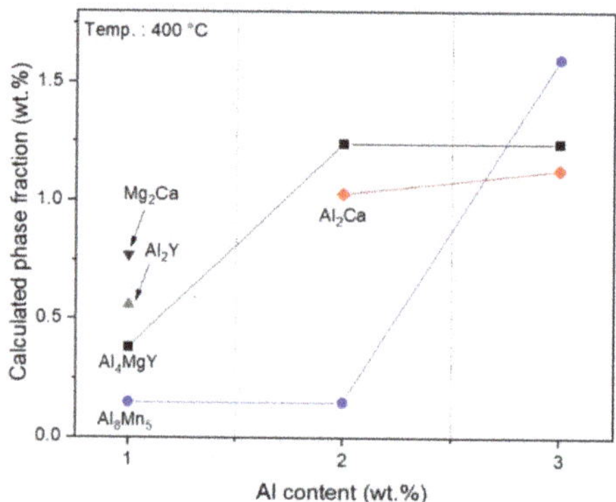

Figure 4. Calculated phase fraction of the investigated alloys at 400 °C as a function of Al content.

The SEM images of the annealed AZXW1000, 2000, and 3000 sheets are demonstrated in Figure 5. The chemical compositions analyzed by SEM-EDS are listed in Table 1 for the secondary phases marked with yellow arrows. The stoichiometry of the particles B, E, and C corresponds to Al_2Ca and $(Mg,Al)_2Ca$ phases, respectively, and the particle D to Al_2Y phase. The particles observed in the AZXW1000, 2000, and 3000 sheets, from the SEM-EDS, does not follow the thermodynamic calculation, and the formation of the Al_2Ca, Mg_2Ca, as well as $(Mg,Al)_2Ca$, are not clearly distinguished depending on the Al content. The formation of Ca-containing phases, Al_2Ca, Mg_2Ca, and $(Mg,Al)_2Ca$ mainly depends on the Ca/Al ratio. That is, the increase of Ca/Al ratio induces the formation of Mg_2Ca and $(Mg,Al)_2Ca$ phases rather than Al_2Ca phase [29], as shown in Figure 4. The chemical composition of the points A and F consisted of Al, Mn, and Y, do not match the equilibrium phases, i.e., Al_8Mn_5, Al_2Y, or Al_4MgY. Structural similarity among these phases [30,31] and solidification in non-equilibrium result in the formation of non-equilibrium phases. Additionally, the high affinity [32] of Al to Mn and Y is considered to agglomerate Al, Mn, and Y, and form the non-equilibrium phase.

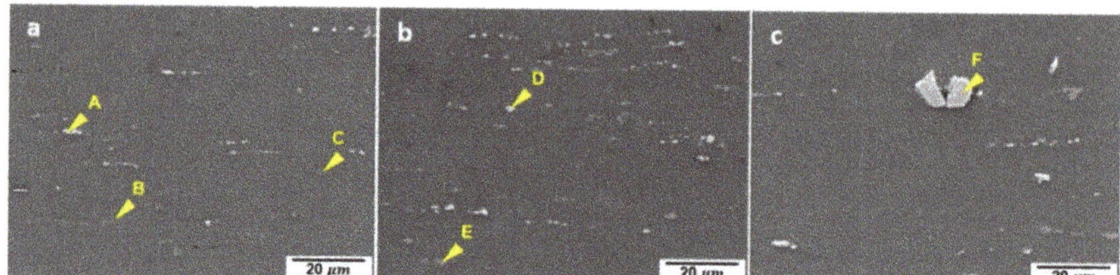

Figure 5. Secondary electron micrographs of the annealed (**a**) AZXW1000, (**b**) AZXW2000, and (**c**) AZXW3000 alloy sheets, respectively. The energy-dispersive X-ray spectroscopy results on the second phases marked by arrows A to G are given in Table 1.

Table 1. Chemical composition (in at.%) of the second phases in the annealed AZXW1000, 2000, and 3000 sheets indicated by yellow arrows in Figure 5, analyzed by SEM-EDS.

Point	Al	Zn	Mn	Ca	Y	Mg
A	35.4	0.7	16.1	0.1	5.7	42.3
B	7.3	1.1	0.07	4.7	0.03	86.9
C	11.0	1.5	0.1	10.9	0.1	76.4
D	49.1	1.2	0.4	0.5	23.5	25.3
E	44.5	0.9	0.09	21.6	0.1	32.8
F	57.7	0.5	29.8	0.1	10.5	1.4

As the Al content increases from 1 wt.% to 3 wt.%, the volume fraction and size of the secondary phases also increase, as shown in Figure 4, as well as the point A and F in Figure 5. It indicates that the high amount of Al is likely to consume the Ca, Y, and Mn solutes by the formation of the secondary phases during the thermo-mechanical treatments. On the contrary, in the alloys with low Al content the other alloying elements can be dissolved in the matrix without forming secondary phases. The amounts of the Ca solutes dissolved in the matrix are plotted as a function of the Al content in Figure 6. The experimental values and the calculated values using by Pandat software confirm that the amount of the dissolved Ca solutes increases with decreasing the Al content. Especially, the dissolved Ca solutes sharply increases by reducing the Al content to 1 wt.%. The differences between the amount of the calculated and the measured Ca solutes arise from the restricted resolution of the SEM-EDS and the formation of the non-equilibrium phases, like Al-Mn-Y, that are not expected from the thermodynamic calculation. It is important to mention that the amount of the Mn and Y solutes dissolved in the matrix also increase as Al content decreases, but the increment is much smaller in comparison to that of the Ca solutes.

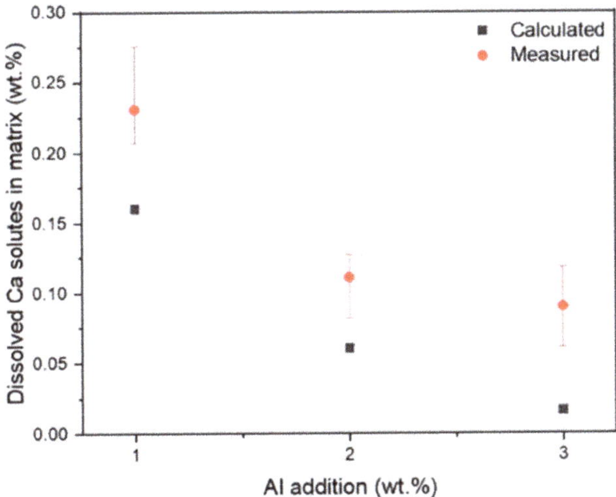

Figure 6. Amount of the dissolved Ca solutes in matrix according to the Al content at 400 °C, comparing the experimentally measured values by SEM-EDS and the calculated amounts by Pandat software.

To investigate the recrystallization behavior and texture evolution during the annealing treatment, EBSD analysis was performed on the sheets annealed to various times. Inverse pole figure maps (IPF) of the examined sheets annealed at 400 °C for 0 s (as-rolled), 10 s, and 30 s are presented in Figure 7 with the corresponding (0001) pole figures. The EBSD measuring points with the confidence index (CI) higher than 0.08, without an additional data clean-up, are shown in the IPF maps, and the non-indexed areas with black color correspond to either highly deformed region or secondary phases. The as-rolled

AZXW1000, 2000, and 3000 sheets have a strong texture with the basal poles mostly tilted to the RD. The as-rolled AZXW1000 sheet shows a distinct broadening of the basal poles towards the TD, in comparison to that of the as-rolled AZXW2000 and AZXW3000 sheets. Accordingly, a large number of grains having the tilted basal pole to the TD (green grains) are found in the as-rolled AZXW1000 sheet. Misorientation angle distribution plots of the as-rolled AZXW1000, 2000 and 3000 sheets are shown in Figure 8. All of the examined sheets show three peaks at around 38°, 56°, and 86°, and the peaks correspond to secondary, contraction and extension twins, respectively [33]. Interestingly, the fraction of the extension twin in the as-rolled AZXW1000 sheet is higher than that in the as-rolled AZXW2000 and AZXW3000. It can be mentioned that the AZXW1000 sheet have higher extension twinning activities than that of the AZXW2000 and AZXW3000 sheets, which matches to the distinct broadening of the basal poles towards the TD. Meanwhile, a higher fraction of the secondary twin is found in the AZXW3000, in comparison to that of the AZXW1000 and AZXW2000 sheets.

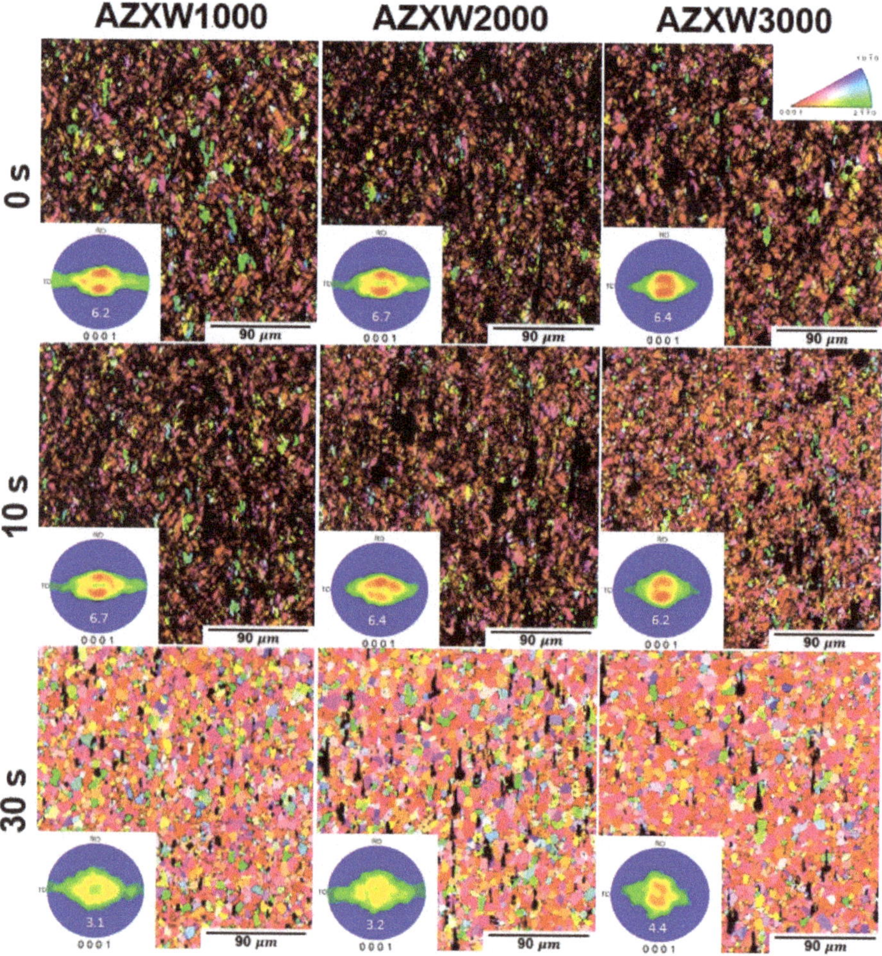

Figure 7. Inverse pole figure maps of the AZXW1000, 2000, and 3000 alloy sheets annealed for 0 s (as-rolled), 10 s, and 30 s.

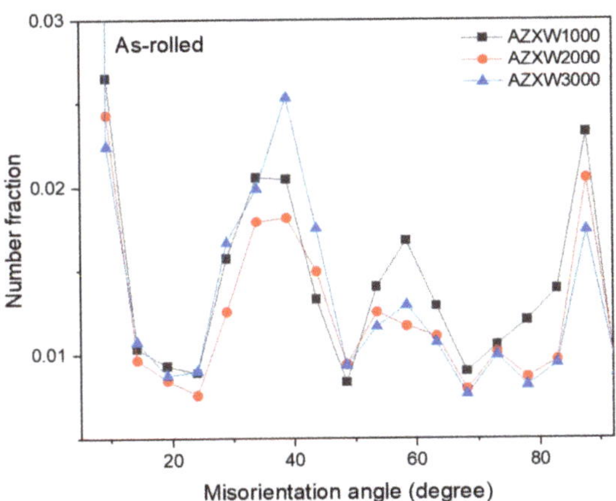

Figure 8. Misorientation angle distribution of the as-rolled AZXW1000, 2000 and 3000 sheets calculated from the IPF maps in Figure 7.

A few papers reported that the secondary twin is an origin of strain localization and shear band formed in the as-rolled sheet [33,34]. The secondary twins were commonly observed in the shear bands with a high density of dislocations. The basal planes reoriented by secondary twins are favorable for the basal slip, and dislocation slip abundantly occurs and piles up around the twin boundaries, resultantly then, bring out stress concentration and distortion of the twin regions [34]. The as-rolled microstructure of the AZXW3000 sheet, Figure 3c, shows the distinct strain localization region developed across more grains in comparison to that of the as-rolled AZXW1000 and AZXW2000 sheets, Figure 3a,b. It arises from the highest fraction of secondary twins due to the stress concentration in the AZXW3000 sheet.

As the recrystallization proceeds during the annealing treatment for 10 s, the basal pole distribution of the investigated sheets broaden to the RD and TD. Even though the maximum pole densities of the sheets annealed for 10 s are barely changed, the basal pole densities weaken uniformly along all sample directions without a preferred weakening at a certain sample direction. After 30 s of annealing, the AZXW1000 and AZXW2000 sheets show further a uniform texture weakening accompanying the basal pole broadening towards the RD and TD. On the contrary, the weakening of the tilted basal pole to the RD is relatively less in the annealed AZXW3000 sheet for 30 s, resulting in a higher texture intensity.

The fraction of the recrystallized grains for the investigated sheets along the annealing time are shown in Figure 9, indicating distinct recrystallization kinetics of each alloy sheet. The recrystallized grains are defined by the grain orientation spread (GOS) less than 1°. The fraction of the recrystallized grains in the as-rolled AZXW1000, 2000, and 3000 sheets are close to 0, and the recrystallization of the three alloy sheets is almost complete after the annealing treatment for 30 s as shown in Figure 7. Comparing the fractions of the recrystallized grains in the annealed sheets for 10 s, i.e., at the beginning stage of the recrystallization, it is clear that the examined alloy sheets exhibit different recrystallization kinetics. The fraction of the recrystallized grains in the AZXW1000, 2000, and 3000 sheets are 0.17, 0.21, and 0.26, respectively, indicating that the AZXW3000 sheet has faster recrystallization kinetics than the AZXW1000 and AZXW2000 sheets. It is reported that the recrystallization of the Mg alloy sheets containing RE, Ca, and Zn [14,16,23–25] is retarded by solutes segregation at the grain boundaries, stacking faults and twin boundaries, and resultantly, the recrystallization kinetics of the sheets become slow. For example, the ZW01 [16] and Mg-0.8Zn-0.2Ca [14] alloy sheets show slower recrystallization kinetics than the investigated alloy sheets in this study. The amount of the dissolved solutes in the above alloys are calculated by Pandat software at the same temperature, 400 °C, for comparison. The amounts are 0.79 wt.% of Y, 0.05 wt.% of Zn, and 0.2 wt.% of Ca and 0.8 wt.% of Zn, respectively, which are higher than those in the examined sheets of the present study, Figure 6. Again, it is plausible that the recrystallization kinetics is influenced by the amount of the dissolved solutes. Accordingly, the recrystallization kinetics of the AZXW1000 and AZXW2000 alloy sheets are slower than that of the AZXW3000, which corresponds to the larger amount of the dissolved Ca solutes in the alloys with lower Al contents.

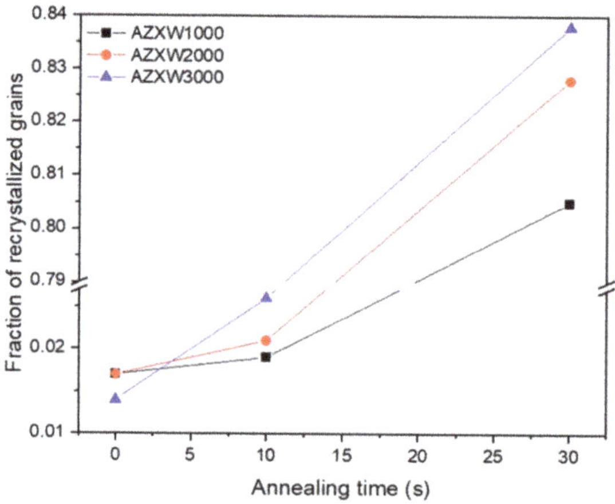

Figure 9. Fraction of the recrystallized grains of the AZXW1000, 2000, and 3000 alloys according to the annealing time calculated from the IPF maps in Figure 7.

The EBSD IPF maps and (0001) pole figures of the annealed sheets for 600 s, which correspond to Figures 1 and 3, are shown in Figure 10. While the AZXW1000 and AZXW2000 sheets maintain the basal poles tilted toward the TD and RD, the AZXW3000 sheet shows a higher concentration of the basal poles at the ND. The tendency of the basal-type texture development, i.e., alignment of the basal poles in the ND, with increasing the annealing time is commonly observed during the recrystallization of the commercial Mg alloy sheets, resulting in basal-type texture. The development of the basal-type texture is effectively hindered by the higher amount of the dissolved solutes, and brings out a more randomized texture with the basal pole tilted towards TD [14].

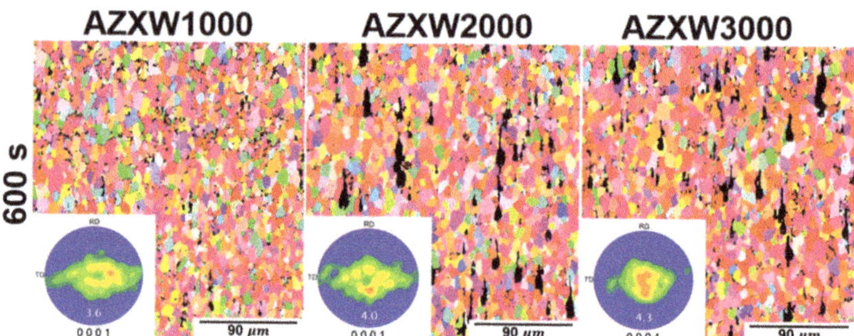

Figure 10. IPF maps and corresponding (0001) pole figures of the AZXW1000, 2000, and 3000 annealed for 600 s.

In addition to the retarded boundary motion by the dissolved solutes, the pinning effect by precipitates is also to be considered. As reported in [35], the thermally stable Y-containing precipitates in the Mg-0.7Ca-0.7Y alloy sheet hinder the grain boundary movement during the homogenization treatment, and contribute to obtaining the fine microstructure. Likewise, the fine particles that are observed in the microstructure of the as-rolled and annealed AZXW1000 and AZXW2000 sheets, Figure 3a,b,d,e, can contribute to retarding the recrystallization. A more detailed analysis regarding the interaction between the precipitates and grain boundary motion will clarify the role of the fine precipitates on the recrystallization kinetics of the examined alloys.

4. Conclusions

The effect of the Al content on the texture evolution and recrystallization behavior of the Mg alloy sheets containing Ca and Y were investigated in this study. The decrease of the Al content in the alloy leads to the increase in the amount of the dissolved alloying elements in matrix, i.e., increase of the Ca solute. The retarded recrystallization and the texture weakening observed in the AZXW1000 and AZXW2000 sheets, in comparison to that of the AZXW3000 sheet, are attributed to the relatively high amount of the solutes. These results clearly show that the formability of the Mg sheet, which contains the alloying elements contribute to the texture weakening like Ca and Y, can be improved in 2~3 times by reducing Al content from 3 wt.% to 1 wt.%.

Author Contributions: Conceptualization, S.J. and S.Y.; Experiments and analyzed the experimental results, S.J.; Writing—original draft preparation, S.J.; Writing—review and editing, S.Y. and S.J.; Project administration, D.L. All authors have read and agreed to the published version of the manuscript.

Funding: This work was partly funded by NST-National Research Council of Science & Technology (No. CRC-15-06-KIGAM), Ministry of Science and ICP (MSIP).

Institutional Review Board Statement: Not applicable.

Informed Consent Statement: Not applicable.

Data Availability Statement: The data presented in this study are available on reasonable request to the corresponding author.

Acknowledgments: The authors gratefully acknowledge Y.M. Kim at Korea Institute of Material Science (KIMS) for a fruitful discussion on the experimental results and non-flammable alloys. The authors are grateful to Y.K. Shin, at Helmholtz-Zentrum Geesthacht (HZG), for supporting the sample preparation and EBSD analysis.

Conflicts of Interest: The authors declare no conflict of interest.

References

1. Lee, S.; Chen, Y.-H.; Wang, J.-Y. Isothermal sheet formability of magnesium alloy AZ31 and AZ61. *J. Mater. Process. Tech.* **2002**, *124*, 19–24. [CrossRef]
2. Kaiser, F.; Bohlen, J.; Letzig, D.; Kainer, K.U.; Styczynski, A.; Hartig, C. Influence of Rolling Conditions on the Microstructure and Mechanical Properties of Magnesium Sheet AZ31. *Adv. Eng. Mater.* **2003**, *5*, 891–896. [CrossRef]
3. Jäger, A.; Lukáč, P.; Gärtnerová, V.; Bohlen, J.; Kainer, K.U. Tensile properties of hot rolled AZ31 Mg alloy sheets at elevated temperatures. *J. Alloys Compd.* **2004**, *378*, 184–187. [CrossRef]
4. Takuda, H.; Yoshii, T.; Hatta, N. Finite-element analysis of the formability of a magnesium-based alloy AZ31 sheet. *J. Mater. Process. Tech.* **1999**, *89–90*, 135–140. [CrossRef]
5. Fuh-Kuo, C.; Tyng-Bin, H. Formability of stamping magnesium-alloy AZ31 sheets. *J. Mater. Process. Tech.* **2003**, *142*, 643–647. [CrossRef]
6. Masoudpanah, S.M.; Mahmudi, R. Effects of rare-earth elements and Ca additions on the microstructure and mechanical properties of AZ31 magnesium alloy processed by ECAP. *Mater. Sci. Eng. A* **2009**, *526*, 22–30. [CrossRef]
7. Shang, L.; Yue, S.; RVerma; Krajewski, P.; Galvani, C.; Essadiqi, E. Effect of microalloying (Ca, Sr, and Ce) on elevated temperature tensile behavior of AZ31 magnesium sheet alloy. *Mater. Sci. Eng. A* **2011**, *528*, 3761–3770. [CrossRef]
8. Bae, G.T.; Bae, J.H.; Kang, D.H.; Lee, H.; Kim, N.J. Effect of Ca addition on microstructure of twin-roll cast AZ31 Mg alloy. *Met. Mater. Int.* **2009**, *15*, 1–5. [CrossRef]

9. Yi, S.; Park, J.H.; Letzig, D.; Kwon, O.D.; Kainer, K.U.; Kim, J.J. Microstructure and mechanical properteis of Ca containing AZX310 alloy sheets produced via twin roll casting technology. In *Magnesium Technology 2016*; Singh, A., Solanki, K., Manuel, M.V., Neelameggham, N.R., Eds.; Spinger: Cham, Switzerland, 2016; pp. 383–387. [CrossRef]
10. Trang, T.T.T.; Zhang, J.H.; Kim, J.H.; Zargaran, A.; Hwang, J.H.; Suh, B.C.; Kim, N.J. Designing a magnesium alloy with high strength and high formability. *Nat. Commun.* **2018**, *9*. [CrossRef]
11. You, B.S.; Kim, Y.M.; Yim, C.D.; Kim, H.S. Oxidation and Corrosion Behavior of Non-Flammable Magnesium Alloys Containing Ca and Y. In *Magnesium Technology 2014*; Alderman, M., Manuel, M.V., Hort, N., Neelameggham, N.R., Eds.; Springer: Cham, Switzerland, 2016; pp. 325–329. [CrossRef]
12. Kim, Y.M.; Yim, C.D.; Kim, H.S.; You, B.S. Key factor influencing the ignition resistance of magnesium alloys at elevated temperatures. *Scripta Mater.* **2011**, *65*, 958–961. [CrossRef]
13. Lee, D.B. High temperature oxidation of AZ31+0.3wt.%Ca and AZ31+0.3wt.%CaO magnesium alloys. *Corros. Sci.* **2013**, *70*, 243–251. [CrossRef]
14. Zeng, Z.R.; Zhu, Y.M.; Xu, S.W.; Bian, M.Z.; Davies, C.H.J.; Birbilis, N.; Nie, J.F. Texture evolution during static recrystallization of cold-rolled magnesium alloys. *Acta Mater.* **2016**, *105*, 479–494. [CrossRef]
15. Bohlen, J.; Wendt, J.; Nienaber, M.; Kainer, K.U.; Stutz, L.; Letzig, D. Calcium and zirconium as texture modifiers during rolling and annealing of magnesium–zinc alloys. *Mater. Charact.* **2015**, *101*, 144–152. [CrossRef]
16. Kim, Y.M.; Mendis, C.; Sasaki, T.; Letzig, D.; Pyczak, F.; Hono, K.; Yi, S. Static recrystallization behaviour of cold rolled Mg-Zn-Y alloy and role of solute segregation in microstructure evolution. *Scripta Mater.* **2017**, *136*, 41–45. [CrossRef]
17. Kurukuri, S.; Worswick, M.J.; Bardelcik, A.; Mishra, R.K.; Carter, J.T. Constitutive Behavior of Commercial Grade ZEK100 Magnesium Alloy Sheet over a Wide Range of Strain Rates. *Metall. Mater. Trans. A* **2014**, *45*, 3321–3337. [CrossRef]
18. Li, Z.H.; Sasaki, T.T.; Bian, M.Z.; Nakata, T.; Yoshida, Y.; Kawabe, N.; Kamado, S.; Hono, K. Role of Zn on the room temperature formability and strength in Mg-Al-Ca-Mn sheet alloys. *J. Alloys Compd.* **2020**, *847*, 156347. [CrossRef]
19. Bian, M.Z.; Sasaki, T.T.; Nakata, T.; Yoshida, Y.; Kawabe, N.; Kamado, S.; Hono, K. Bake-hardenable Mg–Al–Zn–Mn–Ca sheet alloy processed by twin-roll casting. *Acta Mater.* **2018**, *158*, 278–288. [CrossRef]
20. Shi, R.; Miao, J.; Avey, T.; Luo, A.A. A new magnesium sheet alloy with high tensile properties and room-temperature formability. *Sci. Rep.* **2020**, *10*, 10044. [CrossRef]
21. Bachmann, F.; Hielscher, R.; Schaeben, H. Texture Analysis with MTEX–Free and Open Source Software Toolbox. *Solid State Phenom.* **2010**, *160*, 63–68. [CrossRef]
22. Chen, S.-L.; Daniel, S.; Zhang, F.; Chang, Y.A.; Yan, X.-Y.; Xie, F.-Y.; Schmid-Fetzer, R.; Oates, W.A. The PANDAT software package and its applications. *Calphad* **2002**, *26*, 175–188. [CrossRef]
23. Stanford, N.; Barnett, M. Effect of composition on the texture and deformation behaviour of wrought Mg alloys. *Scripta Mater.* **2008**, *58*, 179–182. [CrossRef]
24. Bohlen, J.; Nürnberg, M.R.; Senn, J.W.; Letzig, D.; Agnew, S.R. The texture and anisotropy of magnesium–zinc–rare earth alloy sheets. *Acta Mater.* **2007**, *55*, 2101–2112. [CrossRef]
25. Chino, Y.; Ueda, T.; Otomatsu, Y.; Sassa, K.; Huang, X.; Suzuki, K.; Mabuchi, M. Effects of Ca on Tensile Properties and Stretch Formability at Room Temperature in Mg-Zn and Mg-Al Alloys. *Mater. Trans.* **2011**, *52*, 1477–1482. [CrossRef]
26. Bian, M.Z.; Sasaki, T.T.; Suh, B.C.; Nakata, T.; Kamado, S.; Hono, K. Development of Heat-Treatable High-Strength Mg–Zn–Ca–Zr Sheet Alloy with Excellent Room Temperature Formability. In *Magnesium Technology 2018*; Orlov, D., Joshi, V., Solanki, K.N.N., Eds.; Springer: Cham, Switzerland, 2018; pp. 361–364.
27. Chino, Y.; Huang, X.; Suzuki, K.; Mabuchi, M. Enhancement of Stretch Formability at Room Temperature by Addition of Ca in Mg-Zn Alloy. *Mater. Trans.* **2010**, *51*, 818–821. [CrossRef]
28. Nakata, T.; Li, Z.H.; Sasaki, T.T.; Hono, K.; Kamado, S. Room-temperature stretch formability, tensile properties, and microstructures of precipitation hardenable Mg–6Zn-0.2Ca (mass%) alloy sheets micro-alloyed with Ce or Y. *Mat. Sci. Eng. A* **2020**. [CrossRef]
29. Liang, S.M.; Chen, R.S.; Blandin, J.J.; Suery, M.; Han, E.H. Thermal analysis and solidification pathways of Mg–Al–Ca system alloys. *Mat. Sci. Eng. A* **2008**, *480*, 365–372. [CrossRef]
30. Suzuki, A.; Saddock, N.; Jones, J.; Pollock, T. Structure and transition of eutectic (Mg,Al)2Ca Laves phase in a die-cast Mg-Al-Ca base alloy. *Scripta Mater.* **2004**, *51*, 1005–1010. [CrossRef]
31. Raghavan, V. Al-Ca-Mg (Aluminum-Calcium-Magnesium). *J. Phase Equilib. Diff.* **2010**, *32*, 52–53. [CrossRef]
32. Takeuchi, A.; Inoue, A. Classification of bulk metallic glasses by atomic size difference, heat of mixing and period of constituent elements and its application to characterization of the main alloying element. *Mater. Trans.* **2005**, *46*, 2817–2829. [CrossRef]
33. Guan, D.; Rainforth, W.M.; Gao, J.; Sharp, J.; Wynne, B.; Ma, L. Individual effect of recrystallisation nucleation sites on texture weakening in a magnesium alloy: Part 1- double twins. *Acta Mater.* **2017**, *135*, 14–24. [CrossRef]
34. Zhang, K.; Zheng, J.-H.; Huang, Y.; Pruncu, C.; Jiang, J. Evolution of twinning and shear bands in magnesium alloys during rolling at room and cryogenic temperature. *Mater. Design* **2020**, *193*. [CrossRef]
35. Jo, S.; Whitmore, L.; Woo, S.; Aramburu, A.U.; Letzig, D.; Yi, S. Excellent age hardenability with the controllable microstructure of AXW100 magnesium sheet alloy. *Sci. Rep.* **2020**, *10*, 22413. [CrossRef] [PubMed]

Article

Superplasticity at Intermediate Temperatures of ZK60 Magnesium Alloy Processed by Indirect Extrusion

César Palacios-Trujillo [1], José Victoria-Hernández [2,3], David Hernández-Silva [1,*], Dietmar Letzig [2,3] and Marco A. García-Bernal [4,*]

1   Department of Metallurgical Engineering, Instituto Politécnico Nacional, ESIQIE, UPALM EDIF. 7, 07738 Mexico City, Mexico; cpalaciost0900@alumno.ipn.mx or cpt17_azul@hotmail.com

2   Institute of Material and Process Design, Helmholtz-Zentrum Geesthacht, Max-Planck-Strasse 1, D-21502 Geesthacht, Germany; jose.victoria-hernandez@hzg.de (J.V.-H.); dietmar.letzig@hzg.de (D.L.)

3   Magnesium Innovation Centre MagIC, Helmholtz-Zentrum Geesthacht, Max-Planck-Strasse 1, D-21502 Geesthacht, Germany

4   SEPI, Instituto Politécnico Nacional, ESIME Unidad Ticomán, Av. Ticomán No. 600, Col. San José Ticomán, 07340 Mexico City, Mexico

*   Correspondence: dhernandez@ipn.mx or dhs07670@yahoo.com (D.H.-S.); magarciabe@ipn.mx or magarciabe@gmail.com (M.A.G.-B.)

Abstract: Magnesium alloys usually exhibit excellent superplasticity at high temperature. However, many Mg alloys have poor formation ability near room temperature. Therefore, preparation of Mg alloys with suitable microstructures to show low or intermediate temperature superplasticity is an important goal. In this work, the superplastic behavior at intermediate temperatures of a commercial ZK60 magnesium alloy processed by indirect extrusion was investigated. After extrusion, the alloy showed a refined and homogeneous microstructure with an average grain size of 4 ± 2 μm. Overall texture measurement indicated that the alloy showed a strong prismatic texture with the highest intensity oriented to pole $\langle 10\bar{1}0 \rangle$. A texture component $\langle \bar{1}2\bar{1}1 \rangle$ parallel to the extrusion direction was found; this type of texture is commonly observed in Mg alloys with rare earth additions. Tensile tests were performed at temperatures of 150, 200, and 250 °C at three strain rates of 10^{-2}, 10^{-3}, and 10^{-4} s^{-1}. A very high ductility was found at 250 °C and 10^{-4} s^{-1}, resulting in an elongation to failure of 464%. Based on calculations of the activation energy and on interpretation of the deformation mechanism map for magnesium alloys, it was concluded that grain boundary sliding (GBS) is the dominant deformation mechanism.

Keywords: magnesium alloys; texture; indirect extrusion

Citation: Palacios-Trujillo, C.; Victoria-Hernández, J.; Hernández-Silva, D.; Letzig, D.; García-Bernal, M.A. Superplasticity at Intermediate Temperatures of ZK60 Magnesium Alloy Processed by Indirect Extrusion. *Metals* **2021**, *11*, 606. https://doi.org/10.3390/met11040606

Academic Editor: Daolun Chen

Received: 1 February 2021
Accepted: 20 February 2021
Published: 9 April 2021

Publisher's Note: MDPI stays neutral with regard to jurisdictional claims in published maps and institutional affiliations.

1. Introduction

Magnesium alloys are increasingly used in electronics, automotive, and aerospace industries due to their low density, high specific strength, and excellent machinability [1]. However, the hexagonal crystal structure of magnesium limits its ductility, particularly at low temperatures. Magnesium exhibits only three slip systems at low temperature. Recent studies have shown that the ductility of this alloy can be improved with the reduction of grain size, and this could lead to superplastic behavior [1–5].

Superplasticity refers to the ability of a material to achieve high elongations of at least 200% when a sample is tested in tension [6]. By definition, superplasticity allows some metals and alloys to reach extensive ductility under restricted circumstances such as (1) low strain rates (usually between 10^{-5} and 10^{-3} s^{-1}), (2) a fine and stable microstructure, and (3) a deformation temperature around 0.5 Tm or higher (with Tm being the absolute melting temperature) [7]. The superplastic behavior in magnesium alloys can provide the ability to deform complex parts that are difficult to form [8].

One advantage of Al-free Mg alloys is the exceptional grain-refining ability of Zr. Furthermore, in combination with a severe plastic deformation process such as rolling, extrusion, drawing, equal channel angular pressing (ECAP), and friction stir processing (FSP), it can lead to stable fine grain structures in wrought alloys [9–13]. Among these, extrusion processes have aroused special interest, through which the microstructure can be strongly deformed, resulting in a refined microstructure and a suitable texture [13–16]. Specifically, the indirect extrusion process has shown a good combination of high strength and ductility at room temperature and superplasticity at elevated temperatures [16,17].

The purpose of this work is to investigate the superplastic behavior in a ZK60 commercial magnesium alloy processed by indirect extrusion by means of tensile tests at intermediate temperatures at three different strain rates and to elucidate the deformation mechanisms involved.

2. Materials and Methods

The commercial magnesium ZK60 alloy used in this study was processed by indirect extrusion, and it was provided by the consortium responsible for the MAGFORGE Project, a European Community Research on Forging of Magnesium Alloys [18]. The alloy was provided in the form of round bars of 2 in. diameter. The chemical composition of the alloy is given in Table 1.

Table 1. Alloy composition in weight percentage.

Alloy	Mg	Zn	Zr
ZK60	94.55	5.00	0.45

Samples for microstructural analysis were cut from the bars in order to analyze a plane parallel to the extrusion direction, as shown in Figure 1. Subsequently, they were ground and polished with a solution of OPS (colloidal silica of 0.05 μm), using as lubricant a solution of distilled water, liquid soap, and sodium hydroxide. The chemical attack to reveal the microstructure was carried out with a picric acid solution (150 mL of ethanol, 36 mL of distilled water, 6.5 mL of acetic acid, and 36 g of picric acid). An EBSD measurement was carried out using a field emission gun scanning electron microscope Zeiss Ultra 55 (Carl Zeiss AG, Oberkochen, Germany) equipped with an EDAX-TSL OIM system (EDAX-TSL OIM-7, Ametek-EDAX inc, Draper, UT, USA). The sample used was prepared in the same way as described above. In addition, it was electropolished using a Struers AC2 solution (Struers GmbH, Willich, Germany) at 16 V for 35 s at −25 °C to obtain a surface free of defects, such as deformation from mechanical polishing and scratches. The measurement was also carried out on the extrusion direction (ED)–normal direction (ND) plane. The EBSD measurement was conducted with an acceleration voltage of 15 kV and a step size of 0.3 μm.

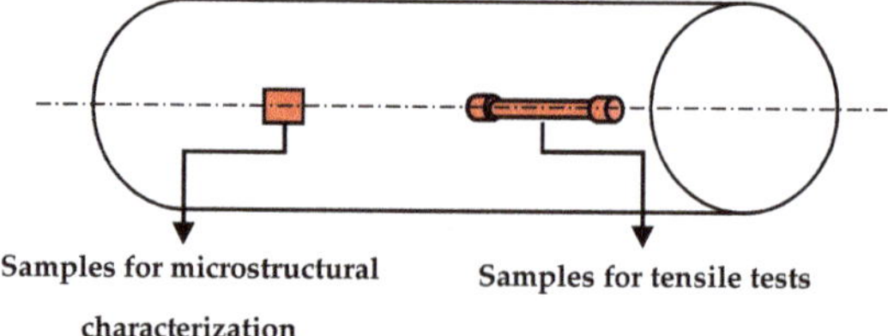

Figure 1. Location and orientation of samples for microstructural characterization and tensile tests.

The global texture was measured by means of a PANalytical X-ray diffractometer (Malvern Panalytical GmbH, Almelo, The Netherlands), using a texture goniometer and a sample in the form of a disc with a diameter of 1 cm. Before measurement, the sample was ground and polished. The results obtained were processed with the software X'Pert Texture (Data collector 2.0, Malvern Panalytical GmbH, Almelo, The Netherlands) to obtain the pole figures.

Tensile test samples were machined from the bars with the load direction parallel to the extrusion direction according to DIN 50125 M6, as shown in Figures 1 and 2. Tensile tests were carried out at intermediate temperatures of 150, 200, and 250 $^\circ$C and strain rates of 10^{-2}, 10^{-3}, and 10^{-4} s^{-1}.

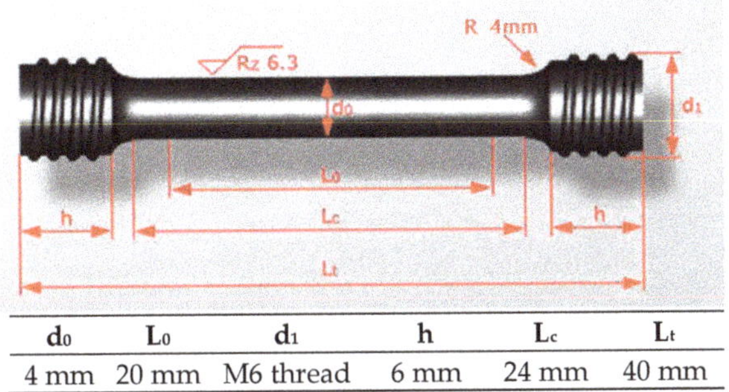

d$_0$	L$_0$	d$_1$	h	L$_c$	L$_t$
4 mm	20 mm	M6 thread	6 mm	24 mm	40 mm

Figure 2. Dimensions of tensile samples according to DIN 50125 M6.

The heating of the tensile test samples was applied by means of an electric furnace, which kept the specimens at a constant temperature within a range of ± 1 $^\circ$C. Before the test, the specimens were preheated to the selected temperature for 10 min in order to have a homogeneous and stable temperature. Once the test specimens had fractured, they were immediately quenched in water.

3. Results and Discussion

3.1. Initial Microstructure and Global Texture

The microstructure of the as-received commercial ZK60 alloy (see Figure 3a) showed recrystallized grains with an average grain size of 4 ± 2 µm measured by the linear intercept method. The microstructure showed additional bands of unrecrystalized grains. Shahzad et al. [17] studied these bands and found that they were zones of unrecrystallized grains formed by Zr segregations. Additionally, an analysis by EDS in SEM revealed particles rich in Zr and Zn. It is possible to consider that such particles belong to the Zr$_2$Zn phase previously identified [19].

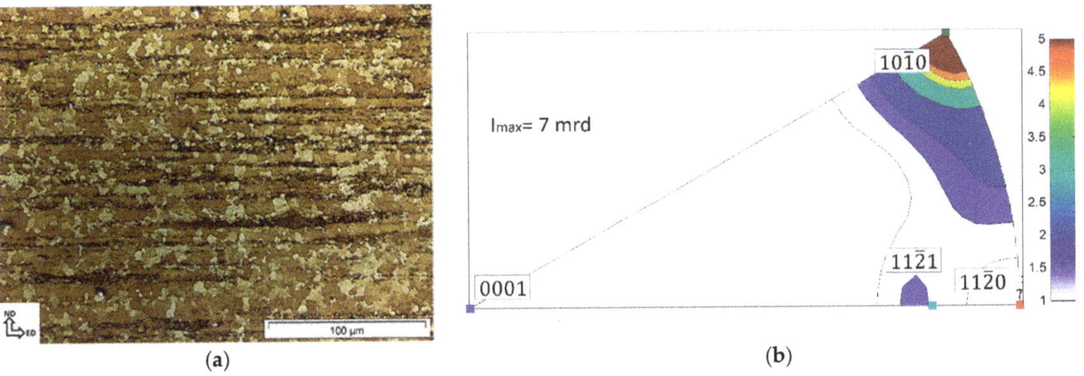

Figure 3. Microstructure of alloy ZK60 as received (**a**) and inverse pole figure (0001) (**b**).

In order to determine the texture of this alloy, pole figures of basal, prismatic, and pyramidal planes were constructed by means of X-ray diffraction. The inverse pole figures

(IPFs) were obtained with the help of the MTEX software (Version 5.3.1, free and open source software toolbox).

The commercial ZK60 alloy showed a strong prismatic texture (see Figure 3b), and the highest intensity was found at the pole $\langle 10\bar{1}0 \rangle$. That means that most of the grains were oriented in such a way that their basal plane was parallel to the extrusion direction.

During extrusion, a texture component $\langle 11\bar{2}1 \rangle$ was developed parallel to the extrusion direction (ED), which caused the grains oriented in this direction to be favorable for basal slip, resulting in an improvement in ductility at room temperature [20]. This type of texture component is not often found in conventional alloys, but it has been seen more frequently in magnesium alloys with additions of rare earth (RE) elements such as Ce, Nd, Gd, and Y [20,21]. However, a minor influence on the mechanical behavior of this texture component in tension was expected, since most of the grains were aligned around the $\langle 10\bar{1}0 \rangle$ pole. On the other hand, it could have contributed to a decrease in the yield asymmetry in samples tested in compression in comparison to the yield stress in tension. However, the analysis of the yield asymmetry is out of the focus of the present work. For more details on the influence of the $\langle 11\bar{2}1 \rangle$ texture component, readers are referred to [22].

3.2. Ductility

Figure 4 shows the engineering stress–strain curve of ZK60 magnesium alloy at room temperature at a strain rate of 10^{-3} s^{-1}. As a result, the material presented a yield stress = 223 MPa, UTS = 290 MPa, and a ductility of ~13%. These data agree with the literature [20].

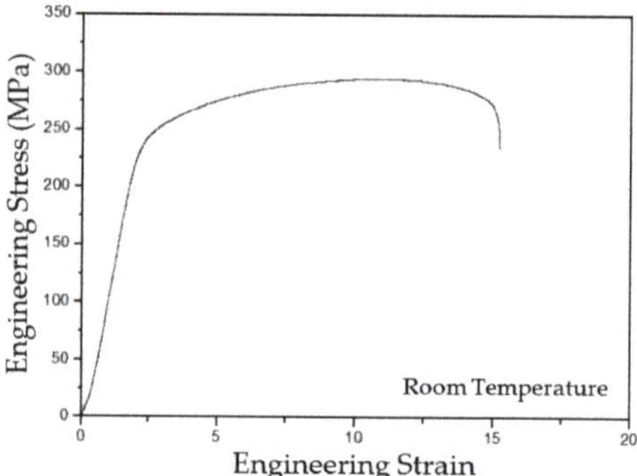

Figure 4. Stress–strain curve at strain rate of 10^{-3} s^{-1} for alloy ZK60.

At a temperature of 150 °C with a strain rate of 10^{-2} s^{-1}, the commercial ZK60 alloy exhibited a good ductility of 49%. As the strain rate decreased, the ductility increased, reaching an elongation to fracture of 89% at the lowest strain rate (see Figure 5). On the other hand, the yield stress and peak stress decreased as the strain rate decreased, while elongation to fracture increased, as shown in Figures 6 and 7. When increasing the test temperature to 200 °C, the behavior was very similar to that observed at the temperature of 150 °C, and at the lowest strain rate tested, an elongation to fracture of 137% was obtained.

Finally, at the highest test temperature used in this work (250 °C) and a strain rate of 10^{-4} s^{-1}, superplastic behavior was observed in this alloy, since it reached 464% of elongation to fracture. Several authors have already reported large percentages of elongation for this alloy. For example, W.J. Kim [23] reported ~1000% with a temperature of 250 °C and strain rate of 10^{-3} s^{-1} and attributed this percentage to a very fine grain size of 1.4 μm obtained by differential speed rolling. H. Watanabe et al. [24] also obtained a refined microstructure with a grain size of 1.4 μm processed by equal channel angular extrusion (ECAE) that reached elongation percentages above 400% at low strain rates and a temperature of 200 °C. The value reached in this study turns out to be quite acceptable, since this alloy was processed by conventional indirect extrusion and deformed at low temperature.

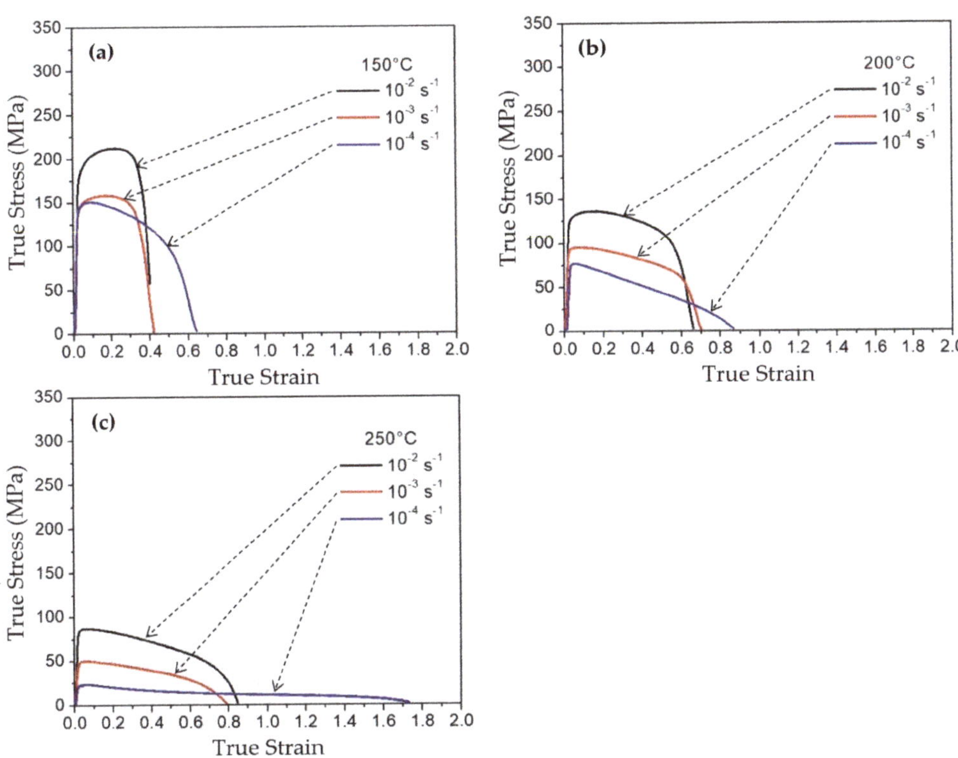

Figure 5. True stress–true strain curves at different temperatures and strain rates: (**a**) 150, (**b**) 200, and (**c**) 250 °C.

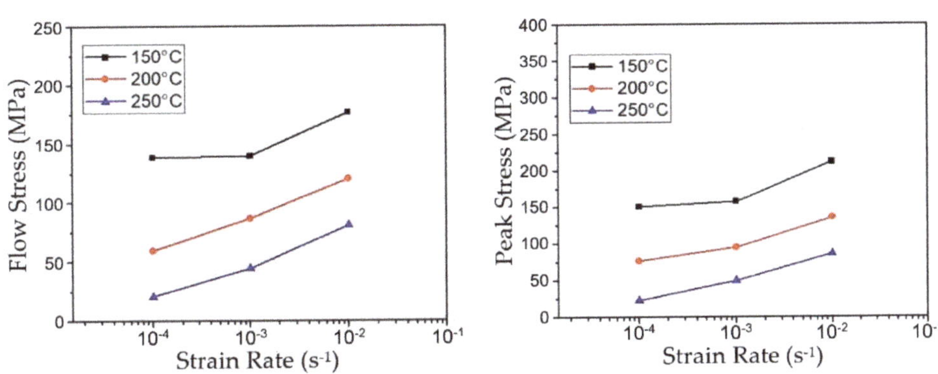

Figure 6. Flow and peak stress as a function of strain rate at different temperatures.

The samples deformed in tension at intermediate temperatures and three different strain rates, as well as at room temperature, are shown in Figure 8. In general, the samples present a ductile fracture. This type of fracture occurs when the material is subjected to a severe or excessive plastic deformation. In addition, this type of fracture can be recognized by the characteristic neck formation in the fractured area; however, this neck was less pronounced at lower strain rates. This is especially noticeable in the sample that reached superplastic-like behavior in which a rather diffuse neck can be observed.

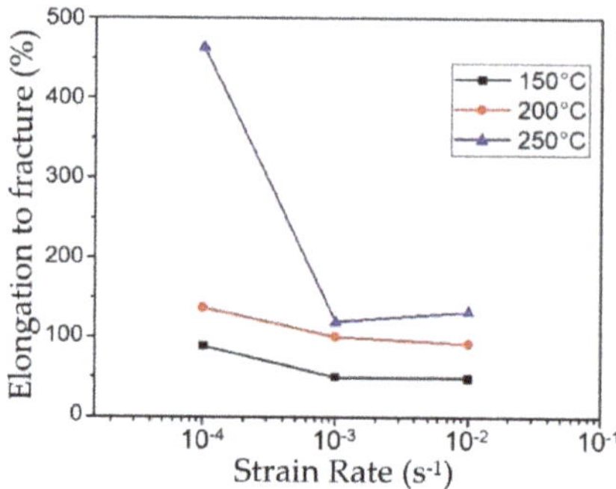

Figure 7. Elongation to fracture as a function of strain rate at different temperatures.

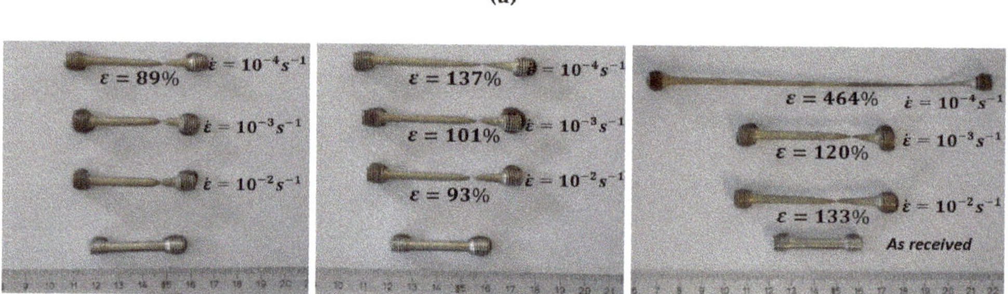

Figure 8. Appearance of tensile specimens after deformation to failure at (**a**) room temperature and at (**b**) 150, (**c**) 200, and (**d**) 250 °C under various strain rates.

3.3. Microstructure after Deformation

The microstructural changes observed after the elongation to failure tests are shown in Figures 9 and 10. The sample tested at room temperature shows a deformed microstructure with some signs of twinning inside the relatively large unrecrystallized grains. This is despite the initial texture in which the majority of grains have a basal-type texture. However, it should be mentioned that the micrograph was taken near the fracture. Thus, some instabilities due to necking could occur, which could have led to the twinning behavior (see Figure 9). Moreover, due to the initial texture with some off-basal texture components, the activation of {10$\bar{1}$2} extension twins was feasible. At elevated temperatures (see Figure 10), the micrographs show a fine-grained structure, especially above 200 °C. At a temperature of 150 °C, the grain was refined as the strain rate decreased. A very different behavior was observed at 200 °C, since the smallest grain size was found at the highest strain rate. It is clearly observed that the dynamic recrystallization process was present as the strain rate decreased, obtaining a homogeneous microstructure at 250 °C and 10^{-4} s^{-1}. Despite the coarser grain size in comparison to samples deformed at higher strain rates at 250 °C, this sample achieved the highest ductility in this study.

Figure 11 exhibits an inverse pole figure map and respective local texture of a sample deformed at 250 °C and a strain rate of 10^{-4} s^{-1}. Despite the fact that a high elongation to fracture was reached and that a significant dynamic recrystallization took place, the deformation texture shared many similarities with the initial texture; i.e., it retained the majority of grains oriented near the $\langle 10\bar{1}0 \rangle$ pole with no obvious development of the $\langle 112\bar{1} \rangle$

texture component. However, an important texture-weakening was observed, most likely because of the combined activity of dynamic recrystallization and the activation of grain boundary sliding (GBS).

Figure 9. Microstructure obtained near the fractured zone after tensile test at room temperature and strain rate of 10^{-3} s^{-1}.

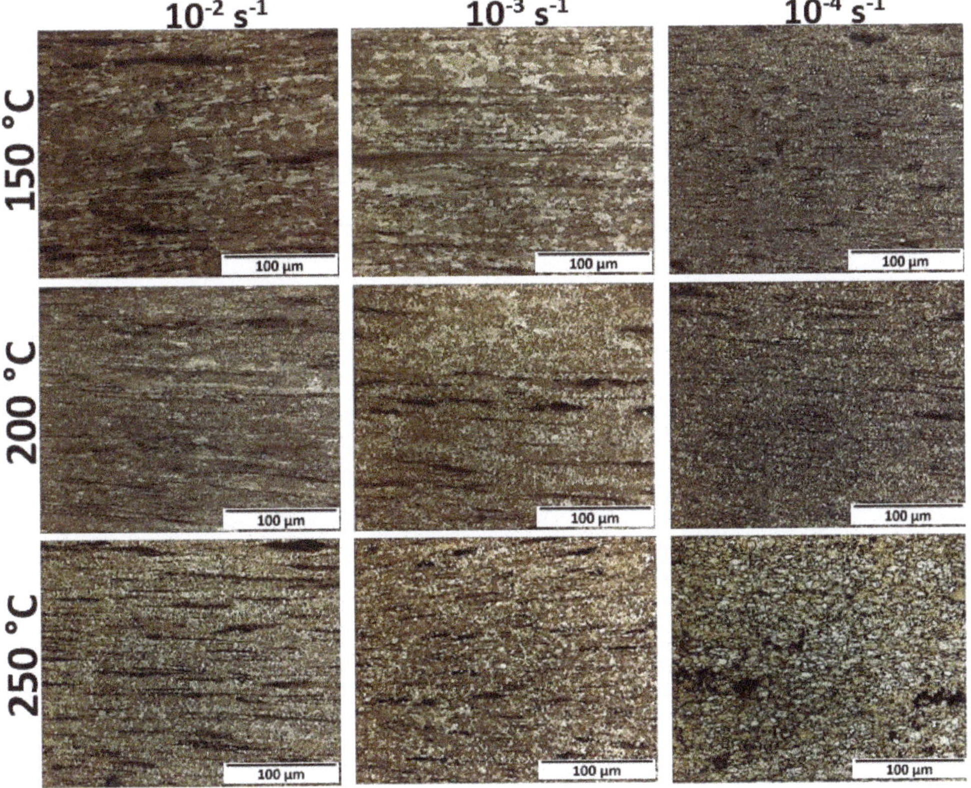

Figure 10. Microstructures obtained near the fractured zone after tensile tests at different temperatures and strain rates.

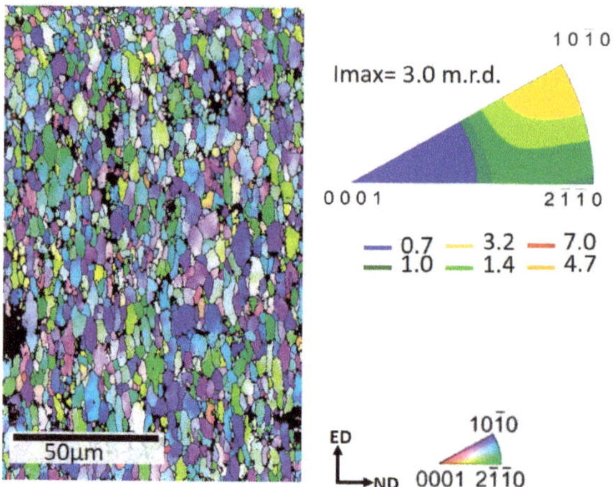

Figure 11. Inverse pole figure map and respective local texture measured near the fracture of a sample deformed at 250 °C and a strain rate of 10^{-4} s^{-1}.

3.4. Zener–Hollomon Parameter

In order to analyze the different values of mechanical properties obtained after the tensile tests, the Zener–Hollomon parameter was calculated for this alloy. The determination of this value (Z) was based on the study carried out by Liu et al. [25], according to the following equation:

$$Z = \dot{\varepsilon} \cdot exp\left(\frac{Q}{RT}\right) = A \cdot (sinh(\alpha\sigma))^n \tag{1}$$

where Z is the temperature-compensated strain rate (i.e., the Zener–Hollomon parameter); $\dot{\varepsilon}$ is the strain rate; Q is the activation energy; R is the gas constant; T is the absolute temperature; and A, α, and n are material constants. It was assumed that the relationship of the Zener–Hollomon parameter and the steady-rate stress were suitable for Equation (1). The values of A, y, and n can be obtained from the plot of $\ln(sinh(\alpha\sigma))$ versus $\ln Z$. However, the values of α and Q must be determined before the values of A and n are fixed. The approximate value of α is determined as follows.

The hot working favored the power law at high stress levels, and the equation is presented as follows:

$$Z \approx A_1 \sigma^{n\prime} \tag{2}$$

The favored equation is the exponent law at low stress levels, which is given by

$$Z \approx A_2 \exp(\beta\sigma) \tag{3}$$

Assuming the value of $n\prime = n$, the approximate value of n can be taken as the slope of the plot of $\ln\sigma$ versus $\ln\dot{\varepsilon}$ at low stress levels. It can be seen by comparing Equation (1) with Equation (3) that $A = A_2/2^n$ and the value of $\alpha = \beta/n$. The values of A_2 and β can be determined according to the σ versus $\ln\dot{\varepsilon}$ plot at high stress levels. It should also be noted that σ can be referred to as the peak stress σ_p [26], which is easier to obtain; therefore, it is the stress used in this study. The value of Q is calculated in the following equation:

$$Q = R\left[\frac{\partial \ln(sinh(\alpha\sigma_p))}{\partial(1/T)}\right]_{\dot{\varepsilon}} \cdot \left[\frac{\partial \ln\dot{\varepsilon}}{\partial \ln(sinh(\alpha\sigma_p))}\right]_T \tag{4}$$

On the right-hand side of the above equation, the first term represents the slope of the $\ln(sinh(\alpha\sigma_p))$ versus $1/T$ plot, called q_1, while the second term represents the reciprocal value of inclination of the $\ln(sinh(\alpha\sigma_p))$ versus $\ln\dot{\varepsilon}$ plot, called $1/q_2$.

The exact values of A and n are obtained from the plot $\ln\left(\sinh\left(\alpha\sigma_p\right)\right)$ versus $\ln Z$, being the relationship of σ_p and Z as follows:

$$\ln(Z) = \ln(A) + n \cdot \ln\left(\sinh\left(\alpha\sigma_p\right)\right) \tag{5}$$

In this way, the approximate values of n and β are estimated from Figure 12a,b, respectively. So, the value of α is 0.0056. On the other side, q_1 and $1/q_2$ are obtained from Figure 12c,d, respectively. As a result, the average value of Q is 466.1 kJ/mol. Now, substituting Q, R, T, and $\dot{\varepsilon}$ in Equation (1), the values of $\ln Z$ can be obtained, which are in the range of 50 to 250. Finally, the exact values of A and n are obtained from the $\ln\left(\sinh\left(\alpha\sigma_p\right)\right)$ versus $\ln Z$ plot (see Figure 12d), where n is 84.81 and $\ln A$ is 58.52.

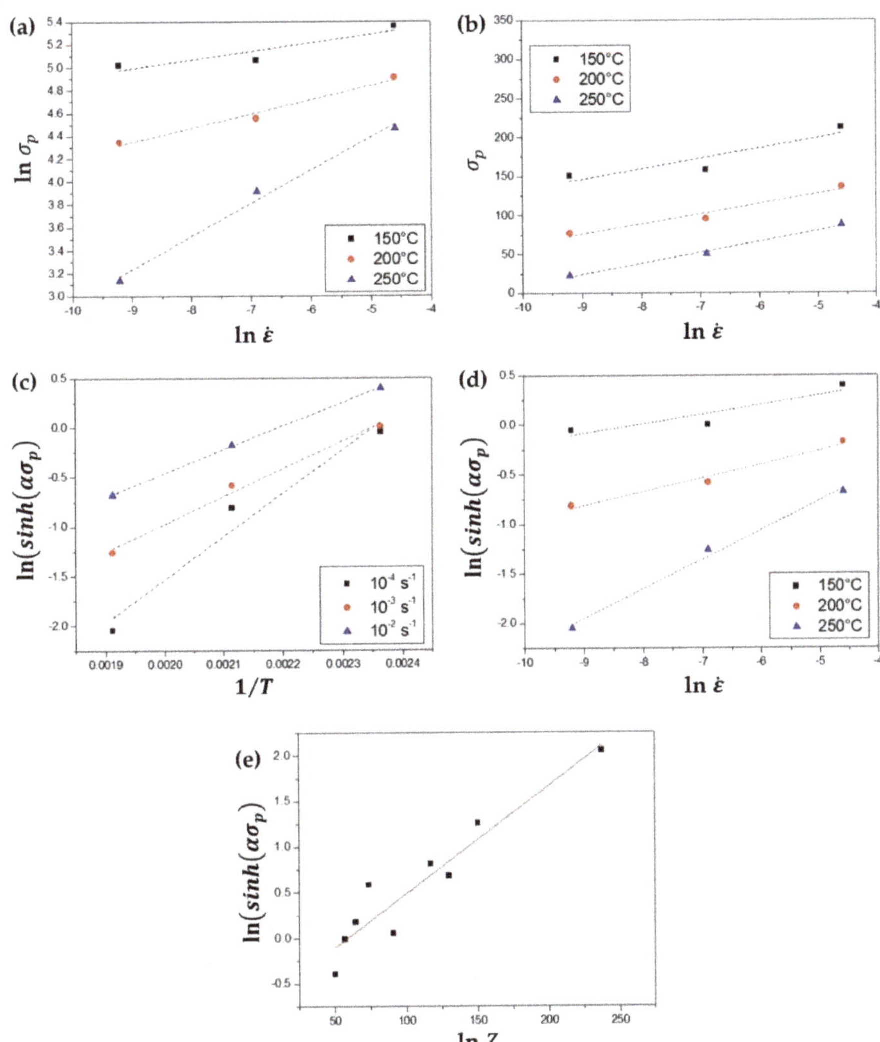

Figure 12. Schematic of the dependence of the peak stress on temperature and strain rate. Plots of (**a**) $\ln\sigma_p$ versus $\ln\dot{\varepsilon}$, (**b**) σ_p versus $\ln\dot{\varepsilon}$, (**c**) $\ln(\sinh(\alpha\sigma_p))$ versus $1/T$, (**d**) $\ln(\sinh(\alpha\sigma_p))$ versus $\ln\dot{\varepsilon}$, and (**e**) $\ln(\sinh(\alpha\sigma_p))$ versus $\ln Z$.

The calculation of the activation energy allows an estimation of the deformation mechanism that predominates during the deformation under different conditions of strain rate and temperature. Based on the obtained average activation energy of 466.1 kJ/mol through the range of strain rates and temperature, the dominant deformation mechanism in the present alloy is grain boundary sliding (GBS) [27].

Another way to establish the dominant deformation mechanism involved in this alloy is through the method developed by Kim et al. [3,4].

Several criteria must be met to expect the material to behave superplastically [28], and this alloy meets some of the requirements. For example, it has a fine and stable microstructure at the end of the tensile test at 250 °C and 10^{-4} s^{-1} with an average grain size of ~4 μm, and there is the presence of second-phase particles that inhibit grain growth. On the other hand, superplastic materials generally exhibit high values of strain rate sensitivity "m" during stress–strain tests [27–29], which is expressed by the following equation [8,28,30]:

$$\sigma = k\dot{\varepsilon}^{m} \tag{6}$$

where σ is the true flow stress, k is a constant, and $\dot{\varepsilon}$ is the strain rate. Metals normally exhibit $m < 0.2$ and superplastic alloys commonly have values of $m > 0.33$. However, the commercial ZK60 alloy of this study that obtained 464% elongation atfracture has a value of $m = 0.29$.

Bussiba et al. [31] reported an elongation of 220% at a strain rate of 1×10^{-5} s^{-1} in a ZK60 alloy with a value of $m = 0.2$; they concluded that grain boundary sliding (GBS) was the dominant deformation mechanism.

H. Watanabe [24] demonstrated a superplastic behavior in a conventional ZK60 alloy by reporting an elongation greater than 400% during tension in the temperature range from 200 to 225 °C and strain rates from 2.6×10^{-6} to 3×10^{-5} s^{-1} with values of $m = 0.5$.

For the determination of the deformation mechanism that predominates during deformation, a deformation mechanism map (DMM) created by Kim et al. [3] was used, since this is valid for magnesium alloys in general, according to the author mentioned. The calculation of the activation energy allows for an estimation of the deformation mechanism that predominates during the deformation under different conditions of strain rate and temperature.

It was observed that the deformation mechanisms that were activated during tensile tests for ZK60 alloys were dislocation glide and grain boundary sliding (GBS), depending on the test conditions. As the stress decreased (at lower strain rates), the grain boundary sliding (GBS) was activated.

Although the DMM created by Kim et al. [3] did not correspond to that of the ZK series alloys, it provided a close estimate for magnesium-based alloys in general.

It can be established that as ln Z decreases, the elongation to fracture will increase with a concomitant decrease of the peak stress (see Figure 13). A similar case was reported by Victoria-Hernández et al. [32]. For an AZ31 alloy with a ductility above ~400%, ln Z was 19, while for an AZ61 alloy with a ductility above 500%, ln Z had a value of 28—much lower values when compared to the alloys studied in this work. On the other hand, the maximum stress values are much higher in this work when compared to the study by Victoria-Hernández et al. [32], due to the strengthening effect of the alloying elements where Zr played an important role in controlling the grain size even after the occurrence of dynamic recrystallization.

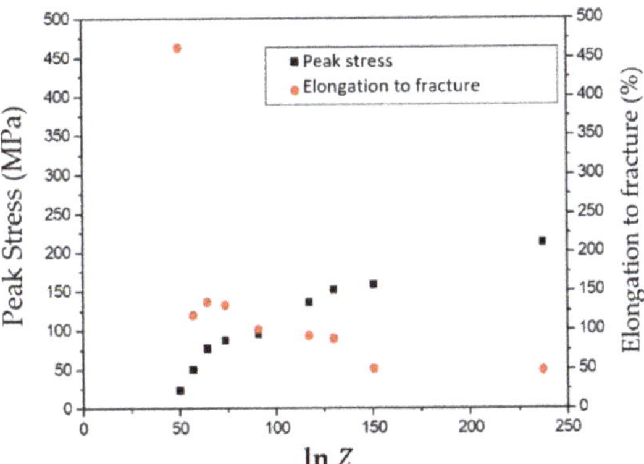

Figure 13. Tensile strength and elongation to failure versus $\ln Z$ plot.

4. Conclusions

A homogeneous and refined microstructure with a grain size of 4 ± 2 μm was obtained after conventional thermomechanical treatment and attributed to the amount of zirconium. The overall texture results show the development of the conventional basal texture where most of the basal planes are aligned parallel to ED. The texture component $\langle 1\bar{2}\bar{1}1 \rangle$ was observed parallel to the extrusion direction; this type of texture is commonly observed in Mg alloys with rare earth (RE) additions. However, it did not seem to play a significant role during further tensile deformation at room temperature and after tensile tests at elevated temperatures. There was no strengthening of the aforementioned pole, even in the case of a sample deformed under a superplastic regime.

The commercial ZK60 alloy showed excellent ductility at 250 °C at three different strain rates, especially at the slowest strain rate (10^{-4} s^{-1}). This alloy was able to present a superplastic behavior, as it achieved a peak elongation of 464%, which is explained due to the combined effect of dynamic recrystallization and the likely activation of GBS. The latter mechanism was believed to be dominant during tests at 250 °C temperature and slow strain rates. As calculated, the activation energy indicates the feasible activation of this mechanism. Furthermore, the texture-weakening effect found in the sample deformed under a superplastic regime suggests the dominant activation of this deformation mechanism. By increasing the strain rate or reducing the temperature, the dominant deformation mechanism seemed to change to dislocation glide.

Author Contributions: Conceptualization, J.V.-H. and C.P.-T.; methodology, J.V.-H. and C.P.-T.; formal analysis, C.P.-T. and D.H.-S.; investigation, C.P.-T.; writing—original draft preparation, M.A.G.-B.; writing—review and editing, M.A.G.-B. and D.H.-S.; visualization, M.A.G.-B..; supervision, D.H.-S.; project administration, D.L. and J.V.-H.; funding acquisition, D.L. All authors have read and agreed to the published version of the manuscript.

Funding: This research received no external funding.

Data Availability Statement: Data supporting reported results can be shared on a reasonable request to corresponding authors.

Acknowledgments: C.P.-T. is grateful for the fruitful discussions with J. Bohlen during his research stay at HZG. C.P.-T. also acknowledges the economic support (Becas Mixtas) of CONACyT-Mexico for his research stay in Germany. The technical support of Y. Shin and A. Reichert at HZG is gratefully acknowledged.

Conflicts of Interest: The authors declare no conflict of interest.

References

1. Zheng, M.; Wu, K.; Liang, M.; Kamado, S.; Kojima, Y. The effect of thermal exposure on the interface and mechanical properties of Al18B4O33w/AZ91 magnesium matrix composite. *Mater. Sci. Eng. A* **2004**, *372*, 66–74. [CrossRef]
2. Panicker, R.; Chokshi, A.; Mishra, R.; Verma, R.; Krajewski, P. Microstructural evolution and grain boundary sliding in a superplastic magnesium AZ31 alloy. *Acta Mater.* **2009**, *57*, 3683–3693. [CrossRef]
3. Kim, W.-J.; Chung, S.; Chung, C.; Kum, D. Superplasticity in thin magnesium alloy sheets and deformation mechanism maps for magnesium alloys at elevated temperatures. *Acta Mater.* **2001**, *49*, 3337–3345. [CrossRef]
4. Kim, W.J.; Park, J.D.; Yoon, U.S. Superplasticity and superplastic forming of Mg–Al–Zn alloy sheets fabricated by strip cast-ing method. *J. Alloys Compd.* **2008**, *464*, 197–204. [CrossRef]
5. Watanabe, H.; Fukusumi, M. Mechanical properties and texture of a superplastically deformed AZ31 magnesium alloy. *Mater. Sci. Eng. A* **2008**, *477*, 153–161. [CrossRef]
6. Watanabe, H.; Fukusumi, M.; Somekawa, H.; Mukai, T. Texture and mechanical properties of superplastically deformed magnesium alloy rod. *Mater. Sci. Eng. A* **2010**, *527*, 6350–6358. [CrossRef]
7. Raghavan, K.S. Superplasticity. *Bull. Mater. Sci.* **1984**, *6*, 689–698. [CrossRef]
8. Padmanabhan, K.A.; Vasin, R.A.; Enikeev, F.U. *Superplastic Flow, Phenomenology and Mechanics*, 1st ed.; Springer: Berlin/Heidelberg, Germany, 2001.
9. Zhou, H.; Ye, B.; Wang, Q.; Guo, W. Uniform fine microstructure and random texture of Mg–9.8Gd–2.7Y–0.4Zr magnesium alloy processed by repeated-upsetting deformation. *Mater. Lett.* **2012**, *83*, 175–178. [CrossRef]
10. Valive, R.Z.; Langdon, T.G. Principles of equal-channel angular pressing as a processing tool for grain refinement. *Prog. Mater. Sci.* **2006**, *51*, 881–981. [CrossRef]
11. Yu, Y.; Kuang, S.; Chu, D.; Zhou, H.; Li, J.; Li, C. Microstructure and Low-Temperature Superplasticity of Fine-Grain ZK60 Magnesium Alloy Produced by Equal-Channel Angular Pressing. *Met. Microstruct. Anal.* **2015**, *4*, 518–524. [CrossRef]
12. García-Bernal, M.A.; Mishra, R.S.; Verma, R.; Hernández-Silva, D. Influence of friction stir processing tool design on microstructure and superplastic behavior of Al-Mg alloys. *Mater. Sci. Eng. A* **2016**, *670*, 9–16. [CrossRef]
13. Zhanga, T.; Cuia, H.; Cuia, X.; Chenb, H.; Zhaoc, E.; Changd, L.; Pana, Y.; Fenga, R.; Zhaie, S.; Chai, S. Effect of addition of small amounts of samariumon microstructural evolution and mechanical properties enhancement of an as-extruded ZK60 magnesium alloy sheet. *J. Mater. Res. Technol.* **2020**, *9*, 133–141. [CrossRef]
14. Zhou, W.; Lin, J.; Dean, T.A.; Wang, L. Feasibility studies of a novel extrusion process for curved profiles: Experimentation and modelling. *Int. J. Mach. Tools Manuf.* **2018**, *126*, 27–43. [CrossRef]
15. Zhou, W.; Yu, J.; Lin, J.; Dean, T.A. Manufacturing a curved profile with fine grains and high strength by differential velocity sideways extrusion. *Int. J. Mach. Tools Manuf.* **2019**, *140*, 77–88. [CrossRef]
16. Kim, B.; Baek, S.; Lee, J.; Park, S. Enhanced strength and plasticity of Mg–6Zn–0.5Zr alloy by low-temperature indirect extru-sion. *J. Alloys Compd.* **2017**, *706*, 56–62. [CrossRef]
17. Shahzad, M.; Wagner, L. Microstructure development during extrusion in a wrought Mg–Zn–Zr alloy. *Scr. Mater.* **2009**, *60*, 536–538. [CrossRef]
18. Sillekens, W.H.; Letzig, D. The MAGFORGE Project: European Community research on forging of magnesium alloys. In Proceedings of the 7th International Conference on Magnesium Alloys and their Applications, Dresden, Germany, 6–9 November 2006.
19. Friedrich, H.E.; Mordike, B.L. *Magnesium Technology, Metallurgy, Design Data, Applications*; Springer: Berlin/Heidelberg, Germany, 2006.
20. Jiang, M.; Xu, C.; Nakata, T.; Yan, H.; Chen, R.; Kamado, S. Rare earth texture and improved ductility in a Mg-Zn-Gd alloy after high-speed extrusion. *Mater. Sci. Eng. A* **2016**, *667*, 233–239. [CrossRef]
21. Luo, A.A.; Mishra, R.K.; Sachdev, A.K. High-ductility magnesium–zinc–cerium extrusion alloys. *Scr. Mater.* **2011**, *64*, 410–413. [CrossRef]
22. Cano-Castillo, G.; Victoria-Hernández, J.; Bohlen, J.; Letzig, D.; Kainer, K.U. Effect of Ca and Nd on the microstructural development during dynamic and static recrystallization of indirectly extruded Mg–Zn based alloys. *Mater. Sci. Eng. A* **2020**, *793*, 139527. [CrossRef]
23. Kim, W.J.; Kim, M.J.; Wang, J.Y. Superplastic behavior of a fine-grained ZK60 magnesium alloy processed by high-ratio dif-ferential speed rolling. *Mater. Sci. Eng. A* **2009**, *527*, 322–327. [CrossRef]
24. Watanabe, H.; Mukai, T.; Ishikawa, K.; Higashi, K. Low temperature superplasticity of a fine-grained ZK60 magnesium alloy processed by equal-channel-angular extrusion. *Scr. Mater.* **2002**, *46*, 851–856. [CrossRef]
25. Liu, J.; Cui, Z.; Li, C. Modelling of flow stress characterizing dynamic recrystallization for magnesium alloy AZ31B. *Comput. Mater. Sci.* **2008**, *41*, 375–382. [CrossRef]
26. McQueen, H.; Ryan, N. Constitutive analysis in hot working. *Mater. Sci. Eng. A* **2002**, *322*, 43–63. [CrossRef]
27. Frost, H.J.; Ashby, M.F. *Deformation Mechanism Maps: The Plasticity and Creep of Metals and Ceramics*; Pergamon Press: Oxford, UK, 1982.
28. Sherby, O.D.; Wadsworth, J. Superplasticity—Recent advances and future directions. *Prog. Mater. Sci.* **1989**, *33*, 169–221. [CrossRef]

29. Olguín-González, M.; Hernández-Silva, D.; García-Bernal, M.; Sauce-Rangel, V. Hot deformation behavior of hot-rolled AZ31 and AZ61 magnesium alloys. *Mater. Sci. Eng. A* **2014**, *597*, 82–88. [CrossRef]
30. Edington, J.W.; Melton, K.N.; Cutler, C.P. Superplasicity. *Prog. Mater. Sci.* **1976**, *21*, 61–170. [CrossRef]
31. Bussiba, A.; Ben Artzy, A.; Shtechman, S.; Ifergan, M. Kupiec. Grain refinement of AZ31 and ZK60 Mg alloys — towards superplasticity studies. *Mater. Sci. Eng. A* **2001**, *302*, 56–62. [CrossRef]
32. Victoria-Hernández, J.; Yi, S.; Letzig, D.; Hernández, D.; Bohlen, J. Microstructure and texture development Mg-Al-Zn alloys during tensile testing at intermediate temperatures. *Acta Mater.* **2013**, *61*, 2179–2193. [CrossRef]

Article

Fatigue Properties of AZ31B Magnesium Alloy Processed by Equal-Channel Angular Pressing

Ryuichi Yamada [1,*], Shoichiro Yoshihara [2] and Yasumi Ito [1]

1   Graduate Faculty of Interdisciplinary Research Faculty of Engineering, Mechanical Engineering (Mechanical Engineering), University of Yamanashi, 4-3-11 Takeda, Kofu-shi 400-8511, Japan; yasumii@yamanashi.ac.jp
2   Department of Engineering and Design, Shibaura Institute of Technology, 3-9-14 Minato-ku, Tokyo 108-8548, Japan; yoshi@shibaura-it.ac.jp
*   Correspondence: ryamada@yamanashi.ac.jp; Tel.: +81-55-220-8091

Abstract: A stent is employed to expand a narrowed tubular organ, such as a blood vessel. However, the persistent presence of a stainless steel stent yields several problems of late thrombosis, restenosis and chronic inflammation reactions. Biodegradable magnesium stents have been introduced to solve these problems. However, magnesium-based alloys suffer from poor ductility and lower than desired fatigue performance. There is still a huge demand for further research on new alloys and stent designs. Then, as fundamental research for this, AZ31 B magnesium alloy has been investigated for the effect of equal-channel angular pressing on the fatigue properties. ECAP was conducted for one pass and eight passes at 300 °C using a die with a channel angle of 90°. An annealed sample and ECAP sample of AZ31 B magnesium alloy were subjected to tensile and fatigue tests. As a result of the tensile test, strength in the ECAP (one pass) sample was higher than in the annealed sample. As a result of the fatigue test, at stress amplitude σ_a = 100 MPa, the number of cycles to failure was largest in the annealed sample, medium in the ECAP (one pass) sample and lowest in the ECAP (eight passes) sample. It was suggested that the small low cycle fatigue life of the ECAP (eight passes) sample is attributable to severe plastic deformation.

Keywords: magnesium alloy; fatigue; equal-channel angular pressing; grain refinement; S–N curve; stent

Citation: Yamada, R.; Yoshihara, S.; Ito, Y. Fatigue Properties of AZ31B Magnesium Alloy Processed by Equal-Channel Angular Pressing. *Metals* **2021**, *11*, 1191. https://doi.org/10.3390/met11081191

Academic Editor: Sangbong Yi

Received: 30 June 2021
Accepted: 22 July 2021
Published: 26 July 2021

Publisher's Note: MDPI stays neutral with regard to jurisdictional claims in published maps and institutional affiliations.

1. Introduction

Ischaemic heart disease is the world's biggest killer, accounting for a combined 9.4 million deaths in 2016 [1]. This disease has remained the leading cause of death glob-ally in the last 15 years [1]. A stent, as a countermeasure against ischaemic heart disease, is employed to expand a narrowed tubular organ, such as a blood vessel. A stent is a tiny tube that a doctor places in an artery or duct to help keep it open and restore the flow of bodily fluids in the area. Stents help relieve blockages and treat narrow or weakened arteries. Despite this, the persistent presence of a metallic stent yields several problems of late thrombosis, restenosis and chronic inflammation reactions. Conventionally, an austenite-based stainless steel that is a general stainless steel, such as SUS316, SUS316 L, etc., is used for the material of a stent for vasodilatation [2]. Once the stainless steel stent is implanted, it will remain in an artery permanently because of its excellent corrosion resistance. In some cases, restenosis may occur [3]. Restenosis is when too much tissue grows around the stent. This could narrow or block the artery again. In general, a stent should be removed within 6 to 9 months. Stent removal surgery is expensive and physically demanding for the patient. Biodegradable magnesium stents have been introduced to solve these problems. There is still a huge demand for further research on new alloys and stent design [4,5]. Ideally, implanted stents can maintain their mechanical integrity during the healing of the vessel wall and then dissolve after healing. The mechanical strength and properties of magnesium are suitable for biodegradable implants, especially for stent application. Magnesium is bi-ocompatible because it is essential for several biological reactions and as a co-factor for enzymes. However, magnesium degradation is accelerated in chloride-abundant en-vironments, such as human body fluids. Therefore, magnesium must be modified to improve its corrosion resistance. In addition, magnesium-based alloys suffer from poor ductility [6] and lower than desired fatigue performance. Further, the great success of stents in treating cardiovascular disease is undermined by their long-term fatigue failure. Pulsatile pressure and repetitive mechanical forces within the coronary artery may result in fatigue fracture after stent implantation, particularly in patients with complex coronary disease [7]. It has also been reported that stents implanted near the heart have a higher probability of fatigue failure. In this research group, it has been found that the corrosion rate of magnesium alloy becomes larger under pulsating flow [8]. If the fatigue properties can be improved, there is a possibility of reducing the corrosion rate. These aspects make fatigue properties an important attribute of cardi-ovascular stents. In this study, we focused on the possibility of improving the strength and fatigue properties simultaneously by grain refinement. Then, as fundamental re-search for this, AZ31 B magnesium alloy has been investigated for the effect of equal-channel angular pressing on tensile and fatigue properties.

2. Materials and Methods

2.1. Specimens

The chemical composition (in mass %) of the AZ31 B magnesium alloy made by MACRW Co., Ltd. (Shizuoka, Japan) used in this study is listed in Table 1. The AZ31 B alloy was hot extruded to a rod with a diameter of 6 mm and then cut into pieces with a length of 60 mm. ECAP was conducted on the as-extruded material through a die with an internal angle φ of 90° between the vertical and horizontal channels and a curvature angle ψ of 90° (Figure 1). As a lubricant, we used molybdenum disulphide (MoS_2) manufactured by Sumico Lubricant Co., Ltd. (Tokyo, Japan). Conditions of ECAP are presented in Table 2. ECAP was conducted for 1 pass and 8 passes, which creates an equivalent strain of 0.91 during one passage through the die as shown in Equation (1) [9].

$$\varepsilon_N = \left(\frac{N}{\sqrt{3}} \right) [2cot\left\{ \left(\frac{\varphi}{2} \right) + \left(\frac{\psi}{2} \right) \right\} + \psi cosec\left\{ \left(\frac{\varphi}{2} \right) + \left(\frac{\psi}{2} \right) \right\}] \tag{1}$$

ε_N is the strain of the material after ECAP for N passes and N is extrusion pass number. For comparison, the annealed sample for AZ31 B consisted of hold for 1 h at 450 °C, furnace cooling. ECAP-8 was used because the maximum number of passes that would not cause cracking under the conditions of this experiment was 8 passes. Furthermore, ECAP-8 p samples were found to have inferior fatigue properties compared to annealed samples [10]. ECAP-1 p was also added to investigate the effect of reducing the number of passes for ECAP. Three distinct kinds of test pieces, i.e., ECAP-1 p, ECAP-8 p and annealed, had commonly equiaxed microstructures (Figure 2). The average grain sizes of annealed and ECAP-8 p samples were confirmed by optical microscope observation to be 40 μm and 6–7 μm, respectively. ECAP-1 p had a bimodal structure, and it was determined that accurate grain size could not be calculated from the average grain size. It was the optical microscope structure in which coarse grains of 30–40 μm and fine grains of less than 10 μm were mixed.

Table 1. Composition of the specimens in mass %. Bai. in Mg is an abbreviation for balance.

	Al	**Zn**	**Mn**	**Si**	**Fe**	**Mg**
AZ31 B	3.14	1.11	0.33	0.02	0.001	Bal.

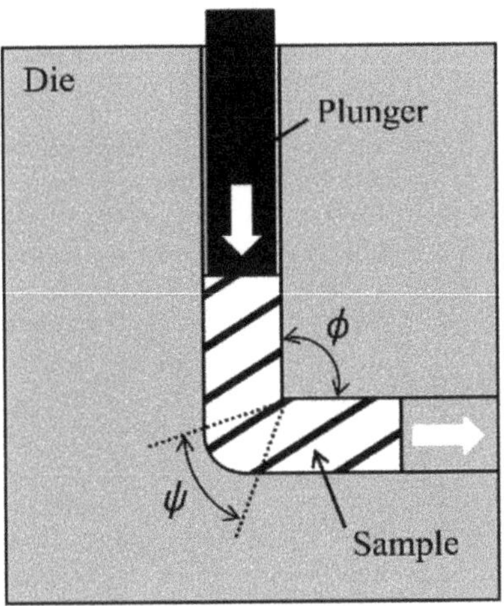

Figure 1. Schematic diagram of ECAP processing.

Table 2. Composition of the specimens in mass%.

Processing Temperature	300 °C
Processing rate	3 mm/min
Hole diameter of die	6 mm
Channel angle φ, ψ	90°, 90°
Number of passes (Rout Bc)	1 pass 8 passes

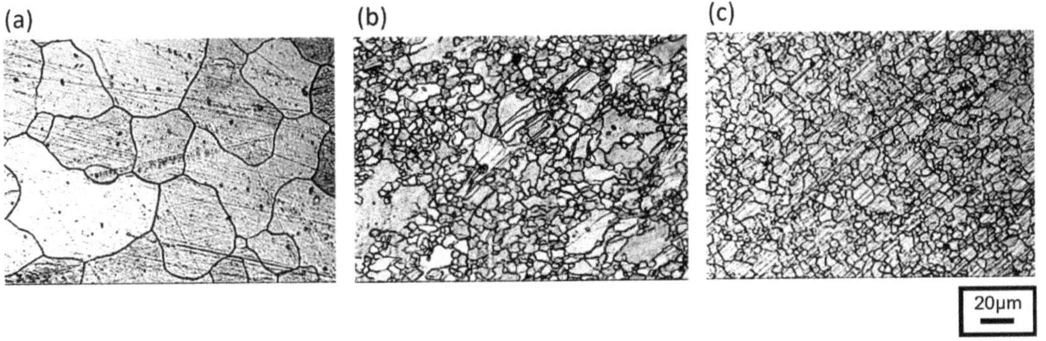

Figure 2. Optical micrographs of AZ31 B alloy. (**a**) Annealed, (**b**) ECAP-1 p and (**c**) ECAP-8 p.

2.2. Experimental Procedure

From the hot-extruded rod or ECAPed sample, tensile and fatigue test pieces were produced by machining. The gauge length and diameter of tensile and fatigue test pieces were 15 mm and 3 mm, respectively. The round-rod test pieces were turned on a lathe and manually mechanically polished and buffed to a mirror finish. The tensile test was carried out on one specimen of each type at an initial strain rate of 1.5×10^{-3} s^{-1} at room temperature. ECAP and tensile tests were performed by an Autograph AG-250 kNG, a precision universal tensile tester manufactured by SHIMADZU CORPORATION (Kyoto,

Japan). Fatigue test was carried out on one specimen of each type in a sinusoidal stress wave of a frequency of 30 Hz with a stress ratio of R (= K_{min}/K_{max}) = 0.1 at room temperature, followed by fractographic observation by a scanning electron microscope (SEM, JSM-7100 F by JEOL Ltd., Tokyo, Japan). The fatigue testing machine was an ElectroPuls E10000 Linear-Torsion all-electric dynamic test instrument manufactured by INSTRON Japan (Kanagawa, Japan). Residual stress measurement was carried out by X-ray diffraction (XRD, SmartLab by Rigaku Corporation, Tokyo, Japan) on the gripping surface of the test piece after the fatigue test. XRD was employed to measure the residual stress of the rods, in which Cr Kα radiation was used as the X-ray source. X-ray generator was attached by a collimator, which could carry out the incident beam of 1 mm spot sizes.

3. Experimental Results

3.1. Tensile Test

Figure 3 shows nominal stress vs. strain curves for the three specimens. Strength in the ECAP-1 p sample was higher than in the annealed sample. The tensile strength of ECAP-1 p was 321 MPa, which was slightly improved by about 1.2 times compared to the annealed sample. In contrast, elongation to failure was larger in the ECAP-8 p sample than in the annealed sample. In spite of processing more, ECAP-8 p had lower tensile strength and lower yield strength than ECAP-1 p. The elongation to failure of ECAP-8 p was 23%, which was improved by about 1.4 times compared to the annealed sample. Tensile properties obtained from tensile tests for the three specimens are summarized in Table 3.

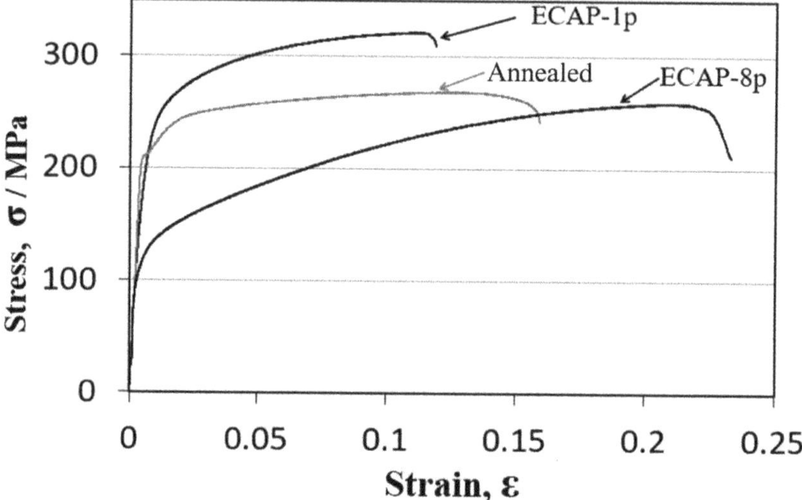

Figure 3. Nominal stress vs. strain curves for annealed, ECAP-1 p and ECAP-8 p samples of AZ31 B alloy.

Table 3. Tensile properties (ultimate tensile strength σ_u, yield strength σ_y, elongation to failure δ) of AZ31B alloy.

	Tensile Strength σ_u MPa	Yield Strength σ_y MPa	Elongation δ %
AZ31B Annealed	269	217	16
AZ31B ECAP-1p	321	200	12
AZ31B ECAP-8p	259	126	23

3.2. Fatigue Test

As a result of the fatigue test in Figure 4, at stress amplitude σ_a = 100 MPa number of cycles to failure was largest in the annealed sample, medium in the ECAP-1 p sample and lowest in the ECAP-8 p sample. At stress amplitude σ_a = 80 MPa, the number of cycles to failure was largest in the ECAP-1 p sample, medium in the annealed sample and lowest in the ECAP-8 p sample. It was confirmed that grain refinement improved fatigue life at low stress amplitude. When the stress amplitude σ_a exceeds 80 MPa, the annealed sample has the best fatigue properties and the ECAP-8 p sample has the worst fatigue properties, suggesting that among tensile properties, strength is more effective in improving fatigue properties than ductility. And since the fatigue properties of the annealed sample are superior to those of the ECAP-1 p sample, it is suggested that increasing the yield strength rather than the tensile strength is effective in improving the fatigue properties.

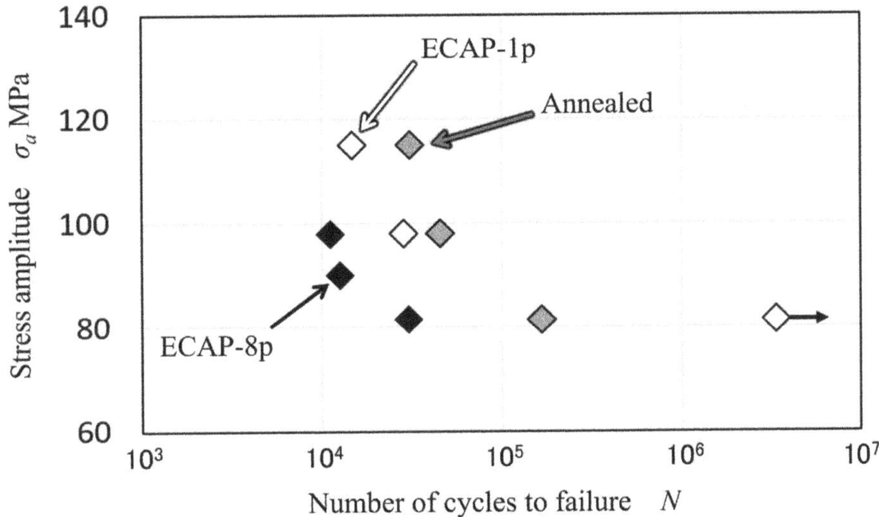

Figure 4. S–N curves for annealed, ECAP-1 p and ECAP-8 p samples of AZ31 B alloy.

Figure 5 shows the SEM images of fracture surface at stress amplitude σ_a = 100 MPa. From this, many fatigue crack initiation sites were located at each fracture surface. In all fracture surfaces, the crack growth area extends radially from the starting point of the fracture indicated by the arrow to the area delineated by the dashed line. In ECAP-1 p and ECAP-8 p, which are shown in Figure 5b,c, the crack growth areas from different fracture initiation points overlap each other. In ECAP-8 p, which had a large number of passes, the number of starting points for fracture was four, which was larger than the two in ECAP-1 p, suggesting early fatigue fracture. In the annealed sample, the starting points of the destruction overlap, however they are not completely connected. Therefore, it is inferred that the fatigue life of ECAP-1 p and ECAP-8 p was reduced by the sudden decrease in the cross section of the parallel section of the fatigue specimen when the crack growth areas overlapped each other. Although it is qualitative, the crack growth area ratio is larger in the annealed and ECAP-1 p sample than in ECAP-8 p. Therefore, it is possible that the crack growth area ratio of the fracture surface increases with the superior fatigue property.

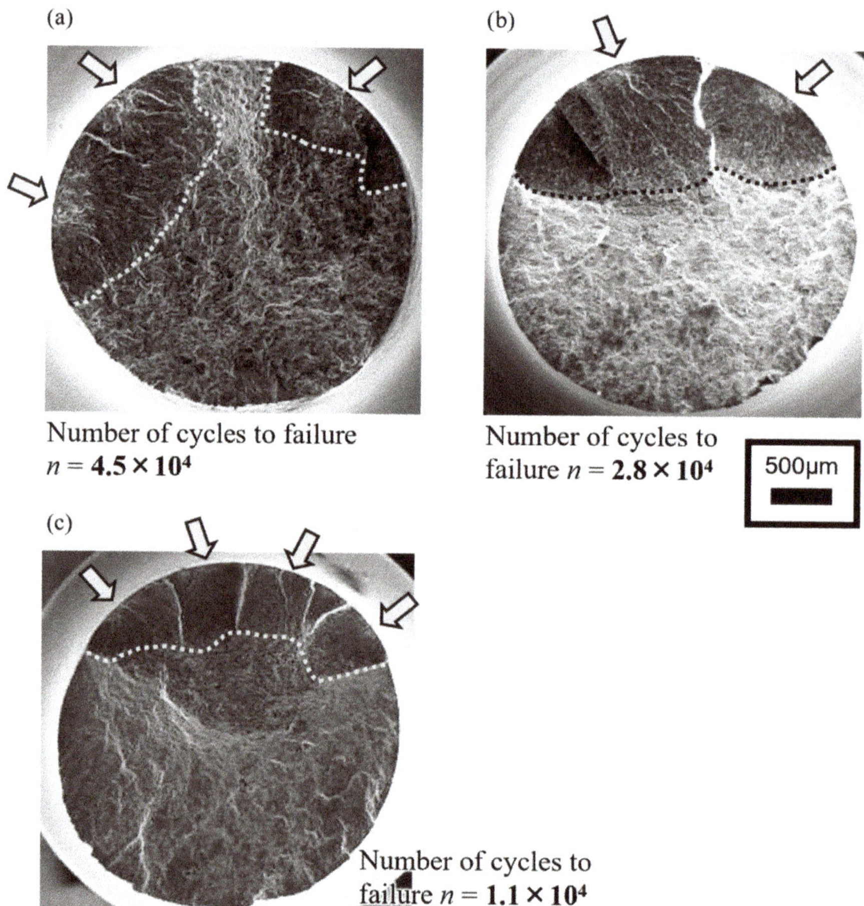

Figure 5. SEM images of fracture surfaces of annealed, ECAP-1 p and ECAP-8 p samples of AZ31 B alloy at stress amplitude σ_a = 100 MPa. (**a**) Anneal, (**b**) ECAP-1 p, (**c**) ECAP-8 p.

Figure 6 is a SEM image which enlarged fatigue crack initiation sites; fracture surfaces are found mainly to consist of transgranular quasi-cleavage. Annealed and ECAP-1 p samples showed a fracture surface with severe surface undulations as if plastic deformation had occurred to a large extent, which could be seen as twinning. No twinning-like structures were observed in ECAP-8 p. It is inferred that the twinning deformation at the starting point of fracture in the annealed and ECAP-1 p samples resist fatigue fracture and the fatigue life is longer than that of ECAP-8 p.

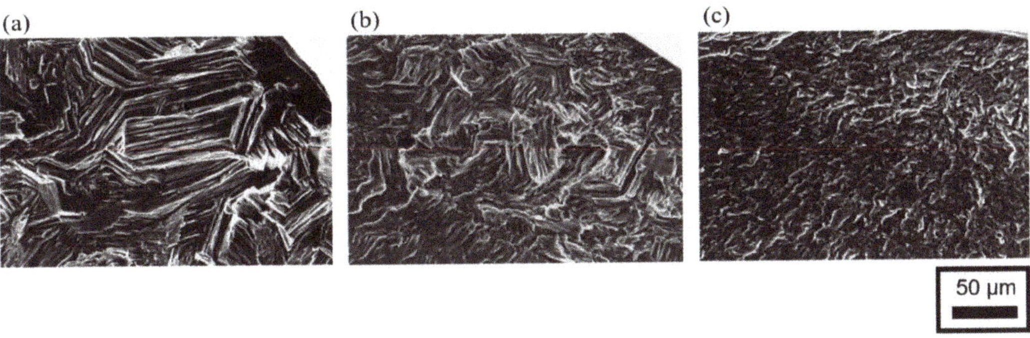

Figure 6. High-magnification SEM images of fracture surfaces of annealed, ECAP-1 p and ECAP-8 p samples of AZ31 B alloy at stress amplitude σ_a = 100 MPa. (**a**) Anneal, (**b**) ECAP-1 p, (**c**) ECAP-8 p.

4. Discussion

In a related study, it was found that the fatigue properties of Mg–Al alloys decreased as the number of passes in ECAP increased [11]. Although it is AZ80 magnesium alloy, the fatigue properties of the six passes material decreased, and the two passes material was superior in fatigue properties [11]. It has been reported that the fatigue properties of AZ31 magnesium alloy, which is close in composition, were improved by ECAP processing, but the number of passes was small, four passes [12]. Therefore, the decrease in the fatigue properties of ECAP-8 p in the present experimental results may be due to the increase in the number of passes. Residual stress measurement was performed to consider the cause of the deterioration of fatigue properties due to ECAP processing at high stress amplitude (Figure 7). For all three specimens, the residual compressive residual stress after the fatigue test was measured. The annealed sample seems to have had little residual stress before the fatigue test. Since the compressive stress value is lower in the annealed sample, it can be inferred that ECAP-1 p and ECAP-8 p had tensile stress generated by ECAP processing. ECAP-1 p and ECAP-8 p showed tensile residual stress, which increased with the number of passes, and ECAP-8 p was considered to have the earliest fatigue failure. It has also been reported that both strength and elongation can be increased by artificial aging heat treatment after ECAP processing in AZ91 magnesium alloy [13]. As future work, it is necessary to verify whether artificial aging of ECAP-1 p can exceed the fatigue properties of annealed samples. The mechanical properties of ECAP samples with varying number of passes between two and seven should also be continued to be investigated.

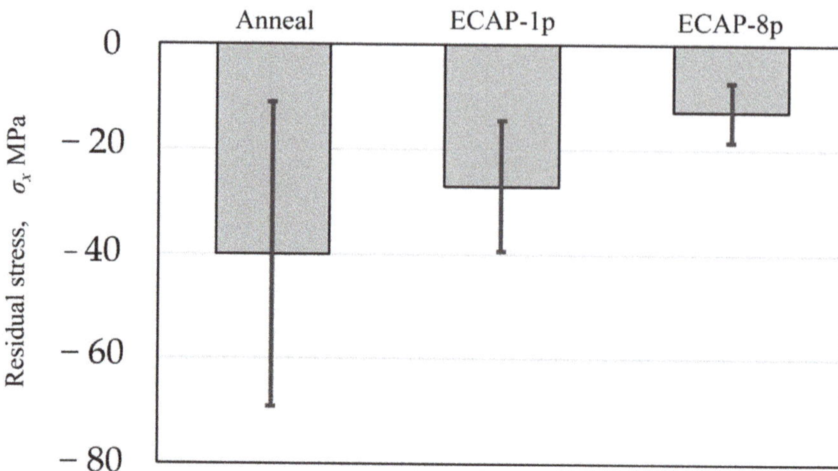

Figure 7. Residual stress of annealed, ECAP-1 p and ECAP-8 p samples of AZ31 B alloy measured by XRD after fatigue test at stress amplitude σ_a = 100 MPa.

5. Conclusions

The effects of varying the number of passes on tensile and fatigue properties of AZ31 B magnesium alloy under the present experimental conditions were investigated when equal-channel angular pressing was performed. Tensile strength in ECAP-1 p was higher than in the annealed sample. In contrast, elongation to failure was larger in the ECAP-8 p sample than in the annealed sample. From the fatigue test, at stress amplitude σ_a = 100 MPa, number of cycles to failure was largest in the annealed sample, medium in the ECAP-1 p sample and lowest in the ECAP-8 p sample. At stress amplitude σ_a = 80 MPa, number of cycles to failure was largest in the ECAP-1 p sample and medium in the annealed sample. It was found that grain refinement by ECAP was effective in improving fatigue life depending on the value of stress amplitude. Strength is more effective than ductility in improving fatigue properties, and it is considered important to increase the yield strength. It was suggested that the low fatigue life of the ECAP-8 p sample is attributable to tensile residual stress in severe plastic deformation.

Author Contributions: Conceptualization, R.Y. and S.Y. and Y.I.; methodology, R.Y.; writing—original draft preparation, R.Y.; supervision, R.Y. and S.Y. and Y.I.; project administration, R.Y.; funding acquisition, S.Y. All authors have read and agreed to the published version of the manuscript.

Funding: This research was funded by JSPS KAKENHI, grant number JP19K04095.

Institutional Review Board Statement: Not applicable.

Informed Consent Statement: Not applicable.

Conflicts of Interest: The authors declare no conflict of interest.

References

1. WHO Fact Sheets. *The Top 10 Causes of Death*; WHO: Geneva, Switzerland, 2018.
2. Ikeuchi, K. Mechanical properties of stents. *Jpn. J. Artif. Organs* **2009**, 46–48.
3. Karas, S.P.; Santoian, E.C.; Gravanis, M.B. Restenosis following coronary angioplasty. *Clin. Cardiol.* **1991**, *14*, 791–801. [CrossRef] [PubMed]
4. Peng, K.; Qiao, A.; Ohta, M.; Putra, N.K.; Cui, X.; Mu, Y.; Anzai, H. Structural Design and Analysis of a Nonel Biodegradable Zinc Alloy stent. *Comput. Modeling Eng. Sci.* **2018**, *117*, 17–28. [CrossRef]
5. Menown, I.B.; Noad, R.; Garcia, E.J.; Meredith, I. The platinum chromium element stent platform: From alloy, to design, to clinical practice. *Adv. Ther.* **2010**, *27*, 129–141. [CrossRef] [PubMed]

6. Vinogradov, A. Effect of severe plastic deformation on tensile and fatigue properties of fine-grained magnesium alloy ZK60. *J. Mater. Res.* **2017**, *32*, 4362–4374. [CrossRef]
7. Hao, P.; Enoki, M.; Sakurai, K. Finite Element Analysis of Tensile Fatigue Behavior of Coronary Stent. *Mater. Trans.* **2012**, *53*, 959–962. [CrossRef]
8. Ohmura, Y. Thesis on Monks, Integrated Graduate School of Medicine, Engineering, and Agricultural Sciences. Master's Thesis, University of Yamanashi, Yamanashi, Japan, 2015; pp. 63–70.
9. Iwahashi, Y.; Horita, Z.; Nemoto, M.; Wang, J.; Langdon, T.G. Principle of equal-channel angular pressing for the processing of ultra-fine grained materials. *Scr. Mater.* **1996**, *35*, 143–146. [CrossRef]
10. Yamada, R.; Orii, T.; Hiyagawa, M.; Yasui, H.; Yoshihara, S.; Ito, Y. Effect of grain refinement on mechanical properties in AZ31B magnesium alloy by Equal-Channel Angular Pressing. In Proceedings of the 136th Conference of Japan Institute of Light Metals, Toyama, Japan, 11 May 2019; pp. 77–78.
11. Fereshteh-Saniee, F.; Akbaripanah, F.; Kim, H.K.; Mahmudi, R. Effects of extrusion and equal channel angular pressing on the microstructure, tensile and fatigue behaviour of the wrought magnesium alloy AZ80. *Fatigue Fract. Eng. Mater. Struct* **2012**, *35*, 1167–1172. [CrossRef]
12. Zhu, R.; Wu, Y.; Ji, W. Low-cycle Fatigue Properties of an Ultrafine-grained Magnesium Alloy Processed by Equal-channel Angular Pressing. *J. Wuhan Univ. Technol. Mater. Sci. Ed.* **2012**, *27*, 1029–1032. [CrossRef]
13. Yuan, Y.; Ma, A.; Jiang, J.; Lu, F.; Jian, W.; Song, D.; Zhu, Y.T. Optimizing the strength and ductility of AZ91 Mg alloy by ECAP and subsequent aging. *Mater. Sci. Eng. A* **2013**, *588*, 329–334. [CrossRef]

MDPI
St. Alban-Anlage 66
4052 Basel
Switzerland
Tel. +41 61 683 77 34
Fax +41 61 302 89 18
www.mdpi.com

Metals Editorial Office
E-mail: metals@mdpi.com
www.mdpi.com/journal/metals